Agriculture in America
1622 - 1860

Americana to 1860: Four Bibliographies of Printed Works in the Collections of Three Philadelphia Libraries

Volume I: American Philanthropy, 1731 - 1860

Volume II: Agriculture in America, 1622 - 1860

AGRICULTURE IN AMERICA
1622 - 1860

Printed Works in the Collections of

The American Philosophical Society
The Historical Society of Pennsylvania
The Library Company of Philadelphia

Compiled by
Andrea J. Tucher

GARLAND PUBLISHING, INC.
NEW YORK & LONDON
1984

Library of Congress Cataloging in Publication Data

Tucher, Andrea J.
Agriculture in America, 1622–1860.

(Americana to 1860 : four bibliographies of printed works in the collections of three Philadelphia libraries ; v. 2)
Includes index.
1. Agriculture—United States—History—Sources—Bibliography—Union lists. 2. Agriculture—United States—Bibliography—Union lists. 3. Agriculture—Bibliography—Union lists. 4. Bibliography—Early printed books—Union lists. 5. Catalogs, Union—Pennsylvania—Philadelphia. I. American Philosophical Society. II. Historical Society of Pennsylvania. III. Library Company of Philadelphia. IV. Title. V. Series: Americana to 1860 ; v. 2.
Z5075.U5T83 1984 016.63'0973 83-49300
[S441]
ISBN 0-8240-8967-7 (alk. paper)

Printed on acid-free, 250-year-life paper
Manufactured in the United States of America

Table of Contents

Foreword

The Andrew W. Mellon Foundation in 1977 seeking to increase the scholarly usefulness of the collections of the American Philosophical Society, The Historical Society of Pennsylvania, and The Library Company of Philadelphia, made a grant to be administered by the first institution for the benefit of all three. Whitfield J. Bell, Jr., then Librarian of the Philosophical Society, James E. Mooney, Director of the Historical Society, and Edwin Wolf, 2d, Librarian of the Library Company determined that bibliographies of printed materials in the three institutions on American philanthropy, education, natural history, and agriculture should be prepared from the earliest period to 1860. Cornelia S. King was selected to compile bibliographies on the first two categories and Andrea J. Tucher on the second two.

Work got underway in 1978 at the Library Company and continued until 1981, when the search of the collections of all three institutions was completed. In all over 25,000 cards covering the four bibliographies have been typed, proofed, and coded for specially created computer-related programs. _Agriculture in America, 1622-1860_ is the second volume to be published; the other two will follow shortly.

We would like to acknowledge the generous support of The Andrew W. Mellon Foundation that has made this collaborative effort possible, and also specially thank Andrea J. Tucher for the skillful and dedicated manner in which she has compiled this most valuable bibliography.

February 1984

Edward C. Carter II
Librarian
American Philosophical Society

Preface

This catalogue is an alphabetical listing of those printed works issued before 1861 which relate to agriculture in the continental United States and which are held in any of three institutions: the library of the American Philosophical Society; the library of The Historical Society of Pennsylvania; and The Library Company of Philadelphia.

The volume is intended to serve as a resource guide for scholarly researchers. I have taken particular care to prise out those ephemeral items -- broadsides, trade catalogues, prospectuses, circular letters, petitions -- that so easily lie concealed in the corners of large collections. My method of attack included the reading of the entire Library Company card catalogue and the relevant portions of the shelf lists of the Historical Society and the Philosophical Society, as well as a sweep of the shelves themselves and a careful plumbing of literally thousands of pamphlet boxes and volumes. As a further aid to the scholar, I have compiled two appendixes. The first is a subject and institution index with authors' names listed in brief form and with key-word titles; the second is a chronological rearrangement with similarly abbreviated entries.

Perhaps the most difficult task I encountered in compiling this catalogue was deciding exactly what categories of items should be included under the heading of American agriculture. Some fine judgments and distinctions have been necessary to keep the volume a manageable size. A well-meaning but unwary step into the quagmire of agricultural tariff debate, for example, would have made impossible a retreat with anything less than a finding-list for Federalist economics; an earnest attempt to include all almanacs that contain any agricultural news at all would have duplicated most of Drake. The following guidelines have been used. Always included are works on raising, breeding, using, or processing crops or livestock (including grains, fruits, vegetables, landscape gardens, vines, decorative plants or flowers; poultry, sheep, pigs, horses, cattle, fish, bees, silkworms, rabbits; dairying, milling, brewing, distilling), as well as items dealing with farriery and veterinary medicine, soil culture, the application of chemistry or geology, and rural architecture. Also included without question are materials concerning agricultural education, publications of agricultural societies, addresses before such societies, and periodicals devoted to specifically agricultural pursuits. Catalogues or prospectuses of nurserymen, manufacturers of agricultural equipment, feed or fertilizer companies, and land salesmen also appear. Categories in which judgments of relevance were made are: works on agricultural commerce, not included if their chief concerns are political or economic; emigrants' guides, listed only if they are addressed primarily to farmers and include notices of farm lands or farm practices; and almanacs, rejected unless they are intended specifically and chiefly for farmers. Not listed at all are twilight books; disbound articles; most federal government documents and separately-printed laws without a named personal author; and facsimile reprints or photostat copies. Items dealing with slave labor in agriculture have already been described in the Historical Society/Library Company catalogue of Afro-Americana, 1553-1906 (1973), and thus do not appear here. The researcher's attention is also directed to the companion volume to this work, Natural History in America, 1609-1860, which lists general, non-prescriptive descriptions of flora and "natural advantages."

Each entry contains the following information:

1. Author, authors, or corporate author when known. Authors' names are always given in full (most of the three institutions' cataloguing having been done well before AACR2), with dates when ascertainable. Square brackets enclose a name that does not appear on the title-page of the work in question, but the Note records any appearance elsewhere in the work. Corporate authors whose names include a geographical location are entered under that location. The Chester County Agricultural Society is listed as CHESTER CO. [PA.]. AGRICULTURAL SOCIETY; but compare AMERICAN SILK SOCIETY, Baltimore, or HIBBERT AND BUIST, Nurserymen, Philadelphia. Entries for land companies include the location of their lands, which is not necessarily the location of their offices.

2. Title, following the title-page exactly in punctuation and initial capitalization. Transcriptions include as much of the title as is practical or helpful; ellipses mark omissions of text but not of authors' names or honorifics if such omissions do not affect the sense of the text. For the sake of clarity, any square brackets that appear in titles are quietly transcribed as parentheses.

3. Place of publication, in modern (Charleston, not Charles-Town), English form.

4. Publishers' and/or printers' names exactly as they appear in the imprint, except that addresses are omitted without ellipses; additional printers named on the title-page verso or in the colophon are added in square brackets. Curtailment of a long list of publishers' names is indicated by "*et al.*"

5. Date of publication, in Arabic numerals; inferred or supplied dates appear within square brackets.

6. Physical description. Because this is a union catalogue of a particular group of items, not a bibliography of ideal copies, descriptions in all cases have been based on the copy or copies in hand, although every effort has been made to ascertain whether these copies are complete. Cases in which the copy in hand is known to be anomalous or defective in some way are clearly distinguished, in that the holding library's symbol is prefixed to that description and the proper or putative configuration is given in parentheses. The following aspects are included:

a. Pagination. Two pagination strategies have been employed. For those items whose final page number (or the final page number of each separately numbered series) represents the actual page count, only those numbers are given and followed by the abbreviation "p.": xvi, 161 p. For those items whose numbering does not reflect an accurate page count, the abbreviation "Pp." is prefixed to an enumeration of the various series: Pp. [1-4], [9]-54. In each case, unprinted blank leaves or a final blank verso are not noted, but unnumbered pages with printed matter are noted in square brackets. An engraved title leaf appearing in addition to the letterpress title is not included in the pagination, but rather is considered illustrative matter (see b. below). Advertisement leaves, numbered or not, are included but distinguished by the parenthetical comment "adv." (for any general commercial material) or "pub. adv." (for advertisements by the publishing firm itself).

b. Illustrative matter. Counts are recorded of all plates, maps, tables, charts, or diagrams that are not included in the pagination, with particular note taken of any that are folding, colored, or tinted. Frontispieces and added engraved title-leaves are counted separately from other plates. Figures, cuts, or drawings appearing in the text or included in the page count are described by the term "illustrations."

c. Unprinted wrappers or paper covers are not noted, but any copy with a printed wrapper has a note to that effect. "Wrappers" is intended to mean that the copy has both a front and a back wrapper; "wrapper" is used to refer to the front wrapper only.

7. Notes, on an ad hoc basis. These include information on content, attribution, publishing history, or bibliographical points, whenever such information is deemed necessary, helpful, or merely interesting. Particular effort has been made to distinguish among editions, states, or issues of the copies in hand. As a rule, manuscript notations are not described unless they are useful in establishing date, author, or other obscure points. The expression "signed at end," unless qualified with "in MS.," should be taken to mean a letterpress name.

Two major exceptions from the above rules should be noted. Periodicals, because of frequent changes in their titles, authors, publishers, and even places of publication, have been given only nominal description, and no effort has been made to detail such changes. However, the issues and numbers held by each library have been fully listed, even including, for the sake of continuity, issues published after 1860. While all entries are by title, editors who appear consistently are given cross-reference notes. And agricultural almanacs, whose titles are perhaps their least important aspect, are all listed in the separate Almanac section (numbers 27 to 655 in the alphabetical listing) by place of publication, and chronologically within place.

Extensive use has been made of cross-references. As a general rule, addresses before societies by a named author are entered under the personal author's name, but in every such case, the society is given a "See also" note. In other words, listed under the heading for each institution are the names of any speaker who addressed it. Conversely, a second or third joint author, a speaker whose address was appended to a general society report, or the subject of a biographical sketch or critical commentary, all have references to the institution or main author. Nurseries or other firms that undergo changes of proprietor or name have notes to that effect, and societies with similar names (the four South Carolina agricultural societies, for instance) are distinguished by date of founding.

My thanks go to the Andrew W. Mellon Foundation for funding the project, and to the staffs of the three libraries who generously lent us their card catalogue drawers, their stacks, their typewriters, and their expertise. I am especially grateful to Edwin Wolf 2nd; to Mary Dunn and Joan Poole of the Temple University Word Processing Center, who handled a nervewracking manuscript with care and dispatch; to the Special Collections Department at Temple and particularly to the Contemporary Culture Collection, for providing me with a congenial spot in which to read proof; and to the many friends and fellow bookpeople who helped with information, advice, and inspiration along the way. And a laurel wreath goes to my colleague, Cornelia S. King, who alone dealt with the welter of prepublication detail after I had left Philadelphia.

Location Symbols & Abbreviations

Location Symbols

APS - In the collection of The American Philosophical Society, Philadelphia
HSP - In the collection of The Historical Society of Pennsylvania, Philadelphia
LCP - In the collection of The Library Company of Philadelphia
HSP in LCP - In the collection of The Historical Society of Pennsylvania, but housed in The Library Company of Philadelphia

Abbreviations

½ t.p. - half title
adv. - advertisement
brs. - broadside
cap. title - caption title
col. - colored
colop. - colophon
contrib. auth. - contributing author
ded. - dedication, dedicatory
ed. - editor, edited
edn. - edition
eng. - engraved
fold. - folding
frag. - fragment, fragmentary
front. - frontispiece
illus. - illustrated, illustration(s)
incl. - including
jt. auth. - joint author
no. - number
p(p). - pages(s)
pdcl. - periodical
pl(s). - plate(s)
prelim(s). - preliminary page(s)
pt(s). - part(s)
pub. - published, publisher
pub. adv. - publisher's advertisement
sep. - separate
tbl. - table
t.p. - title page, title leaf
vol(s). - volume(s)

Reaping Machines in Eden: An Introductory Essay

The history of American agricultural literature before the Civil War is a chapter in the conflict between life and art. Art took the early ascendency with its portrayal of the New World as a newfound Eden, a hothouse-cum-menagerie-cum-treasury bursting with exotic natural luxuries like spices, silk, and gems. Life moved in later but robustly with straightforward instruction in the prosy tasks of reaping and mowing, constructing beehives, brewing small beer, and curing catarrh in horses. And when Life finally and decisively succeeded in entrenching reapers and catarrh in Eden, and heaving out the ginger and pearls, Art simply encamped in Arcadia instead.

Life was little in evidence for nearly the first half of the period covered by this catalogue. Many of the earliest works are agricultural only in the optative and imperative moods; they were written by Englishmen who were convinced that the colonies were a goldmine -- or, perhaps, that the colonies had a moral duty to be a goldmine -- that could produce the luxuries like wine and olives as well as the necessities like hemp on which the mother country had been paying such high import duties. While the settlers themselves were preoccupied with sweating their tobacco, burying fishheads in their cornhills, and figuring out how to eat a pumpkin, Englishmen wrote cheery tracts that airily promised rich rewards to anyone who would just ship out and grow mulberry trees. One of the most ecstatic of these works is the earliest item listed in this catalogue, a tract written in 1622 by John Bonoeil, which described the culture of silkworms and included a puff for pomegranates, oranges, wine grapes, and almonds. The piece was prefaced by a letter from King James I himself, directing the settlers to stop dallying with tobacco, which "brings with it many disorders and inconueniences," and to turn their attention forthwith to producing silk, a "rich and solid Commodity." Bonoeil gave detailed and foolproof instructions, even suggesting scenting the warming-pots in the worms' hothouses with "Rosemary, or Tyme, Roses, or other sweet smels," for "those sweet smels are very agreeable and pleasing to the Wormes." Kingly interest in the project remained high; the supply ships from England often carried pounds of "silkworme seed" and even some experienced silk culturists to help tend the worms. But Sir Edwin Sandys soon had to report back that the settlers could not be weaned from their disorderly tobacco.

This first failure did not discourage the men with the west in their eyes. Only the stoniest-hearted potential adventurer could have resisted the thrilling pictures painted in Edward Williams's two promotional tracts of 1650 about Virginia. Williams portrayed the colony as the "rich-bosomed Sister" to Persia, whose latitude she shared, and the possessor of "an equality in her plenty of Mulberries, Silke, and Gums, Vines, Maiz, Rice, and all sorts of Graine." Even China, the richest and most mysterious empire in the world, was not Virginia's superior, for both produced flax, furs, hemp, and pearls, and nothing could prevent the American cousin from soon challenging China's corner in sugar, indigo, and ginger. Strangely enough, however, the silk, pearls, and ginger did not spring up in Virginia's rich bosom, which continued to put forth chiefly tobacco; but the vision of exotic luxuries continued to excite mercantilists throughout the colonial period, and to flit through many of their works. Even after the Revolution, when British interest faded, the legacy remained; the dream of silk and wine continued for years to inspire a continuation of wishful King Jamesian prose. In the 1830s in New England a resurgence of interest in silk generated dozens of manuals, societies, and periodicals, but very little fabric. And although some farmers in pockets of South Carolina and Ohio did produce some usable wine grapes, winemaking was never a large success in this period. The English critic Mrs. Trollope coolly dismissed America as too immature to produce good wine -- although she was not sanguine about the new nation's readiness for other of life's finer things either. In her revised, 1839 edition of *Domestic Manners of the Americans*, she noted that all the wine she had tasted was "miserable stuff" even though vines grew profusely everywhere; "perhaps," she sniffed, " . . . the champagne and Bordeaux of the Union may appear simultaneously with a Shakspeare, a Raphael, and a Mozart."

Not all of the earliest works were quite so Artistic as those of Bonoeil and Williams. A few colonial Americans were rather more clear-eyed than their enraptured but far-off cousins. Francis Rawle attempted to cheer Delaware Valley residents suffering under the hard-money crisis of the 1720s by pointing out their region's great natural wealth. He touted no commodities that stirred the heart, but he did suggest some that could help to fill the larder or the barn: wheat, barley, and oats; distilled drinks made from apples, peaches, or rye (as well as "decay'd Wines"); horses. Johannes Deigendesch of Germantown, Pennsylvania issued numerous editions of his compendium of medical and traditional lore for the care of horses, a genre that soon became a Pennsylvania Dutch specialty. Jared Eliot, a Connecticut clergyman, produced a series of essays on field husbandry in which nuggets of practical wisdom on drainage or tillage were interspersed with musings on the structure of the prophet Elisha's twelve-yoke plow, and Vergil's faults as a writer on husbandry. Even Eliot succumbed to silk fever, though; he devoted most of his sixth and last essay to the culture of mulberries and the care of silkworms. And Edmund Quincy produced what he called "little more" than a collection from American and European sources on the culture of hemp, but his modesty was unnecessary. While that difficult plant never did become a large trade crop, it was often an important item in local economies. Quincy's description of how to sow, water, harvest, and brake hemp offered a rare piece of practical wisdom that was actually applicable to an American product.

This early paucity of activity in the unartistic area of practical improvement bothered some agriculturists, particularly the wealthy gentlemen planters. They tried in a variety of ways to keep in touch with new practices. They exchanged letters with each other and with European friends; George Washington, for one, had a large agricultural correspondence. They read English books in the hope of gleaning ideas useful for their own circumstances; Thomas Jefferson, an avid bibliophile, had a sizeable collection of standard authors including Tull, Hales, Bradley, and Evelyn. They read and wrote pieces for periodicals like the _Transactions_ of the American Philosophical Society. But rather to their own dismay, the urge for improvement was confined chiefly to these upperclass gentlemen. The small farmer, the dirt farmer, and the pioneer rarely felt compelled to experiment with possible innovations -- or even to consider the possibility that innovations existed at all. The same methods were followed by farmer after farmer, by farmer's son after farmer's son, and many of those methods were wasteful if not positively harmful. Washington criticized them; Jefferson criticized them; European visitors criticized them. A variety of entrenched attitudes lay behind the farmers' inertia.

Influential, of course, was a nearly universal and nonchalant faith in America's plenty. One could always move on to new land if one exhausted the soil of the old. Important with Southern crops was the farmer's need to use labor -- including slave labor -- to best advantage; planting a field with clover or grass might replenish the soil for tobacco in another year but would give the laborers little work to do in that one. Isolation and semi-literacy were other contributing factors. In rural areas, communication was laborious, ideas circulated slowly, the annual almanac was often the only printed matter in evidence, and few farmers would have trusted anything they read in a book anyway. But the problem that annoyed the gentlemen improvers most was the typical farmer's simple and stubborn resistance to change of any sort. Writing somewhat later, in 1889, the historian Henry Adams -- not a farmer, but certainly a gentleman -- in the second chapter of his vast _History of the United States of America_ described the American temper at the beginning of the Republic: ". . . conservatism possessed the world by right. Experience forced on men's minds the conviction that what had ever been must ever be. At the close of the eighteenth century nothing had occurred which warranted the belief that even the material difficulties of America could be removed . . . ." Adams bemoaned the unscientific character of the average turn-of-the-century American's mind, a moan that had been voiced by that American's elite contemporaries as well.

So diffident an author was one key member of this early gentlemanly circle of improvers that he is still, nearly two hundred years later, paying a penalty for his meekness. The retiring disposition of John Beale Bordley, a respectable if obscure Maryland lawyer and judge, led him to publish most of his works anonymously, and some have been consistently misattributed ever since. Some of his descendants were irked for a long time at his character failing; Bordley's great-great-grandson Francis Sims McGrath lamented as late as 1950 in a family history, _Pillars of Maryland_, that his progenitor's lack of self-confidence kept him from doing his part during the Revolution. This disapproving descendant did make mention of Bordley's agricultural labors, but he was clearly making the best of a bad job; in discussing Bordley's major work, _Essays and Notes on Husbandry and Rural Affairs_, he singled out for special praise the strength of the paper it was printed on. But McGrath's lamentation was misguided, and historians who overlook Bordley and his work are missing an important combatant in the struggle to overturn the inert old order. Bordley may not have made a great contribution to American politics, but in American agriculture his role was signal: he was among the very first to make a practical effort to haul Life into Eden.

Bordley was inspired by one great idea: to convince American farmers that they should cultivate the staple necessities of life in order to become economically as well as politically independent. His first pamphlet, _Necessaries; Best Product of Land; Best Staple of Commerce_, published in 1776, chastised his countrymen for their preoccupation with chasing the everlasting phantoms of wine and silk. Those articles could not, of course, feed and clothe needy families, and Bordley urged farmers to concentrate on raising wheat, corn, cattle, wool, and other humdrum but practical commodities. But the pursuit of luxury crops did more than merely threaten to supplant more wholesome products; it also risked subjecting America to a total and devastating economic dependence on Britain. The mother country "is old, and drinks hard, and her constitution is much impaired"; she would, Bordley warned, forbid Americans to sell their products to any other nations, which would keep those prices down, and would then set artificially high the prices of the bread and other staples that the farmers would need to buy just to survive. The only answer for Americans was complete economic self-sufficiency: to eat, drink, and wear only those items that they produced themselves.

These were not particularly radical words. Nonimportation of British goods was already a well-used American weapon against oppressive taxes, although it had been used with varying degrees of heartiness. But Bordley was no tinkling cymbal; he had already set about proving not only that Americans could do without imported goods, but also that they could so so handsomely. Six years previously he had acquired an estate of 1600 acres on Wye Island, on Maryland's Eastern Shore, and had turned the estate into a virtually self-sufficient village. He had a smithy, a carpenter's shop, a spinnery, a ropewalk, a brickyard, a mill, a brewery, a dairy, a granary, an ice-house, a smokehouse; he raised livestock, wheat, cotton, hemp, peaches, plums, nectarines, melons, grapes, almonds, palma christi, and madder. In Philadelphia in 1865, Elizabeth Bordley Gibson published _Biographical Sketches of the Bordley Family_, a family history that was as lyrical as the later descendant's would be rueful. Mrs. Gibson described the judge's "elegant retirement" in his well-appointed suzerainty within a "well-educated and affluent circle of neighbors"; she pictured him, on a typical day in planting season, pottering in his garden, supervising the ongoing work, visiting and doctoring his slaves, dining, savoring a "cool tankard" of home-brewed sangaree garnished with springs of burnet and balm. But this picturesque life did not come without effort, and Mrs. Gibson took plains to convey Bordley's constant labors toward agricultural improvement. Even as a very young man he had often sent to England for the latest machines, the newest books, and the best seeds; on his estate he experimented constantly. And he even devised a way to share his findings with others while still maintaining his diffident distance. As Mrs. Gibson described it,

> as his systems became matured, he published their results: at first on written cards, sent around his neighborhood, then in printed handbills, of which we have two, one called "Country Economy," another "Farmyard Economy;" and in 1783-4, his small essay, entitled "Summary View of Courses of Crops, &c.," which he presented liberally through the country. His object in so doing was to rouse those farmers who lived by the soil to make exertions for themselves . . . .

No written cards or handbills seem to have survived, but the _Summary View_ was only the first of many pamphlets on practical agricultural improvement that Bordley sent out into the countryside. In his effort to rouse the exertions of local farmers, he wrote essays detailing his ideas on sheep, on crop rotations, on cattle, on farm economy, on farm buildings, and on hemp, among others, all of which he had printed up in small editions and all of which he signed only "B." Doubtless the first recipients of the pamphlets, his near neighbors, knew who the modest author was, but as the essays gained wider circulation other readers were not so sure. Thomas Jefferson had several of Bordley's works, but, as E. Millicent Sowerby notes in her catalogue of Jefferson's library, he ordered Bordley's _Sketches on Rotations of Crops_ several years after its first appearance and still did not know the author's name. The Englishman Richard Parkinson at least gave credit to Bordley when he quoted and commented on the _Sketches_ in his 1805 _Experienced Farmer's Tour_, but the American, who had died the year before, was spared the indignity of seeing his name misspelled throughout.

Modern bibliographers are better now on both spelling and attribution, but these small pamphlets still have some secrets to reveal. Many of them went through extensive revision and exist in several different editions or variants previously unrecorded. Ever the improver, Bordley often made corrections and additions in manuscript on copies he gave to friends; many of the copies listed in this catalogue are so annotated. But Bordley apparently also decided now and then that the revisions warranted an entire new printing, and other times, high demand seems to have inspired a reissue. In just the three libraries represented in this catalogue can be found two distinct editions of both _Intimations_ and _Intending to Retire_, the second edition of each with alterations of text; neither Evans nor Bristol notes more than one. One copy of _Hemp_ has a two-page Supplement added at back. Obscurity envelops the circumstances of the publication of the second edition of _Sketches on Rotations_: there are two issues of the title-page, distinguished only by insignificant differences in typography, but on one copy of each issue the date has been changed in manuscript from 1796 to 1797. Priority of issue may well be indeterminable, but the situation suggests that at some point a reissue of the work was required. And since Bordley produced his major work, an anthology of all his pieces under the title _Essays and Notes on Husbandry and Rural Affairs_, during a "painful state of [his] health, and the injurious effects of a long course of opium," he soon felt compelled to issue a 36-page supplement of "Additions and Amendments"; this survives in two recorded copies, one in Philadelphia. Since these revisions were then incorporated into the second edition of _Essays and Notes_, published in 1801, some of the pieces may exist in as many as six versions: first pamphlet edition, first pamphlet edition with Bordley's manuscript revisions, second pamphlet edition, first collected edition in _Essays and Notes_, first collected edition supplemented with additions and amendments, and second collected edition. This series could well allow a far-sighted reader to accompany Bordley throughout all the phases of his experiments in agricultural improvement.

When Bordley issued his small pamphlets with his signature austerely barbered, he puzzled some readers for a while; but when he completely disguised his editorship of two full-length works behind a pair of vague epithets, he ended up confounding nearly everybody for decades. In 1803 he prepared two English works for publication in Philadelphia, where he had moved some years previously with his second wife. In order that "it might not run into a high price," he condensed and paraphrased William Forsyth's _Treatise on the Culture and Management of Fruit_

Trees; modestly, he dubbed the piece an "epitome," and himself as editor only "an American farmer." And throughout an anonymous compilation of selections from English works, Gleanings from the Most Celebrated Books on Husbandry, Gardening, and Rural Affairs, he interspersed, set off from the rest of the text by quotation marks, a number of "remarks and observations by a gentleman of Philadelphia." But since William Cobbett had recently produced a full-length American edition of Forsyth's work, most major bibliographies have credited Cobbett with the Epitome, as well; the convincing Pierce W. Gaines does so in his definitive bibliography of Cobbett, and cites Pearl, Melville, and Sabin as precedents. And few bibliographers have dared even to guess at either the English compiler or the American editor of the mysterious Gleanings.

Evidence of Bordley's editorship of both works abounds, however. Scattered sources, although none with the voice of a Sabin or a Gaines, have accepted the statement by Mrs. Gibson herself that Bordley was responsible for the Epitome, and a recondite internal clue bears out her assertion. The opening Advertisement in the Epitome is in large part an explanation that Forsyth treated "country habitations" in his work differently than did "the author of essays and notes on husbandry . . . . [because] the former was written upon the happening of certain events," namely some destructive fires. "Country Habitations" was of course the title of one of Bordley's pamphlets, the text of which was also included in his Essays and Notes. Perhaps the lowercase initial letters of the title as quoted have misled bibliographers into reading it as a sort of generic description rather than as the name of a specific work. A specific work it clearly was, however -- a specific work whose author was anxious to reconcile its differences with the work he was editing. The section in the Epitome on country habitations includes especially plentiful editorial comment on practices in America.

The only external evidence for Bordley's editorship of Gleanings is perhaps tenuous: the addition, in an unknown hand, of "J.B. Bordley" following the epithet on the title-page of a copy held by the Widener Library at Harvard University. The diligent reader of the Philadelphia gentleman's additions to the text, however, will notice many that can only be Bordleyan. The editor was clearly a man of sturdy republican virtue; under the heading "Beer" is his comment

> Small Beer, the best family drink: Malt and hops are productions of the Farm. Cyder being precarious, Rum, an intemporate [sic] liquor, is unhappily resorted to for want of cyder. The ingredients of Beer, (better than either) are always in the power of the Husbandman.

Similar in tone is the added paragraph on "Diet," which even includes a citation to Bordley's own work:

> Ought to be attended to; and in the country be simple, plain, and neatly dressed, unpoisoned with foreign, harsh ingredients. Spices, salt, condiments, &c. to be moderately admitted. In Essays and Notes on Husbandry are good general rules to the purpose.

Other of the gentleman's remarks refer to conditions in Maryland, to crops that had been attempted at Wye, or to subjects about which Bordley himself had also written. Under "Gin-Seng," for instance, the editor added an account of the Marylander Daniel Preston who had cured himself of asthma by chewing on the plant and drinking tea made from its leaves; a new paragraph commended palma christi as a good course of castor oil, a "very valuable family purge"; and the section on hemp was expanded. The inscriber of Harvard's copy may remain unknown, but his information was assuredly correct.

So earnest an experiment was John Beale Bordley, and so uncommon were his practical efforts to improve the lot of the small farmer, that one is pained to learn of his eventual disillusionment. Until nearly the day of his death, in January of 1804, he continued to work on his writings; the invaluable Mrs. Gibson described him, at a time "when his disease was far advanced upon him," continuing to labor:

> . . . that hope [of helping his country] supported him in perseverance, under the most excruciating pains, ofttimes so torturing that his family would entreat him to desist and take rest, but he would resume his pen again and again, with this and similar expressions: "No! I cannot stop; Americans must learn to know the value of agriculture, and if I can expedite or smooth the way to them, I must do it!"

But by that time his labors were chiefly on behalf of his fellow gentleman farmers. After many years of sending pamphlets out into the countryside,

> he began to fear [wrote Mrs. Gibson] that these efforts failed of producing the durable impression he desired; - it was a subject on which every small farmer thought himself already qualified to judge, or sufficiently informed for his own purpose. The lazy maxim, "'twill do well enough," was a bar to improvement that Mr. Bordley full often complained of . . . .

In 1784-5 he had worked with a group of like-minded city gentlemen to found the Philadelphia Society for Promoting Agriculture -- farmers in his own state being too scattered or too settled into raising tobacco to support such a society -- and he began to spend more and more time with this group, especially after moving to Philadelphia in 1791 and leaving Wye in his son's care. By 1799 he had even ceased to produce his pamphlets for the countryside, and instead spent his time preparing hefty volumes which were undoubtedly bought only by substantial farmers or scientifically-minded gentlemen. The greatest early effort to supplant the silk and wine of Eden with good American hemp and small beer was thus baffled by the inertia of farmers who preferred the familiar disorders of tobacco to either. But for two decades on the Eastern Shore of Maryland, a man so modest that he signed his full name to only one of his many publications -- so shy that his descendant four generations removed still thought of his failing with pique -- proved that American soil and American ingenuity together could produce virtually any commodity that American taste could require.

Although Bordley died too early to have seen much widespread or productive interest in agricultural progress, within just a few years after his death a revolution in attitude was beginning to take place. As the nineteenth century advanced and the new nation grew in both size and confidence, Americans came to realize that the state of agriculture could -- and must -- be improved. New incursions of citydwellers who could not grow their own food had to be fed; new types of land in the west required new methods and crops; new machines caused changes in traditional work practices. Life was striding boldly into Eden -- and practical agricultural literature was coming to life.

By 1860, the terminal point for this catalogue, books were available on every possible aspect of husbandry. From the panoramic inclusiveness of Richard Lamb Allen's American Farm Book, which covered "soils, manures, draining, irrigation, grasses, grain, roots, fruits, cotton, tobacco, sugar cane, rice, and every staple product of the United States, with the best methods of planting, cultivating, and preparation for market," to the singlemindedness of B. Eastwood's treatise devoted entirely to the cranberry; from the pure science of Justus von Liebig's works on agricultural chemistry to the pure domesticity of Louisa Johnson's Every Lady her own Flower Gardener, "addressed to the industrious and economical"; from the earthiness of Samuel Luther Dana's muck manuals to the tenderness of Edward Townley's plea for "humanity to honey bees," the American cultivator could find a volume to fill any need. Periodicals also flourished, streaming through post offices all over the country to bring agricultural news and suggestions to even the most isolated farmers; John S. Skinner's American Farmer, begun in Baltimore in 1819, and the New York State Agricultural Society's Cultivator, issued in Albany starting in 1834, were among the more influential. But for some of the most intriguing and the most revealing agricultural writing, modern readers should turn not to the well-known periodicals, not to the standard treatises, but to the cheap pamphlets, handbooks, merchants' catalogues, and county society publications that also abounded in this period, for in those works are best caught and crystallized the farmer's daily concerns, frustrations, and victories. Readers can pore over the extensive catalogues of nurserymen like the Prince family in Flushing, New York, or the Bartrams near Philadelphia, and make a selection from the "most choice and splendid varieties"; ponder the purchase of Manny's Patent Adjustable Reaping and Mowing Machine for the field, or an Atmospheric Attempering Churn for the dairy from J.H. Bump; or gain inspiration from the list of premiums offered by the Philadelphia Society for Promoting Agriculture at its annual exhibition in Powelton. They can weight the benefits of joining their local country agricultural society, or any of the more specialized societies ranging from the St. Louis Vine & Fruit Growers' Association to the Lebanon County Horse Company For the Detection of Horse Thieves. And if they are thinking of pulling up stakes and moving west, they can study Josiah T. Marshall's Farmer's and Emigrant's Hand-Book for hints on how to clear forest or prairie land, how to plant the most useful crops, how to keep livestock healthy -- and, should adversity strike, how to survive on bread made from wood chips.

Thus Life took over Eden, and works on practical agricultural improvement, once as rare in America as native vintage wines were, became as commonplace as beer. But this new emphasis did not destroy the artistic tradition; even if blissful portrayals of the luxuriant growth of exotics had lost their appeal, or at least their persuasive power, a new kind of idyll soon replaced the old. In the hundreds of speeches and orations delivered across the country at agricultural fairs and society meetings, America became Arcadia, and the American farmer turned into the embodiment of all simple, homely republican virtue.

Listening to speeches in general was a beloved national pastime. In an era when no more than twenty percent of the population lived in urban places and few citizens saw more than a dozen people in the same spot at once, Americans loved to gather in large audiences to hear lengthy orations delivered by the local military hero, the travelling lecturer, or the visiting politician. And the canny politician loved to have the opportunity to heap compliments on the unsung tillers of the soil for their crucial contributions to America's greatness -- and then to reap a double benefit by circulating printed copies of the speech among other readers who would be touched by an agrarian theme. As the nation built more and more railroads and factories, as citizens reluctantly learned to accept hordes of immigrants who were often poorly skilled and poorly washed, as the cities grew dirtier and more frightening, as banks and commercial men gained power, a terrible nostalgia for the passing of a more innocent, more virtuous world gripped much of the population. The apathetic, tradition-bound small farmer who continued to plant tobacco because his grandfather always had, underwent a rosy transformation into the last guardian of a purity that was rapidly dissolving away.

So everyone made agricultural speeches. Stephen Douglas made one. Horace Greeley made one. Lewis Cass made one. Even the urbane Daniel Webster made one in 1840, the year the Whigs magically converted themselves into the keepers of the Jeffersonian flame in William Henry Harrison's Log Cabin campaign. Log cabins became so fashionable among the normally rather aristocratic Whigs that in another famous speech that year, Webster apologized for not having been born in one himself, but quickly pointed out that his elder siblings had been. Some of these agricultural speeches did contain useful information. Webster described some farming practices in England, and Greeley, always the reformer, took time to explain why workmen should be complete masters of their materials and implements. But almost every speech began with a ritualistic disclaimer of the speaker's ability to say anything his audiences had not already figured out for themselves; Webster "did not esteem himself," the paraphrased report of his address revealed, "to be particularly qualified to judge of the subject in all its various aspects and departments ... but the subject had been one of careful observation to him . . . ." And almost every speech ended with a paean to a boyhood home or a long-awaited retirement dwelling. Greeley's conclusion stirred many hearts:

> As for me, long-tossed on the stormiest waves of doubtful conflict and arduous endeavor, I have begun to feel, since the shades of forty years fell upon me, the weary, tempest-driven voyager's longing for land, the wanderer's yearning for the hamlet where in childhood he nestled by his mother's knee, and was soothed to sleep on her breast . . . And so I, in the sober afternoon of life, when its sun, if not high, is still warm, have bought me a few acres of land in the broad, still country, and, bearing thither my household treasures, have resolved to steal from the City's labors and anxieties at least one day in each week, wherein to revive as a farmer the memories of my childhood's humble home . . . .

Some speeches, indeed, contained very little else besides ritual and paean. In 1850, two years after he had lost the presidential election to Zachary Taylor, Lewis Cass addressed the Kalamazoo County Agricultural Society. Since he felt his "incompetence to perform the part of a teacher in the great art of agriculture," he decided instead to present "many general and important considerations, as well moral and statistical, as historical. "Over the course of their presentation, he quoted plentifully from the Bible, referred to his military experience, praised those who worked with their hands, urged the importance of education for understanding the Constitution and the laws, and pitied the strife-torn European nations. In a particularly ringing passage he described his trip to Egypt, and announced that even when standing atop the Great Pyramid of Cheops and looking down at the Nile valley "made fertile by God and not by man," he was still grateful to live in free America, "this best refuge for down trodden humanity." The speech was full of moving reflections and thrilling phrases. And it was printed not in remote Michigan but in the hustling city of Washington.

So in 1860, the end point for this catalogue, agricultural literature was still caught in the struggle between life and art. The earliest writings of the colonial period had concentrated on the land's glorious and infallible capacity to produce silk, gems, wine, and spices, and had virtually ignored the humdrum matters of curing beasts and harvesting fields. Only slowly and laboriously did the concentration shift to the more practical topics that could help the farmer in his increasingly complex task of feeding and clothing a booming nation. But even as the nation boomed, the farmer himself became the revered symbol of the simple rural life that was passing. It may have been impractical, it may have been wasteful and even destructive, but passing it assuredly was -- and that was provender for the artistic tradition. If reaping machines were going to take over Eden, the farmer would be safer in Arcadia.

Andrea J. Tucher

Agriculture in America
1622 - 1860

1. ABBOT, ABIEL, 1770-1828
An Address Delivered Before The Essex [Mass.] Agricultural Society At The Agricultural Exhibition In Danvers, October 17, 1821.
Andover: Printed For The Society By Flagg And Gould, 1822.
14 p. LCP

An Account of East-Florida .... [1766].
See STORK, WILLIAM.

Account Of The Wheat Moth .... 1805.
See EDDOWES, RALPH, 1751-1833.

2. ADAMS, DANIEL, 1773-1864.
The Agricultural Reader, Designed For The Use Of Schools.
Boston: Published By Richardson & Lord; Printed By J.H.A. Frost, 1824.
264 p. LCP

__________, ed.
See also The Medical And Agricultural Register [pdcl.].

An Address To Farmers On The Following Interesting Subjects .... 1796.
See DABNEY, JOHN, 1752-1819, comp.

3. ADLUM, JOHN, 1759-1836.
Adlum On Making Wine.
Georgetown, D.C.: James C. Dunn, Printer, 1826.
36 p. LCP; APS

4. __________.
A Memoir On The Cultivation Of The Vine In America, And The Best Mode Of Making Wine.
Washington: Printed By David And Force, 1823.
142 p. HSP

5. __________.
A Memoir On The Cultivation Of The Vine In America, And The Best Mode Of Making Wine. Second Edition.
Washington: Printed For The Author, By William Greer, 1828.
179, [1] p.; fold. front. LCP (lacking front.); APS

6. "AGRICOLA," A North Carolina Farmer, pseud.
A Series Of Essays On Agriculture & Rural Affairs; In Forty-Seven Numbers.
Raleigh: Printed & Published By Joseph Gales, 1819.
Pp. [1-4], [3]-223, [1-3]; illus. HSP in LCP

7. The Agricultural Museum: Designed To Be A Repository Of Valuable Information To The Farmer ....
Georgetown, Ca. [i.e. D.C.].
APS: Vol. I, nos. 1-24 (July 4, 1810-June 19, 1811)

8. Agriculturist's And Manufacturer's Magazine: Or The Friend Of National Industry.
Philadelphia.
HSP: Vol. 1, no. 2 (Jan. 8, 1820)

9. ALBANY AGRICULTURAL WORKS, Albany, N.Y.
Annual Circular Of Emery Brothers, Sole Proprietors Of The Albany Agricultural Works, Corners of Hamilton, Liberty and Union Streets. Warehouse And Seed Store, Removed To No. 52 State St., ... Albany, N.Y.
Albany: Weed, Parsons & Company, Printers, 1856.
[16] p.; illus. LCP (wrappers)

10. __________.
Annual Descriptive Catalogue Of Agricultural Implements, Horticultural Tools, And Field, Grain And Garden Seeds, For Sale At The Albany Agricultural Warehouse, and Seed Store, ... By Horace L. Emery, Successor To Luther Tucker, 1848.
Albany, N.Y.: Printed By C. Van Benthuysen, [1848].
[32] p.; illus. LCP (wrappers)

ALBANY CO. [N.Y.]. AGRICULTURAL SOCIETY.
See [EATON, AMOS, 1776-1842, and THEODRIC ROMEYN BECK, 1791-1855]. A Geological Survey .... Under The Direction of .... [1820].

ALBANY [N.Y.]. INSTITUTE.
See NEW YORK. SOCIETY FOR THE PROMOTION OF AGRICULTURE, ARTS AND MANUFACTURES [founded 1791]. Merged in 1824 with the above institution.

ALBEMARLE [VA.]. AGRICULTURAL SOCIETY.
See
MADISON, JAMES, 1751-1836. Address Delivered Before .... 1818.
STEVENSON, ANDREW, 1784-1857. Address Of ... Delivered Before .... 1847.

Albertus Magnus, oder Der lange verborgene ....
See HOHMAN, JOHANN GEORG, d. ca. 1846.

11. ALLAIRE, JULIEN-PIERRE, 1742-1816, et al.
Rapport Sur Les Voyages De M. Michaux Pour La Naturalisation Des Arbres Forestiers Des États-Unis, Fait à la Société d'Agriculture du département de la Seine, dans la séance du 1er. fevrier 1809.
Paris: De L'Imprimerie De Madame Huzard, 1809.
11 p. Note t.p verso: Extrait des Mémoires de la Société d'Agriculture, tome XI. LCP; APS

12. [ALLARDICE, ROBERT BARCLAY, 1779-1854].
Agricultural Tour In The United States And Upper Canada, With Miscellaneous Notices, By Captain Barclay Of Ury [pseud.].
Edinburgh: William Blackwood & Sons [Printed By John Stark (½ t.p.verso)], 1842.
xxiii, [1], 181, [1], [2 (pub. adv.)] p. LCP

13. ALLEGHENY COLLEGE, Meadville, Pa.
Report Of A Committee Of Trustees Of Allegheny College On The Manual Labor System. Adopted, and ordered to be printed; Oct. 7th, 1833.
Meadville, Pa.: Jos. C.G. Kennedy, Printer, 1833.
16 p. The labor proposed is chiefly agricultural. LCP; APS; HSP

14. ALLEN & NEEDLES, Oil merchants, Philadelphia.
Remarks On The Origin And Application Of Guano. For Sale By Allen & Needles, Dealers In Oil And Guano .... [wrapper title]
[Philadelphia: 1849?].
16 p. LCP (wrapper)

15. ALLEN, ISAAC JACKSON.
Address Of President Isaac J. Allen, Of The Farmers' College, Delivered September 27th, 1855, During The Exhibition Of The Cincinnati Horticultural Society, With A Preface Giving A Brief History Of The Institution; And A Description Of Some Of The Prominent Features Of The Exhibition. By Geo. Graham, Secretary ....
Cincinnati: Gazette Co. Steam Printing House, 1855.
34 p. LCP (wrappers)

16. ALLEN, JOHN FISKE.
The Culture of the Grape. Embracing Directions For The Treatment Of The Vine, In The Northern States Of America, In The Open Air, And Under Glass Structures, With And Without Artificial Heat.
Boston: Dutton And Wentworth, Printers, 1847.
55 p.; illus. LCP

17. __________.
A Practical Treatise On The Culture And Treatment Of The Grape Vine: Embracing Its History, With Directions For Its Treatment, In The United States ... Second Edition - Enlarged.
Boston: Printed By Dutton And Wentworth, 1848.
[4], 247 p.; front. LCP

18. __________.
A Practical Treatise On The Culture And Treatment Of The Grape Vine
... Third Edition - Enlarged And Revised.
New York: C.M. Saxton, Agricultural Book Publisher [S.W. Benedict, Stereotyper and Printer (t.p. verso)], 1853.
330, [1]+11 (pub. adv.) p.; front.; 3 pls. HSP in LCP

19. ALLEN, LEWIS FALLEY, 1800-1890.
Address Delivered Before The New-York State Agricultural Society, At The Capitol, In The City of Albany, On the Evening of the 18th January, 1849. Published By Order Of The Assembly.
Albany: Weed, Parsons & Co., Public Printers, 1849.
36 p. LCP

20. __________.
Rural Architecture. Being A Complete Description Of Farm Houses, Cottages, And Out Buildings, Comprising Wood Houses, Workshops, Tool Houses, Carriage And Wagon Houses, Stables, Smoke And Ash Houses, Ice Houses, Apiary Or Bee House, Poultry Houses, Rabbitry .... Together With Lawns, Pleasure Gardens And Parks; The Flower, Fruit And Vegetable Garden .... Also, The Best Method Of Conducting Water Into Cattle Yards And Houses. Beautifully Illustrated.
New York: C.M. Saxton, Agricultural Book Publisher [Buffalo, N.Y.: Stereotyped by Jewett, Thomas And Co. (t.p. verso)], 1853.
Pp. [i]-xv, [1], [13]-378 incl. 11 pls.; front.; illus. LCP

21. ALLEN, RICHARD LAMB, 1803-1869.
The American Farm Book; Or Compend Of American Agriculture; Being A Practical Treatise On Soils, Manures, Draining, Irrigation, Grasses, Grain, Roots, Fruits, Cotton, Tobacco, Sugar Cane, Rice, And Every Staple Product Of The United States, With The Best Methods Of Planting, Cultivating, And Preparation For Market. Illustrated By More Than 100 Engravings.
New York: C.M. Saxton [C.W. Benedict, Stereotyper (t.p. verso)], 1849.
325, [1], [10 (pub. adv.)] p.; front.; 1 pl; illus. LCP

22. __________.
The American Farm Book; Or Compend Of American Agriculture ....
New York: C.M. Saxton [C.W. Benedict, Stereotyper (t.p. verso)], 1854.
325, [1], 10 (pub. adv.) p.; front.; illus. LCP

23. __________.
The American Farm Book: Or Compend Of American Agriculture ....
New York: C.M. Saxton, Barker & Co.; San Francisco: H.H. Bancroft & Co. [C.W. Benedict, Stereotyper (t.p. verso)], 1860.
325, [1], 10 (pub. adv.) p.; front.; illus. LCP

24. __________.
Domestic Animals. History And Description Of The Horse, Mule, Cattle, Sheep, Swine, Poultry, And Farm Dogs. With Directions For Their Management, Breeding, Crossing, Rearing, Feeding, And Preparation For A Profitable Market. Also, Their Diseases, And Remedies. Together With Full Directions For The Management Of The Dairy.
New York: Published For The Author, By C.M. Saxton, 1848.
Pp. [1-2], [5]-227, [1], [1-8 (pub. adv.)]; illus. LCP

25. ALLEN, SAMUEL CLESSON, 1772-1842.
An Address Delivered At Northampton, Before The Hampshire, Franklin & Hampden [Mass.] Agricultural Society, October 27, 1830. Published By Request Of The Society.
Northampton: T. Watson Shepard, Printer, 1830.
32 p. LCP (wrapper)

26. ALLEN, STEPHEN MERRIL, 1819-1894.
Fibrilia, Or Flax Cotton: Three Addresses, Before The Legislative Societies Of Massachusetts And Rhode Island [i.e. Mass. Soc. for Promoting Agriculture and the R.I. Soc. for the Encouragement of Domestic Industry], And The Class On Agriculture At Yale College. [wrapper title]
Boston: Wright & Potter, Printers, 1860. [wrapper imprint]
11, [1], 15, [1], 12, 4 p. With sep. t.p. for last 2 sections. Also incl. (last 4 p.): Cultivation Of Flax For Making Fibrilia ... LCP (wrapper)

ALMANACS.

All farmers' or agricultural almanacs, regardless of title, are listed here according to their primary place of imprint, and chronologically within place.

_________. ALBANY.

27. State Of New-York Agricultural Almanack For The Year ... 1824 .... The Agricultural Selections by J. Buel. Published Under The Patronage Of The Board of Agriculture.
Albany: Sold by Daniel Steele; Packard & Van Benthuysen, Printers, [1823].
[48] p.; illus. HSP

28. State Of New-York Agricultural Almanack For The Year ... 1825 .... The Agricultural Selections By J. Buel. Published Under The Patronage Of The Board Of Agriculture.
Albany: Sold by Daniel Steele; Packard & Van Benthuysen, Printers, [1824].
[48] p.; illus. HSP

29. Prynne's Almanac For 1841, Adapted To The Use Of The Merchant, Mechanic and Farmer. Calculated For The Meridian Of Albany ... By Arthur Prynne, Esq.
Albany: Published By Erastus H. Pease; Printed By J. Munsell, [1840].
[24] p.; illus. HSP

30. The Cultivator Almanac, For The Year 1845. By Luther Tucker ....
Albany: Published By E. Van Schaack, [1844].
[32] p.; illus. T.p. headed: No. 2. HSP

31. The Cultivator Almanac For The Year 1847. By Luther Tucker ....
Albany: Published By Erastus H. Pease, Bookseller; Stereotyped and Printed by C. Van Benthuysen and Co., [1846].
[32] p.; illus. T.p. headed: Vol. 1.) 1847. (No. 4. HSP

32. The People's Almanac And Farmers' Calendar For 1859 .... By Joel Munsell .... Astronomical Calculations for the Latitude and Longitude of Albany ....
Albany: Munsell & Rowland, [1858].
[36] p.; illus. T.p. headed: Vol. 1. No. 4. LCP

33. The Illustrated Annual Register Of Rural Affairs And Cultivator Almanac, For The Year 1860 .... By J.J. Thomas ....
Albany, N.Y.: Luther Tucker & Son; New York: A.O. Moore & Co., 1860.
Pp. [1]-20, [229]-332, [1-20 (adv.)]; illus. HSP (wrappers)

_________. ALLENTOWN, PA.

34. Der Neue Pennsylvanische Stadt-und Land-Calender, Auf das Jahr ... 1823 .... Mit besonderm Fleiss nach dem Pennsylvanischen und angranzenden Staaten Horizont und Nordhohe berechnet. Zum erstenmal herausgegeben.
Allentown, (Pennsylv.): gedruckt und zu haben bey Heinrich Ebner und Comp., [1822].
[36] p.; illus. HSP in LCP; LCP

35. Der Neue Pennsylvanische Stadt-und Land-Calender, Auf das Jahr ... 1824 .... Zum zweytenmal herausgegeben.
Allentown, (Pennsylv.): gedruckt und zu haben bey Heinrich Ebner und Comp., [1823].
[36] p.; illus. HSP in LCP (lacking last leaf); LCP

36. Der Neue Pennsylvanische Stadt-und Land-Calender, Auf das Jahr ... 1825 .... Zum drittenmal herausgegeben.
Allentown, (Pennsylv.): gedruckt und zu haben bey Heinrich Ebner und Comp., [1824].
[36] p.; illus. HSP in LCP (3rd leaf frag.; illus. cut out); LCP (lacking 1st 3 and last 3 leaves)

37. Der Neue Pennsylvanische Stadt-und Land-Calender, Auf das Jahr ... 1826 .... Zum viertenmal herausgegeben.
Allentown, (Penn.): gedruckt und zu haben bey Heinrich Ebner und Comp., [1825].
[36] p.; illus. HSP in LCP

38. Der Neue Pennsylvanische Stadt-und Land-Calender, Auf das Jahr ... 1827 .... Zum funftenmal herausgegeben.
Allentown, (Penn.): gedruckt und zu haben bey Heinrich Ebner und Comp., [1826].
HSP in LCP: [34] p. only (of 36; lacking last leaf); illus.

39. Der Neue Pennsylvanische Stadt-und Land-Calender, Auf das Jahr ... 1828 .... Zum sechstenmal herausgegeben.
Allentown, (Penn.): gedruckt und zu haben bey Heinrich Ebner und Comp., [1827].
[36] p.; illus. HSP in LCP; LCP (lacking pp. 31-32)

40. Der Neue Pennsylvanische Stadt-und Land-Calender, Auf das Jahr ... 1829 .... Zum siebenten Mal herausgegeben.
Allentown, (Penn.): gedruckt und zu haben bey Heinrich Ebner und Comp., [1828].
[36] p.; illus. LCP

_________. BALTIMORE.

41. The Farmers' Almanack, For The Year ... 1808 .... Calculated For The Meridians Of Maryland, Pennsylvania, Virginia, Kentucky And Tennessee .... By Father Abraham.
Baltimore: Printed And Sold By Warner & Hanna, [1807].
[36] p. HSP

42. The Farmer's Almanack, For The Year ... 1810 .... By Father Abraham.
Baltimore: Printed & Sold By Warner & Hanna, [1809].
[48] p. LCP; HSP

43. The Farmr's [sic] Almanack, For The Year ... 1811 .... By Father Abraham.
Baltimore: Printed & Sold By Warner & Hanna, [1810].
HSP: [46] p. only (of 48; lacking last leaf)

44. The Farmer's Almanac, For The Year ... 1814 .... By Father Abraham.
Baltimore: Printed & Sold By William Warner, [1813].
[48] p. HSP

45. The Farmer's Almanac For The Year ... 1815 .... By Father Abraham. Astronomical Calculations By John Sharp.
Baltimore: Printed & Sold By William Warner, [1814].
[48] p. HSP

46. The Farmer's Almanac For The Year ... 1816 .... By Father Abraham. Astronomical Calculations By John Sharp.
Baltimore: Printed & Sold By William Warner, [1815].
[48] p. HSP

47. The Farmer's Almanac For The Year ... 1818 .... By Father Abraham. Astronomical Calculations By John Sharp.
Baltimore: Printed & Sold By William Warner, [1817].
[36] p. HSP

48. The Farmer's Almanack, For The Year ... 1820 .... By Father Abraham.
Baltimore: Printed & Sold by William Warner, [1819].
[36] p. LCP

49. City And Country Almanac, For The Year 1836 .... Calculated for meridian of Virginia, Maryland and Pennsylvania, By Charles F. Egelmann.
[Baltimore:] Printed By Richard J. Matchett, [1835].
[24] p.; illus. HSP

50. The Farmer's Calendar, For The Year ... 1838 .... Calculated for the Meridian of Virginia, Maryland, and Pennsylvania. By Charles F. Egelmann.
Baltimore: Cushing & Sons, [1837].
[32] p.; illus. HSP

51. The Farmers' Almanac For 1840 .... By Charles F. Egelmann.
Baltimore: Published By Matchett & Neilson; Matchett & Neilson, Printers, [1839].
[30] p.; illus. HSP

52. The Farmer's Calendar, For The Year ... 1840 .... By Charles F. Egelmann.
Baltimore: Cushing & Sons; S. Sands, printer, [1839].
[32] p.; illus. HSP

53. The Farmer's Calendar, For The Year ... 1842 .... By Charles F. Egelmann.
Baltimore: Cushing & Brother; S. Sands, printer, [1841].
[32] p.; illus. HSP

54. The Citizens' And Farmers' Almanac, For The Year ... 1843 .... Carefully calculated for the Horizon of Maryland, Pennsylvania and Virginia .... Original Calculation By Henry Frost .... No. 13.
Baltimore: published and sold by John T. Hanzsche, [1842].
[32] p. LCP

55. The Farmer's Calendar, For The Year ... 1843 .... By Henry Frost ....
Baltimore: Cushing & Brother; S. Sands, printer, [1842].
[32] p.; illus. HSP

56. The Farmer's Calendar, For The Year ... 1844 .... By Henry Frost ....
Baltimore: Cushing & Brother; S. Sands, printer, [1843].
[32] p.; illus. HSP

57. The American Farmer's Almanac. 1846.
Baltimore: Published By J. B. Keller, [1845].
36 p.; illus. Fuller title p. 3: ... Original Calculations by Edward Hagerty .... T.p. headed: For The Middle And Western States. LCP

58. Farmer's Almanac, For 1846 .... Carefully calculated for the meridians of Maryland, Virginia, Delaware, Pennsylvania and Ohio .... By Edward Hagerty ....
Baltimore: Published By Owen, Kurtz & Co., [1845].
[32] p.; illus. HSP

59. The Farmer's Calendar, For The Year ... 1847 .... By Chas. F. Eaglemann.
Baltimore: Cushing & Brother; Printed by Samuel Sands, [1846].
[32] p.; illus. HSP

60. The Farmer's Calendar, For The Year ... 1848 .... By Chas. F. Eaglemann ....
Baltimore: Cushing & Brother; Printed by Samuel Sands, [1847].
[32] p.; illus.

61. [The American Farmer's Almanac 1854.]
[Philadelphia, New York, Boston and Baltimore: Published annually, by Fisher & Brother (p. 3)], [1853].
HSP: [32] p. only (of 36; lacking t.p. and last leaf). Fuller title p. 3 begins: Fisher & Brother's American Farmer's Almanac ....

62. The American Farmer's Almanac 1855.
Baltimore: J.C. Hines & Co., [1854].
[36] p.; illus. Fuller title p. 3: Fisher & Brother's American Farmer's Almanac .... At bottom of p. 3: Published annually, by Fisher & Brother, Philadelphia, New York, Boston and Baltimore. HSP

63. The Farmer's Calendar For The Year ... 1855 .... By Chas. F. Eaglemann ....
Baltimore: John Cushing & Co; Sands & Mills, Printers, [1854].
[32] p.; illus. HSP

64. The American Farmer's Almanac 1856.
Baltimore: J.C. Hines & Co. Booksellers & Stationers, [1855].
[36] p.; illus. Fuller title p. 3 begins: Fisher & Brother's American Farmer's Almanac ....; at bottom of p. 3: Published annually, by Fisher & Brother, Philadelphia, New York, Boston and Baltimore. LCP

__________. BARNSTABLE, MASS.
See The (Old) Farmer's Almanack ... 1848; 1852; Boston. Wrappers have Barnstable imprint.

__________. BLAIRSVILLE, PA.

65. The Agricultural Almanack, For The Year ... 1828 .... Calculated By John Armstrong.
Blairsville, Pa.: Published And Sold By James F. M'Carty; Zanesville, Ohio: William Davis, [1827].
[36] p. HSP

__________. BOSTON.

66. The Farmer's Almanack, Calculated On A New And Improved Plan, For the Year ... 1800 .... Fitted to the Town of Boston .... By Robert B. Thomas.
Boston: Printed by Manning & Loring, For John West, Proprietor of the Copy-Right, And for Sale at his Bookstore ...; Sold also by the Booksellers in Salem, Newburyport ..., [1799].
HSP: [42] p. only (of 48; lacking pp. 31-32, 41-42, and last leaf); illus. T.p. headed: No. VIII.

67. Farmer's Almanack ... For the Year .... 1801 .... By Robert B. Thomas.
Boston: Printed by Manning & Loring, For John West ..., [1800].
HSP: [46] p. only (of 48; lacking pp. 35-36); illus. T.p. headed: No. IX.

68. The Farmer's Almanack ... For The Year ... 1804 .... By Robert B. Thomas.
Boston: Printed for John West ..., [1803].
[48] p.; illus. T.p. headed: No. XII. HSP in LCP

69. The Farmer's Almanack ... For The Year ... 1805 .... By Robert B. Thomas.
Boston: Printed for John West ...; Munroe & Francis, Printers, [1804].
[42] p. Drake 3662 gives various paginations. HSP

70. The Farmer's Almanack ... For The Year ... 1806 .... By Robert B. Thomas.
Boston: Printed for John West ...; David Carlisle, Printer, [1805].
LCP: [42] p. only (of 48; lacking first 2 leaves)

71. The Farmer's Almanack ... For The Year ... 1807 .... By Robert B. Thomas.
Boston: Printed for John West ...; David Carlisle, Printer, [1806].
[48] p.; illus. T.p. headed: No. XV. LCP

72. The Farmer's Almanack ... For The Year ... 1816 .... By Robert B. Thomas.
Boston: Printed For West And Richardson ...; T.W. White, Printer, [1815].
[48] p.; illus. T.p. headed: Number XXIV. HSP

73. The Farmer's Almanack ... For The Year ... 1817 .... By Robert B. Thomas.
Boston: Printed For West And Richardson ...; T.W. White, Printer, [1816].
[48] p.; illus. T.p. headed: Number XXV. LCP

74. The Farmer's Almanack ... For The Year ... 1819 .... By Robert B. Thomas.
Boston: Printed By J.H.A. Frost For West, Richardson & Lord ..., [1818].
[48] p.; illus. T.p. headed: Number XXVII. HSP

75. The Farmer's Almanack ... For The Year ... 1820 .... By Robert B. Thomas.
Boston: Printed By J.H.A. Frost, For West, Richardson & Lord ..., [1819].
[48] p.; illus. T.p. headed: Number XXVIII. HSP

76. The Farmer's Almanack ... For The Year ... 1821 .... By Robert B. Thomas.
Boston: Printed By J.H.A. Frost, For Richardson & Lord ..., [1820].
[48] p.; illus. T.p. headed: Number XXIX. HSP

77. The Farmer's Almanack ... For The Year ... 1822 .... By Robert B. Thomas.
Boston: Printed By J.H.A. Frost, For Richardson & Lord ..., [1821].
HSP: [46] p. only (of 48; lacking last leaf); illus. T.p. headed: Number XXX.

78. The Farmer's Almanack ... For The Year ... 1824 .... By Robert B. Thomas.
Boston: Printed By J.H.A. Frost, For Richardson & Lord ..., [1823].
[48] p.; illus. T.p. headed: Number XXXII. HSP

79. The Farmer's Almanack ... For The Year ... 1826 .... By Robert B. Thomas.
Boston: Published By Richardson & Lord ...; J.H.A. Frost, Printer, [1825].
[48] p.; illus. T.p. headed: Number XXXIV. HSP

80. The Farmer's Almanack ... For The Year ... 1827 .... By Robert B. Thomas.
Boston: Published By Richardson & Lord ...; J.H.A. Frost, Printer, [1826].
[48] p.; illus. T.p. headed: Number XXXV. HSP

81. The Farmer's Almanack ... For The Year ... 1828 .... By Robert B. Thomas.
Boston: Published By Richardson And Lord ...; J.H.A. Frost, Printer, [1827].
[52] p.; illus. Last 4 p. pub. adv. T.p. headed: Number XXXVI. HSP

82. The New England Farmer's Almanack, For The Year ... 1828. By Thomas Green Fessenden, Editor of the New England Farmer.
Boston: Published By John B. Russell, (Proprietor of the copy-right); And Sold ... By Bowles & Dearborn, Booksellers and Stationers; Hartford, Con.: Sold also ... by O.D. Cooke & Son; Brattleborough, Vt.; Concord, N.H.: Isaac Hill, et al.; John B. Russell, Printer, [1827].
[40] p. LCP

83. The Farmer's Almanack ... For The Year ... 1829 .... By Robert B. Thomas.
Boston: Published By Richardson And Lord ...., [1828].
[48] p.; illus. T.p. headed: Number XXXVII. HSP

84. The New England Almanack, For The Year ... 1829. By Thomas G. Fessenden ....
Boston: Published By John B. Russell ...; And sold ... By Bowles & Dearborn, et al., [1828].
44, [6 (adv.)] p. LCP

85. The Farmer's Almanack, ... For The Year ... 1830 .... By Robert B. Thomas.
Boston: Published By Richardson, Lord And Holbrook ...; Lyman Thurston & Co. Stereotypers, [1829].
LCP: [46] p. only (of 48; lacking pp. 41-42); illus. T.p. headed: Number XLV.

86. The New England Farmer's Almanac. For The Year ... 1832. By Thomas G. Fessenden ....
Boston: Published By Carter, Hendee & Babcock ...; Sold also by John B. Russell, et al.; I.R. Butts, Printer, [1831].
45, [3 (adv.)], [12 (pub. adv.)] p. T.p. headed: No. V. LCP

87. The (Old) Farmer's Almanack, Calculated On A New And Improved Plan, For The Year ... 1832 .... By Robert B. Thomas.
Boston: Richardson, Lord & Holbrook ..., [1831].
[60] p.; illus. Last 12 p. pub. adv. T.p. headed: Number XL. HSP

88. The New England Farmer's Almanac. For the Year ... 1833. By Thomas G. Fessenden.
Boston: Published by Carter, Hendee & Co.; Sold Also By John B. Russell, et al.; I.R. Butts, Printer, [1832].
45, [7] p. T.p. headed: No. VI. LCP

89. The (Old) Farmer's Almanack ... For The Year ... 1833 .... By Robert B. Thomas.
Boston: Carter, Hendee And Co. ..., [1832].
[56] p.; illus. Last 8 p. pub. adv. T.p. headed: Number XLI. HSP

90. The (Old) Farmer's Almanack ... For The Year ... 1834 .... By Robert B. Thomas.
Boston: Published And Sold By Carter, Hendee And Co. ..., [1833].
HSP: 46 (of 48; lacking pp. 47-48), 12, 7 [i.e. 12; pp. 6-12 misnumbered (pub. adv.)] p.; illus. T.p. headed: Number XLII. Incl. (12-p.section): Intemperance Illustrated And Exemplified. HSP

91. The (Old) Farmer's Almanack ... For The Year ... 1835 .... By Robert B. Thomas.
Boston: Published And Sold By Carter, Hendee And Co. ...; Print of Kane & Co., [1834].
[48] p.; illus. T.p. headed: Number XLIII. HSP

92. The Farmer's Almanack ... For The Year ... 1836 .... By Robert B. Thomas.
Boston: Published And Sold By Carter, Hendee And Co. ..., [1835].
[48] p.; illus. T.p. headed: Number XLIV. HSP

93. The Farmer's Almanack ... For The Year ... 1837 .... By Robert B. Thomas.
Boston: Published And Sold By Charles J. Hendee ..., [1836].
[48] p.; illus. T.p. headed: XLV. HSP

94. The Farmer's Almanack ... For The Year ... 1839 .... By Robert B. Thomas.
Boston: Published And Sold By G.W. Palmer & Co. ..., [1838].
[48] p.; illus. T.p. headed: Number XLVII. LCP; HSP

95. The Farmer's Almanack ... For The Year ... 1840 .... By Robert B. Thomas.
Boston: Published And Sold By Jenks & Palmer ..., [1839].
[48] p.; illus. T.p. headed: Number XLVIII. LCP; HSP

96. The Farmer's Almanack ... For The Year ... 1841 .... By Robert B. Thomas.
Boston: Published And Sold By Jenks & Palmer ..., [1840].
[48] p.; illus. T.p. headed: Number XLIX. LCP (wrapper); HSP (wrappers)

97. The Farmer's Almanack ... For The Year ... 1842 .... By Robert B. Thomas.
Boston: Published And Sold By Jenks & Palmer ..., [1841].
48 p.; illus. T.p. headed: Number L. Wrapper title: The (Old) Farmers Almanac ...; wrapper imprint: Worcester: Spooner & Howland, Plain And Ornamental Job Printers. HSP (wrappers)

98. The Farmer's Almanack ... For The Year ... 1843 .... By Robert B. Thomas.
Boston: Published And Sold By Jenks & Palmer ..., [1842].
48 p.; illus. T.p. headed: Number LI. HSP

99. The Farmer's Almanack ... For The Year ... 1846 .... By Robert B. Thomas.
Boston: Published And Sold By Jenks & Palmer ..., [1845].
48 p.; illus. T.p. headed: Number LIV. LCP; HSP

100. The (Old) Farmer's Almanack ... For The Year ... 1848 .... By Robert B. Thomas.
Boston: Published And Sold By Jenks, Palmer & Co. ..., [1847].
48 p.; illus. T.p. headed: Number Fifty-Six. Wrapper title: The Farmer's Almanac ....; wrapper imprint: Barnstable: Sold by O.M. Hinckley, Bookseller & Apothecary. HSP (wrappers)

101. The (Old) Farmer's Almanack ... For The Year ... 1849 .... By Robert B. Thomas.
Boston: Published And Sold By Jenks, Palmer & Co. ..., [1848].
48 p.; illus. T.p. headed: Number Fifty-Seven. Wrapper title: The Farmers' Almanac ....; wrapper imprint: Benjamin B. Mussey & Co., Booksellers, Publishers, And Stationers; Samuel N. Dickinson & Co., Printers. HSP (wrappers)

102. The (Old) Farmer's Almanack ... For The Year ... 1850 .... By Robert B. Thomas.
Boston: Published And Sold By Jenks, Palmer & Co. ..., [1849].
48 p.; illus. T.p. headed: Number Fifty-Eight. Wrapper imprint: Sold By William J. Reynolds And Co., Boston. LCP (wrappers)

103. The (Old) Farmer's Almanack ... For The Year ... 1851 .... By Robert B. Thomas.
Boston: Published And Sold By Jenks, Palmer & Co. ..., [1850].
48 p.; illus. T.p. headed: Number Fifty-Nine. Wrapper imprint: Providence, R.I.: Mather & Burr. HSP (wrappers)

104. The (Old) Farmer's Almanack ... For The Year ... 1852 .... By Robert B. Thomas.
Boston: Published By Jenks, Hickling & Swan ..., [1851].
48 p.; illus. T.p. headed: Number Sixty. Wrapper imprint: Barnstable, Mass.: Sold By O.M. Hinckley. HSP (wrappers)

105. The (Old) Farmer's Almanack ... For The Year ... 1853 .... By Robert B. Thomas.
Boston: Published By Jenks, Hickling & Swan ..., [1852].
48 p.; illus. T.p. headed: Number Sixty-One. Wrapper imprint: Worcester: Sold By John Keith & Co. (Successors To J. Grout,) Booksellers, Stationers, And Blank Book Manufacturers. LCP (wrappers); HSP (wrappers)

106. The (Old) Farmer's Almanack ... For The Year ... 1854 .... By Robert B. Thomas.
Boston: Published By Jenks, Hickling & Swan ..., [1853].
48 p.; illus. T.p. headed: Number Sixty-Two. HSP

107. The (Old) Farmer's Almanack ... For The Year ... 1855 .... By Robert B. Thomas.
Boston: Published By Jenks, Hickling & Swan ..., [1854].
48 p.; illus. T.p. headed: Number Sixty-Three. Wrapper imprint: Boston: B.B. Mussey & Company. HSP (wrappers)

108. The (Old) Farmer's Almanack ... For The Year ... 1856 .... By Robert B. Thomas.
Boston: Published By Hickling, Swan & Brown, [1855].
48 p.; illus. T.p. headed: Number Sixty-Four. HSP (wrappers)

109. The (Old) Farmer's Almanack ... For The Year ... 1857 ....
Boston: Published By Hickling, Swan & Brown ... [Stereotyped By Hobart And Robbins (colop.)], [1856].
46, [2 (pub. adv.)] p.; illus. T.p. headed: Number Sixty-Five. LCP; HSP (wrappers)

110. The (Old) Farmer's Almanack ... For The Year ... 1858 .... By Robert B. Thomas.
Boston: Published By Hickling, Swan & Brewer ..., [1857].
48 p.; illus. T.p. headed: Number Sixty-Six. HSP (wrappers)

111. The (Old) Farmer's Almanack, Calculated On A New And Improved Plan, For The Year ... 1859 .... Fitted For Boston .... By Robert B. Thomas.
Boston: Published By Hickling, Swan & Brewer, [1858].
48 p.; illus. T.p. headed: No. 67. Wrapper imprint: New Bedford: Chas. Taber & Co. With adv. leaves at front, back, and interleaved. HSP (wrappers)

112. The (Old) Farmer's Almanack ... For The Year ... 1860 .... By Robert B. Thomas.
Boston: Published By Hickling, Swan & Brewer ..., [1859].
46, [2 (pub. adv.)] p.; illus. T.p. headed: Number Sixty-Eight. Wrapper imprint: Boston: Crosby, Nichols & Company. HSP (wrappers)

113. The (Old) Farmer's Almanack ... For The Year ... 1861 .... By Robert B. Thomas.
Boston: Published By Swan, Brewer & Tileston ..., [1860].
48, [4 (pub. adv.)] p.; illus. T.p. headed: Number Sixty-Nine. Wrapper imprint: Chase, Nichols & Hill. HSP (wrappers)

__________. CANANDAIGUA, N.Y.

114. The Farmer's Diary, Or Western Almanac, For The Year 1836 .... Calculated for the Meridian of Canandaigua, and will serve for the western part of New-York and Pennsylvania ....
Canandaigua [N.Y.]: Printed And Sold By Morse & Harvey, [1835].
[24] p.; illus. LCP

__________. CHARLESTON.

115. Hoff's Agricultural and Commercial Almanac, Calculated For The States Of Georgia And The Carolinas; For the year ... 1811 .... Calculated By Joshua Sharp.
Charleston: Printed by J. Hoff, and sold at his ... Bookstore; Sold also by W. Aikin, S. & G. Robertson, T. Flemming, W. Walton & Co., et al., [1810].
[48] p. HSP

116. Bason's Country Almanack, For 1821 .... Calculated for North and South Carolina, By Joshua Sharp.
Charleston: Published By W.P. Bason; And Sold by Thomas Fleming & Co.; Boyce & Johnson; Goodman & Weyman; C.S. Simonton & Co.; et al., [1820].
[36] p. on 24 leaves. HSP

117. Bason's Country Almanack, For 1822 .... By Joshua Sharp.
Charleston: Published By W.P. Bason; And Sold by Joseph S. & Joseph Fleming; Joseph T. Weyman; Ker Boyce; Samuel M'Cleary, et al., [1821].
[36] p. on 36 leaves. HSP

118. Miller's Planters' & Merchants' Almanac, For The Year ... 1847 .... Calculated by David Young, for the States of Carolina & Georgia ....
Charleston, S.C.: Published and Sold ... by A.E. Miller; Printed by Miller & Brown, [1846].
[48] p. T.p. headed: (Third Edition.) HSP

119. Miller's Planters' & Merchants' Almanac, For The Year ... 1859 .... The Astronomical Calculations by Samuel H. Wright ... For The States Of Carolina And Georgia ....
Charleston, S.C.: Published and Sold ... by A.E. Miller; Printed by A.E. Miller, [1858].
[60] p. T.p. headed: Third Edition. HSP

__________. CHESTERFIELD, N.H.

120. The Farmer's, Mechanic's And Gentleman's Almanack, And Farmer's Calendar .... Fitted To The Latitude And Longitude Of The City Of Boston .... By Nathan Wild.
Chesterfield, N.H.: Published By The Author; J. Metcalf, Printer - Wendell, Mass., [1825].
[48] p. LCP

__________. CLAREMONT, N.H.

121. The New-England Farmer's Almanac, For The Year ... 1845 .... By Truman Abell.
Claremont, N.H.: Published and sold at the Claremont Bookstore; Lempster: Sold also ... by the Author; Windsor: Preston Merrifield, et al., [1844].
[48] p.; illus. T.p. headed: No. XXXI. HSP

122. The New-England Farmer's Almanac, For The Year ... 1848 .... By Truman Abell.
Claremont, N.H.: Published and sold by the Claremont Manufacturing Company; Lempster: Sold also ... by the Author; Windsor: P. Merrifield, et al., [1847].
[48] p.; illus. T.p. headed: No. XXXIV. HSP

__________. CONCORD, N.H.

123. Farmer's And Scholar's Almanack, For The Year ... 1830 .... Fitted as to the latitude and longitude, so as to answer for all the New-England States. By Dudley Leavitt ....
Concord: Published And Sold By Jacob B. Moore, [1829].
[48] p. T.p. headed: Leavitt's. HSP

124. The New-England Farmer's And Scholar's Almanack, On An Improved Plan. For The Year ... 1833 .... By Dudley Leavitt ....
Concord: Published And Sold By Horatio Hill And Co. ...; Hill And Barton, Printer, [1832].
[48] p.; illus. T.p. headed: Leavitt's. HSP

125. The New-England Almanack ... For The Year 1840 .... By Dudley Leavitt ....
Concord, N.H.: Published By Marsh, Capen, Lyon, And Webb ..., [1839].
40, [8] p.; illus. T.p. headed: Number XLIV. HSP

126. The New England Almanack ... For The Year 1844 .... By Dudley Leavitt ....
Concord, N.H.: Published By John F. Brown ..., [1843].
48 p.; illus. T.p. headed: Number XLVIII. HSP

127. The New-England Farmer's Almanack ... For The Year ... 1847 .... By Dudley Leavitt ....
Concord, N.H.: Published By John F. Brown ..., [1846].
48 p.; illus. T.p. headed: No. LI. HSP

__________. DANBURY, CT.

128. The Farmer's Diary: Or The United States Almanack, For The Year ... 1792 .... Calculated for the Meridian of Danbury, in the State of Connecticut .... By Joseph Leland, Philom.
Danbury: Published By Douglas And Ely, [1791].
HSP in LCP: [34] p. only (of 36; lacking last leaf)

129. The Farmer's Diary: Or The United States Almanack, For The Year ... 1793 .... By Joseph Leland, Philom.
Danbury: Published By Douglas & Ely, [1792].
HSP in LCP: [22] p. only (of 24; lacking pp. 21-22)

130. The Farmer's American Almanac, For the Year ... 1796 ... Calculated for the Meridian of Danbury ... but may serve indifferently for all the adjacent States .... By Andrew Beers, Philom.
Danbury: Printed and Sold, by Ely & Nichol, [1795].
[24] p. HSP in LCP

131. The Farmer's American Almanac, For the Year ... 1798 .... By Andrew Beers, Philom.
Danbury: Printed and Sold ... By Douglas & Nichols, [1797].
34, [2] p. HSP in LCP

132. The Farmer's American Almanac, for the Year ... 1799 .... Calculated for the Meridian of Danbury ... but may serve for the adjacent States .... By Andrew Beers, Philom.
Danbury: Printed & sold ... By Douglas & Nichols, [1798].
HSP in LCP: [12] p. only (of 36; lacking last 12 leaves)

__________. DOYLESTOWN, PA.

133. Miner's Agricultural And Miscellaneous Almanac, For Pennsylvania & New Jersey, For The Year ... 1817 .... The Calculations carefully made by William Collom ....
Doylestown: Printed by Asher Miner, [1816].
[36] p. LCP (lacking last leaf); HSP

134. Miner's Agricultural Almanac, For The Year ... 1818 .... Carefully Calculated By William Collom.
Doylestown: Printed by Asher Miner, [1817].
[36] p. HSP

135. Miner's Agricultural Almanac, For The Year ... 1820 .... Carefully calculated for the Latitude and Meridian of Philadelphia - by William Collom.
Doylestown, Penn.: Printed by Asher Miner, [1819].
[36] p. HSP

136. Miner's Agricultural Almanac, For The Year ... 1821 .... Carefully calculated ... by William Collom.
Doylestown, Penn'a: Printed by Asher Miner, [1820].
[36] p. LCP; HSP

137. Miner's Agricultural Almanac, For The Year ... 1824 .... Carefully calculated ... by Wm. Collom ....
Doylestown, Penn'a: Printed by Asher Miner, [1823].
[36] p. HSP

138. Miner's Agricultural Almanac, For The Year ... 1824 .... Carefully calculated ... by Wm. Collom ....
Doylestown, Penn'a: Printed by Asher Miner, [1823].
HSP: [34] p. only (of 36; lacking last leaf)

__________. EASTON, PA.

139. The Northampton Farmer's Almanac For the Year ... 1824 .... Calculated for the Meredian (sic) of Pennsylvania & New Jersey.
Easton: Printed by Henry & William Hutter, [1823].
[36] p. LCP; HSP (lacking pp. 31-34)

140. The Northampton Farmer's Almanac For the year ... 1825 ....
Easton: Printed by Henry And William Hutter, [1824].
[36] p. HSP in LCP; LCP; HSP

141. The Northampton Farmer's Almanac For the year ... 1826 .... Second Edition.
Easton: Printed by Henry and William Hutter, [1825].
[48] p. LCP

142. The Northampton Farmers' Almanac For The Year ... 1827 ....
Easton, Pennsylvania: Printed by Henry and William Hutter, [1826].
[30] p.; Drake 11557 also has copies with 17 and 18 leaves. HSP

143. The Farmers And Mechanicks Almanack, For The Year ... 1835 .... By Charles Frederick Egelmann.
Easton, Pa.: Published By G.W. Mentz & Son, For Henry Hammann, [1834].
34, [2] p.; illus. LCP

__________. ERIE, PA.

144. The Cultivator Almanac, For The Year 1846. By Luther Tucker ....
Erie, Pa.: Published By T. Moorhead, Jr., Agent for the Cultivator, [1845].
[32] p.; illus. T.p. headed: No. 3. HSP

__________. GALVESTON, TEXAS.

145. The Texas Almanac, For 1857, With Statistics, Historical and Biographical Sketches, &c., Relating To Texas.
Galveston: Prepared, Printed And Published By Richardson & Co., At The News Office, 1856.
159 p. LCP

146. The Texas Almanac, For 1858; Giving The Annual Statistics Of The State ....
Galveston: Prepared, Printed and Published by Richardson & Co., 1857. 192, [122 (adv.)] p.; 5 pls. LCP

__________. HAGERSTOWN, MD.

147. Der neue Nord-Americanische Stadt und Land Calender, Auf das Jahr Christi 1797 .... Nach dem Marylandischen Horizont und Nordhöhe berechnet .... Zum Erstenmal herausgegeben.
Hagerstown: gedruckt und zu haben bey Johann Gruber, [1796].
LCP: [28] p. incl. wrapper; lacking 2 [?] leaves at end.

148. Der neue Nord-Americanische Stadt und Land Calender, Auf das Jahr Christi 1798 .... Zum Zweytenmal herausgegeben.
Hagerstown: gedruckt und zu haben bey Johann Gruber, [1797].
[32] p. incl. wrapper. LCP (wrapper)

149. Der neue, Nord-Americanische Stadt und Land Calender, Auf das Jahr Christi 1799 .... Zum Drittenmal herausgegeben.
Hagerstown: gedruckt und zu haben bey Johann Gruber, [1798].
HSP in LCP: [34] p. only (of 36 incl. wrapper; lacking wrapper).

150. Der neue Nord-Americanische Stadt und Land Calender, Auf das Jahr Christi 1802 .... Zum Sechstenmal herausgegeben.
Hagerstown: gedruckt und zu haben bey Johann Gruber, [1801].
HSP in LCP: [40] p. only (of 42 incl. wrapper; lacking last leaf)

151. Der neue Nord-Americanische Stadt und Land Calender, Auf das Jahr ... 1805 ....
Hagerstown: gedruckt fur Jacob D. Dietrich's Bücher-Stohr, [1804].
[42] p. incl. wrapper. HSP in LCP (wrapper); LCP (lacking wrapper)

152. Der neue Nord-Americanische Stadt und Land Calender, Auf das Jahr ... 1806 .... Zum zehntenmal herausgegeben.
Hagerstown: Gedruckt fur Jacob D. Dietrich's Bücher-Stohr, [1805].
HSP in LCP: [40] p. only (of 42 incl. wrapper; lacking last leaf)

153. Der neue Nord-Americanische Stadt und Land Calender, Auf das Jahr ... 1807 .... Zum eilftenmal [sic] herausgegeben.
Hagerstown, Washington County, Maryland: gedruckt und zu haben bey Johann Gruber, [1806].
[42] p. incl. wrapper. LCP (wrapper)

154. Der neue Nord-Americanische Stadt und Land Calender, Auf das Jahr ... 1807 .... Zum funftenmal herausgegeben.
Hagerstown: gedruckt und zu haben bey Jacob D. Dietrich, [1806].
[48] p. incl. wrapper. HSP in LCP (wrapper)

155. Der neue Nord-Americanische Stadt und Land Calender, Auf das Jahr ... 1808 .... Zum zwölftenmal herausgegeben.
Hagerstown, Washington County, Maryland: gedruckt und zu haben bey Johann Gruber, [1807].
[42] p. incl. wrapper. HSP in LCP (wrapper)

156. Der neue Nord-Americanische Stadt und Land Calender, Auf das Jahr ... 1809 .... Zum dreyzehntenmal herausgegeben.
Hagerstown, Washington County, Maryland: gedruckt und zu haben bey Johann Gruber, [1808].
[42] p. incl. wrapper. HSP in LCP (wrapper); LCP (wrapper)

157. Der neue Nord-Americanische Stadt und Land Calender, Auf das Jahr ... 1810 .... Zum vierzehntenmal herausgegeben.
Hagerstown, Washington County, Maryland: gedruckt und zu haben bey Johann Gruber, [1809].
HSP in LCP: [40] p. only (of 42 incl. wrapper; lacking pp. 33-34)

158. Der neue Nord-Americanische Stadt und Land Calender, Auf das Jahr ... 1811 .... Zum fünfzehntenmal herausgegeben.
Hagerstown, Washington County, Maryland: gedruckt und zu haben bey Johann Gruber, [1810].
[42] p. incl. wrapper; illus. HSP in LCP (wrappper)

159. Der neue Nord-Americanische Stadt und Land Calender, Auf das Jahr ... 1814 .... Zum achtzehntenmal herausgegeben.
Hagerstown, Washington County, Maryland: gedruckt und zu haben bey Johann Gruber, [1813].
[42] p. incl. wrapper. HSP in LCP (wrapper)

160. Der neue Nord-Americanische Stadt und Land Calender, Auf das Jahr ... 1815 .... Zum neunzehntenmal herausgegeben.
Hagerstown, Maryland: gedruckt und zu haben bey J. Gruber und D. May, [1814].
[40] p. incl. wrapper. HSP in LCP (wrapper)

161. The American Farmers' Almanack, For The Year ... 1822 .... Arranged After The System Of The German Calendars .... Calculated for the Meridian of Maryland and the adjacent states ....
Hagerstown: Printed and sold by John Gruber and Daniel May, [1821].
[36] p. incl. wrappers. LCP (wrappers); HSP (wrappers)

162. American Farmers' Almanack, For The Year ... 1823 .... By Charles F. Egelmann.
Hagerstown: Printed and sold by John Gruber and Daniel May, [1822].
[36] p. incl. wrappers. T.p. headed: No. II. LCP (wrappers)

163. American Farmers' Almanac, For The Year ... 1826 .... By Charles F. Egelmann.
Hagerstown, Md.: Printed and sold by John Gruber and Daniel May, [1825].
HSP: [34] p. only (of 36 incl. wrappers; lacking last leaf). T.p. headed: No. V.

164. American Farmers' Almanac, For The Year ... 1827 .... By Charles F. Egelmann.
Hagerstown: Printed and sold by John Gruber and Daniel May, [1826].
[28] p. incl. wrappers. T.p. headed: No. VI. Drake 2530 lists most copies with 15 leaves. LCP (wrappers)

165. American Farmers' Almanac, For The Year ... 1828 .... By Charles F. Egelmann.
Hagerstown: Printed and sold by John Gruber and Daniel May, [1827].
[30] p. incl. wrappers. T.p. headed: No. VII. LCP (wrappers)

166. American Farmers' Almanac, For The Year ... 1830 .... By Charles F. Egelmann.
Hagerstown, Md.: Printed and sold by John Gruber and Daniel May, [1829].
[44] p. incl. wrapper; illus. T.p. headed: No. IX. Drake 2572 has 13, 15, or 16 leaves. HSP (wrappers)

167. The American Farmers' Almanack, For The Year ... 1833 .... By Charles F. Egelmann.
Hagerstown, Md.: Printed and sold by J. Gruber, [1832].
HSP: [28] p. only (of 30 incl. wrapper; lacking last leaf). T.p. headed: No. XII.

168. The Hagerstown Town and Country Almanack, For The Year ... 1843 .... Carefully calculated for the Horizon of Maryland .... By Charles F. Egelmann ....
Hagerstown, Md.: Printed and sold by John Gruber, [1842].
[30] p. incl. wrappers; illus. T.p. headed: No. XXII. LCP (wrappers); HSP (wrappers)

169. The Hagerstown Town and Country Almanack, For The Year ... 1846 .... By Charles F. Egelmann ....
Hagerstown, Md.: Printed and sold by John Gruber, [1845].
HSP: [28] p. (of 30 incl. wrapper; lacking last leaf); illus. T.p. headed: No. XXV.

170. The Hagerstown Town and Country Almanack, For The Year ... 1849 .... By Charles F. Egelmann ....
Hagerstown, Md.: Printed and sold by John Gruber, [1848].
[30] p. incl. wrappers; illus. T.p. headed: No. XXVIII. HSP (wrappers)

171. The Hagers-City Town and Country Almanack, For The Year ... 1850 .... By Charles F. Eagelmann [sic] ....
Hagerstown, Md.: William Stewart's Edition; By permission of John Gruber, inventor hereof, [1849].
[32] p. incl. wrappers; illus. T.p. headed: No. XXIX. HSP (wrappers)

172. The Hagerstown Town and Country Almanack, For The Year ... 1853 .... Calculations By Charles F. Egelmann ....
Hagerstown, Md.: Printed and sold by John Gruber, [1852].
[30] p. incl. wrappers; illus. T.p. headed: No. XXXII. HSP (wrappers)

173. The Hagerstown Town and Country Almanack, For The Year ... 1857 .... Calculation By Charles F. Egelmann ....
Hagerstown, Md.: Printed and sold by John Gruber; T.G. Robertson, Bookseller and Stationer, [1856].
[32] p. incl. wrappers; illus. T.p. headed: No. XXXVI. HSP (wrappers)

174. The Hagerstown Town and Country Almanack, For The Year ... 1858 .... By Charles F. Egelmann ....
Hagerstown, Md.: T.G. Robertson, Bookseller and Stationer, [1857].
[32] p. incl. wrappers; illus. T.p. headed: No. XXXVII. HSP (wrappers)

175. The Hagerstown Town and Country Almanack, For The Year ... 1859 .... By Charles F. Egelmann ....
Hagerstown, Md.: Printed and sold by John Gruber; T.G. Robertson, Bookseller and Stationer, [1858].
[32] p. incl. wrappers; illus. T.p. headed: No. XXXVIII. HSP (wrappers)

176. The Hagerstown Town and Country Almanack, For The Year ... 1861 .... By Charles F. Egelmann ....
Hagerstown, Md.: Printed and sold by John Gruber; T.G. Robertson, Bookseller and Stationer, [1860].
[32] p. incl. wrappers; illus. T.p. headed: No. XL. HSP (wrappers)

__________. HARRISBURG, PA.

177. The Pennsylvania Almanac, And Rural Economist's Assistant, For The Year ... 1823 .... Carefully calculated for the Meridian of Pennsylvania and the adjoining States, by Charles F. Egelman.
Harrisburg: Printed And Sold By John Wyeth, J.S. Wiestling And C. Gleim, [1822].
[36] p. incl. wrappers. HSP (wrappers)

178. The Pennsylvania Almanac, And Rural Economist's Assistant, For The Year ... 1824 .... by Charles F. Egelman.
Harrisburg: Printed And Sold By John Wyeth, J.S. Wiestling And C. Gleim, [1823].
[36] p. incl. wrappers. HSP (wrappers)

__________. KNOXVILLE, TENN.

179. Farmer And Cottage Gardener's Almanac. For The Year 1858.
Knoxville, Tennessee: J.A. Rayl, Bookseller & Stationer, [1857].
33, [3] p.; illus. HSP

__________. LANCASTER, PA.

180. Der Neue, Gemeinnutzige Landwirthschafts Calender, Auf das Jahr ... 1789 .... Mit sonderbarem Fleiss nach dem Pennsylvanischen und der angrenzenden Staaten Horizont und Nordhohe berechnet. Zum Zweytenmal herausgegeben.
Lancaster: Gedruckt und zu haben bey Albrecht und Lahn, [1788].
[44] p. incl. wrappers. HSP in LCP (wrappers)

181. Der Neue, Gemeinnützige Landwirthschafts Calender, Auf das Jahr ... 1790 .... Zum Drittenmal herausgegeben.
Lancaster: Gedruckt und zu haben bey Albrecht und Lahr, [1789].
HSP in LCP; LCP: [42] p. only (of 44 incl. wrapper; lacking wrapper)

182. Der Neue, Gemeinnützige Landwirthschafts Calender, Auf das Jahr ... 1791 .... Zum Viertenmal herausgegeben.
Lancaster: Gedruckt und zu haben bey Johann Albrecht und Comp., [1790].
HSP in LCP: [40] p. only (of 44 incl. wrapper; lacking wrapper and last leaf).

183. Der Neue, Gemeinnützige Landwirthschafts Calender, Auf das Jahr ... 1792 .... Zum Fünftenmal herausgegeben.
Lancaster: Gedruckt und zu haben bey Johann Albrecht und Comp., [1791].
HSP in LCP: [42] p. only (of 44 incl. wrapper; lacking wrapper).

184. Der Neue, Gemeinnützige Landwirthschafts Calender, Auf das Jahr ... 1793 .... Zum Sechstenmal herausgegeben.
Lancaster: Gedruckt und zu haben bey Johann Albrecht und Comp., [1792].
[44] p. incl. wrapper; illus. HSP in LCP (wrapper)

185. Der Neue, Gemeinnützige Landwirthschafts Calender, Auf das Jahr ... 1794 .... Zum Siebentenmal herausgegeben.
Lancaster: Gedruckt und zu haben bey Johann Albrecht und Comp., [1793].
[44] p. incl. wrapper; illus. HSP in LCP (lacking 1st 3 leaves and last leaf); LCP

186. Der Neue Gemeinnützige Landwirthschafts Calender, Auf das Jahr ... 1796 .... Zum Neuntenmal herausgegeben.
Lancaster: Gedruckt und zu haben bey Johann Albrecht und Comp., [1795].
[44] p. incl. wrapper. HSP in LCP (wrapper); LCP (wrapper)

187. Der Neue, Gemeinnützige Landwirthschafts Calender, Auf das Jahr ... 1797 .... Zum Zehntenmal herausgegeben.
Lancaster: Gedruckt und zu haben bey Johann Albrecht und Comp., [1796].
[44] p. incl. wrapper. HSP in LCP; LCP (lacking wrapper and last leaf).

188. Der Neue, Gemeinnützige Landwirthschafts Calender, Auf das Jahr ... 1798 .... Zum Elftenmal herausgegeben.
Lancaster: Gedruckt und zu haben bey Johann Albrecht und Comp., [1797].
HSP in LCP: [42] p. only (of 44 incl. wrapper; lacking last leaf; wrapper frag.)

189. Der Neue, Gemeinnützige Landwirthschafts Calender, Auf das Jahr ... 1799 .... Zum Zwölftenmal herausgegeben.
Lancaster: Gedruckt und zu haben bey Johann Albrecht und Comp., [1798].
[44] p. incl. wrapper. HSP in LCP (wrapper)

190. Der Neue, Gemeinnützige Landwirthschafts Calender, Auf das Jahr ... 1800 .... Zum Dreyzehntenmal herausgegeben.
Lancaster: Gedruckt und zu haben bey Johann Albrecht und Comp., [1799].
[44] p. incl. wrapper. HSP in LCP (wrapper)

191. Der Neue, Gemeinnützige Landwirthschafts Calender, Auf das Jahr ... 1801 .... Zum Vierzehntenmal herausgegeben.
Lancaster: Gedruckt und zu haben bey Johann Albrecht und Comp., [1800].
[44] p. incl. wrapper. HSP in LCP (wrapper)

192. Der Neue, Gemeinnützige Landwirthschafts Calender, Auf das Jahr ... 1802 .... Zum Fünfzehntenmal herausgegeben.
Lancaster: Gedruckt und zu haben bey Johann Albrecht, [1801].
[44] p. incl. wrapper. HSP in LCP (wrapper)

193. Der Neue, Gemeinnützige Landwirthschafts Calender, Auf das Jahr ... 1803 .... Zum Sechszehntenmal herausgegeben.
Lancaster: Gedruckt und zu haben bey Johann Albrecht, [1802].
[44] p. incl. wrapper. HSP in LCP (wrapper); LCP (wrapper)

194. Der Neue, Gemeinnützige Landwirthschafts Calender, Auf das Jahr ... 1804 .... Zum Siebenzehntenmal herausgegeben.
Lancaster: Gedruckt und zu haben bey Johann Albrecht, [1803].
[44] p. incl. wrapper. HSP in LCP (wrapper)

195. Der Neue, Gemeinnützige Landwirthschafts Calender, Auf das Jahr ... 1805 .... Zum Achtzehntenmal herausgegeben.
Lancaster: Gedruckt und zu haben bey Johann Albrecht, [1804].
[44] p. incl. wrapper. HSP in LCP (wrapper); LCP (wrapper)

196. Der Neue, Gemeinnützige Landwirthschafts Calender, Auf das Jahr ... 1806 .... Zum Neunzehntenmal herausgegeben.
Lancaster: Gedruckt und zu haben bey Johann Albrecht, [1805].
[44] p. incl. wrapper. HSP in LCP (wrapper)

197. Der Neue, Gemeinnützige Landwirthschafts Calender, Auf das Jahr ... 1807 .... Zum Zwanzigstenmal herausgegeben.
Lancaster: Gedruckt und zu haben bey Johann Albrecht, [1806].
[44] p. incl. wrapper. HSP in LCP (wrapper); LCP (wrapper)

198. Der Neue, Gemeinnützige Landwirthschafts Calender, Auf das Jahr ... 1808 .... Zum Ein und Zwanzigstenmal herausgegeben.
Lancaster: Gedruckt und zu haben bey Georg und Peter Albrecht, [1807].
[40] p. incl. wrapper. HSP in LCP (wrapper); LCP (lacking wrapper and last leaf)

199. Der Neue, Gemeinnützige Landwirthschafts Calender, Auf das Jahr ... 1809 .... Zum Zwey und Zwanzigstenmal herausgegeben.
Lancaster: Gedruckt und zu haben bey Georg und Peter Albrecht, [1808].
[40] p. incl. wrapper. HSP in LCP (wrapper)

200. Der Gemeinnützige Landwirthschafts Calender, Auf das Jahr ... 1810 .... Zum Drey und Zwanzigstenmal herausgegeben.
Lancaster: Gedruckt und zu haben bey Anton Albrecht, [1809].
[40] p. incl. wrapper. HSP in LCP (wrapper)

201. Der Gemeinnützige Landwirthsachfts [sic] Calender, Auf das Jahr ... 1811 .... Zum Vier und Zwanzigstenmal herausgegeben.
Lancaster: Gedruckt und zu haben bey Anton Albrecht, [1810].
[40] p. incl. wrapper; illus. HSP in LCP (wrapper); LCP (lacking wrapper); HSP (lacking wrapper and last 2 leaves)

202. Der Gemeinnützige Landwirthschafts Calender, Auf das Jahr .... 1812 .... Zum Fünf und Zwanzigstenmal herausgegeben.
Lancaster: Gedruckt und zu haben bey Anton Albrecht, [1811].
[40] p. incl. wrapper. HSP in LCP (wrapper); LCP (wrapper)

203. Der Gemeinnützige Landwirthschafts Calender, Auf das Jahr ... 1813 .... Zum Sechs und Zwanzigstenmal herausgegeben.
Lancaster: Gedruckt und zu haben bey Anton Albrecht, [1812].
[40] p. incl. wrapper. HSP in LCP (wrapper)

204. Der Gemeinnützige Landwirthschafts Calender, Auf das Jahr ... 1814 .... Zum Sieben und Zwanzigstenmal herausgegeben.
Lancaster: Gedruckt und zu haben bey Anton Albrecht, [1813].
[40] p. incl. wrapper. HSP in LCP (wrapper); LCP (wrapper)

205. Der Gemeinnützige Landwirthschafts Calender, Auf das Jahr ... 1815 .... Zum Acht und Zwanzigstenmal herausgegeben.
Lancaster: Gedruckt und zu haben bey Anton Albrecht, [1814].
[40] p. incl. wrapper. HSP in LCP (wrapper); LCP (wrapper; lacking last leaf)

206. Der Gemeinnützige Landwirthschafts Calender, Auf das Jahr ... 1816 .... Zum Neun und Zwanzigstenmal herausgegeben.
Lancaster: Gedruckt und zu haben bey Anton Albrecht, [1815].
[40] p. incl. wrapper. HSP in LCP (wrapper); LCP (wrapper).

207. Der Gemeinnützige Landwirthschafts Calender, Auf das Jahr ... 1817 .... Zum Dreyssigstenmal herausgegeben.
Lancaster: Gedruckt und zu haben bey Anton Albrecht, [1816].
[40] p. incl. wrapper. HSP in LCP (wrapper)

208. Der Gemeinnützige Landwirthschafts Calender, Auf das Jahr ... 1818 .... Zum 31sigstenmal herausgegeben.
Lancaster: Gedruckt und zu haben bey Anton Albrecht, [1817].
[40] p. incl. wrapper; illus. HSP in LCP (wrapper); LCP (wrapper)

209. Der Gemeinnützige Landwirthschafts Calender, Auf das Jahr ... 1821 .... Zum zwey und dreyssigstenmal Herausgegeben.
Lancaster, (Pennsylvania): gedruckt und zu haben bey William Albrecht, [1820].
[36] p. incl. wrapper; illus. HSP in LCP (wrapper); LCP (wrapper)

210. Der Gemeinnützige Landwirthschafts Calender, Auf das Jahr ... 1822 .... Zum drey und dreyssigstenmal Herausgegeben.
Lancaster, (Pennsylvania): gedruckt und zu haben bey William Albrecht, [1821].
[36] p. incl. wrapper; illus. HSP in LCP (wrapper)

211. Der Gemeinnützige Landwirthschafts Calender, Auf das Jahr ... 1823 .... berechnet, von Carl F. Egelmann. Zum vier und dreyssigstenmal Herausgegeben.
Lancaster, (Pennsylvania): gedruckt und zu haben bey William Albrecht, [1822].
[36] p. incl. wrapper. HSP in LCP (lacking wrapper and last 3 leaves); LCP (wrapper)

212. Der Gemeinnützige Landwirthschafts Calender, Auf das Jahr ... 1824 .... Zum fünf und dreyssigstenmal Herausgegeben.
Lancaster (Pennsylvania): gedruckt und zu haben bey William Albrecht, [1823].
[36] p. incl. wrappers; illus. HSP in LCP (wrappers)

213. Der Gemeinnützige Landwirthschafts Calender, Auf das Jahr ... 1825 .... Zum sechs und dreyssigstenmal Herausgegeben.
Lancaster, (Pennsyvania): gedruckt und zu haben bey William Albrecht, [1824].
[36] p. incl. wrappers; illus. HSP in LCP (wrappers)

214. Der Gemeinnützige Landwirthschafts Calender, Auf das Jahr ... 1826 .... Zum sieben und dreyssigstenmal Herausgegeben.
Lancaster, (Pennsylvania): gedruckt und zu haben bey William Albrecht, [1825].
[36] p. incl. wrappers; illus. HSP in LCP (wrappers); LCP (wrappers)

215. Der Gemeinnützige Landwirthschafts Calender, Auf das Jahr ... 1827 .... Zum acht und dreyssigstenmal Herausgegeben.
Lancaster, (Pennsylvania): gedruckt und zu haben bey William Albrecht, [1826].
[36] p. incl. wrapper; illus. HSP in LCP (wrapper); LCP (wrapper)

216. Der Gemeinnützige Landwirthschafts Calender, Auf das Jahr ... 1828 .... Zum neun und dreyssigstenmal Herausgegeben.
Lancaster, (Pennsylvania): gedruckt und zu haben bey William Albrecht, [1827].
[36] p. incl. wrapper; illus. HSP in LCP; LCP (lacking wrapper and last leaf)

217. Agricultural Almanac, For the Year ... 1829 .... Carefully calculated for the Meridian of Pennsylvania and the adjoining States.
Lancaster: Printed And Sold By John Baer, [1828].
[36] p. incl. wrappers; illus. LCP (wrappers); HSP (wrappers)

218. Der Gemeinnützige Landwirthschafts Calender, Auf das Jahr ... 1829 .... Zum vierzigstenmal Herausgegeben.
Lancaster, (Pennsylvania): gedruckt und zu haben bey William Albrecht, [1828].
[36] p. incl. wrapper; illus. HSP in LCP (wrapper); LCP (wrapper)

219. Agricultural Almanac, For the Year ... 1830 .... Carefully calculated for the Meridian of Pennsylvania and the adjoining States.
Lancaster: Printed and sold by John Bear, [1829].
HSP: [34] p. incl. wrappers; illus. Drake 11657 has most copies with 18 leaves.

220. Der Gemeinnützige Landwirthschafts Calender, Auf das Jahr ... 1830 .... Zum Einundvierzigstenmal Herausgegeben.
Lancaster, (Pennsylvania): gedruckt und zu haben bey William Albrecht, [1829].
[36] p. incl. wrapper; illus. HSP in LCP (wrapper); LCP (lacking wrapper)

221. Der Gemeinnützige Landwirthschafts Calender, Auf das Jahr ... 1831 .... Zum Zweyundvierzigstenmal Herausgegeben.
Lancaster, (Pennsylvania): gedruckt und zu haben bey William Albrecht, [1830].
[36] p. incl. wrapper; illus. HSP in LCP (wrapper); LCP (wrapper)

222. Agricultural Almanac, For the Year ... 1832 ....
Lancaster: Printed and sold by John Bear, [1831].
HSP: [34] p. incl. wrappers; illus. Drake 11749 has most copies with 18 leaves.

223. Der Gemeinnützige Landwirthschafts Calender, Auf das Jahr ... 1832 .... Zum Dreyundvierzigstenmal Herausgegeben.
Lancaster, (Pennsylvania): gedruckt und zu haben bey William Albrecht, [1831].
[36] p. incl. wrapper; illus. HSP in LCP (wrapper); LCP (wrapper)

224. Agricultural Almanac, For the Year ... 1833 ....
Lancaster: Printed and sold by John Bear, [1832].
HSP: [34] p. only (of 36; lacking pp. 5-6); illus.

225. Der Gemeinnützige Landwirthschafts Calender, Auf das Jahr ... 1833 .... Zum Vier und vierzigstenmal Herausgegeben.
Lancaster, (Pennsylvania): gedruckt und zu haben bey William Albrecht, [1832].
[36] p. incl. wrapper; illus. HSP in LCP (wrapper); LCP (lacking wrapper)

226. Agricultural Almanac, For the Year ... 1834 ....
Lancaster: Printed and sold by John Bear, [1833].
[36] p. incl. wrappers; illus. HSP (wrappers)

227. Agricultural Almanac, For the Year ... 1835 ....
Lancaster: Printed and sold by John Bear, [1834].
HSP: [32] p. only (of 36 incl. wrappers; lacking pp. 27-31); illus.

228. Agricultural Almanac, For the Year ... 1836 ....
Lancaster: Printed and sold by John Bear, [1835].
[36] p. incl. wrappers; illus. HSP (wrappers)

229. Agricultural Almanac, For the Year ... 1837 ....
Lancaster: Printed and sold by John Bear, [1836].
[36] p. incl. wrappers; illus. HSP (wrappers)

230. Agricultural Almanac, For the Year ... 1838 ....
Lancaster: Printed and sold by John Bear, [1837].
[36] p. incl. wrappers; illus. HSP (wrappers)

231. Agricultural Almanac, For the Year ... 1839 ....
Lancaster: Printed and sold by John Bear, [1838].
[36] p. incl. wrappers; illus. HSP (wrappers)

232. Agricultural Almanac, For the Year ... 1840 ....
Lancaster: Printed and sold by John Bear, [1839].
[36] p. incl. wrappers; illus. HSP (wrappers)

233. Agricultural Almanac, For the Year ... 1841 ....
Lancaster: Printed and sold by John Bear, [1840].
[36] p. incl. wrappers; illus. HSP (wrappers)

234. Agricultural Almanac, For the Year ... 1842 ....
Lancaster: Printed and sold by John Bear, [1841].
[36] p. incl. wrappers; illus. LCP (wrappers); HSP (wrappers)

235. Agricultural Almanac, For the Year ... 1843 ....
Lancaster: Printed and sold by John Bear, [1842].
[36] p. incl. wrappers; illus. HSP in LCP (wrappers); HSP (wrappers)

236. Agricultural Almanac, For The Year ... 1845 ....
Lancaster: Printed and sold by John Bear, [1844].
[36] p. incl. wrappers; illus. HSP (wrappers)

237. Agricultural Almanac, For The Year ... 1846 ....
Lancaster: Printed and sold by John Bear, [1845].
[36] p. incl. wrappers; illus. HSP (wrappers)

238. Agricultural Almanac, For The Year ... 1847 ....
Lancaster: Printed and sold by John Bear, [1846].
[36] p. incl. wrappers; illus. HSP (wrappers)

239. Agricultural Almanac, For The Year ... 1848 ....
Lancaster: Printed and sold by John Bear, [1847].
[36] p. incl. wrappers; illus. HSP (wrappers)

240. Agricultural Almanac, For The Year ... 1849 ....
Lancaster: Printed and sold by John Bear, [1848].
[36] p. incl. wrappers; illus. HSP (wrappers)

241. Agricultural Almanac, For The Year ... 1850 ....
Lancaster: Printed and sold by John Bear, [1849].
HSP: [32] p. only (of 36 incl. wrappers; lacking wrappers); illus.

242. The Agricultural Almanac, For The Year ... 1852 ....
Lancaster: Printed and sold by John Bear, [1851].
HSP: [28] p. only (of 36 incl. wrappers; lacking front wrapper, t.p., and pp. 17-21); illus.

243. Agricultural Almanac, For The Year ... 1853 ....
Lancaster: Printed and sold by John Bear, [1852].
HSP: [34] p. only (of 36 incl. wrappers; lacking back wrapper); illus.

244. Agricultural Almanac, For The Year ... 1854 ....
Lancaster: Printed and sold by John Bear, [1853].
[36] p. incl. wrappers; illus. HSP (wrappers)

245. Agricultural Almanac, For The Year ... 1855 ....
Lancaster: Printed and sold by John Baer & Sons, [1854].
[36] p. incl. wrappers; illus. HSP (wrappers)

246. Agricultural Almanac, For The Year ... 1856 ....
Lancaster: Printed and sold by John Baer & Sons, [1855].
[36] p. incl. wrappers; illus. HSP (wrappers)

247. Agricultural Almanac, For The Year ... 1857 ....
Lancaster: Printed and sold by John Baer & Sons, [1856].
[36] p. incl. wrappers; illus. HSP (wrappers)

248. Agricultural Almanac, For the Year ... 1858 ....
Lancaster: Printed and sold by John Baer & Sons, [1857].
[36] p. incl. wrappers; illus. HSP (wrappers)

249. Agricultural Almanac, For The Year ... 1859 ....
Lancaster: Printed and sold by John Baer & Sons, [1858].
[36] p. incl. wrappers; illus. HSP (wrappers)

250. Agricultural Almanac, For The Year ... 1860 ....
Lancaster: Printed and sold by John Baer's Sons, [1859].
[36] p. incl. wrappers; illus. HSP (wrappers)

251. Agricultural Almanac, For The Year ... 1861 ....
Lancaster: Printed and sold by John Baer's Sons, [1860].
HSP: [34] p. only (of 36 incl. wrappers; lacking pp. 33-34); illus.

__________. LOCKPORT, N.Y.

252. Tucker & Co.'s National Almanac For the use of Farmers, Planters, Merchants, Mechanics & Families For The Year 1858. Calculated for the Meridian of New York and Charleston .... [wrapper title; cap. title p. 1 continues: Calculated By Samuel H. Wright, Dundee, Yates Co., N.Y.]
Lockport, N.Y.: Published By M.H. Tucker & Co. Successors To George W. Merchant; Buffalo: Stereotyped And Printed By E.R. Jewett & Co., [1857]. [wrapper imprint]
32 p. HSP (wrappers)

__________. LOWELL, MASS.

253. Ayer's American Almanac, For The Use Of Farmers, Planters, Mechanics, And All Families. 1859. Calculated for the four great Meridians of this country .... [wrapper title]
Lowell, Mass., U.S.A.: Published By Dr. J.C. Ayer & Co., [1858]. [wrapper imprint]
36 p.; illus. HSP (wrappers)

__________. MORRISTOWN, N.J.

254. The Citizen And Farmer's Almanac, For The Year ... 1808 .... Fitted To The Latitude And Meridian Of Morristown (N.J.)
Morristown: Printed By Jacob Mann, [1807].
HSP [30] p. only (of 36; lacking last 3 leaves)

255. The Citizen & Farmer's Almanac, For The Year ... 1818 .... By David Young, Philom. ....
Morristown: Printed By Jacob Mann, [1817].
[36] p. HSP

256. The Citizen's & Farmer's Almanac, For The Year ... 1821 .... By David Young, Philom. ....
Morristown: Printed By Jacob Mann, [1820].
HSP: [34] p.; Drake 5342 has most copies with 18 leaves.

257. The Citizen's & Farmer's Almanac, For The Year ... 1823 .... By David Young, Philom. ....
Morristown: Printed By Jacob Mann, [1822].
[36] p. HSP

258. Farmer's Almanac, For The Year 1846 .... Calculated for the Horizon and Meridian of New Jersey ... By David Young, Philom. ....
Morristown, N.J.: Published And Sold ... By William Nevin, [1845].
35, [1] p.; illus. HSP

259. Farmer's Almanac, For The Year ... 1850 .... By David Young, Philom. ....
Morristown, N.J.: Published And Sold ... By William Nevin, [1849].
35, [1] p.; illus. T.p. headed: No. 25. HSP

260. Farmer's Almanac, For The Year ... 1853 .... By David Young, Philom. ....
Morristown, N.J.: Published And Sold ... By William Nevin, [1852].
35, [1] p.; illus. HSP

261. Farmer's Almanac, For The Year ... 1855 .... By Samuel Hart Wright, Dundee, Yates Co., N.Y.
Morristown, N.J.: Published And Sold ... By John R. Runyon, [1854].
35, [1] p.; illus. T.p. headed: No. 30. HSP

__________. NEW BEDFORD, MASS.
See The (Old) Farmer's Almanack .... 1859; Boston. Wrapper has New Bedford imprint.

__________. NEW HAVEN.

262. An Agricultural and Economical Almanack, For The Year ... 1816 ... By Elijah Middlebrook .... Published under the patronage of the Society for promoting Agriculture in the State of Connecticut.
New Haven: Printed By J. Barber, And For Sale At H. Howe's Book-Store, [1815].
[36] p. T.p. headed: No. I. APS

__________. NEW LONDON, CT.

263. The New-England Almanac, And Gentlemen and Ladies' Diary, Enlarged, For the Year ... 1798 .... Fitted to the meridian of New-London .... By Nathan Daboll.
New London: From Springer's Press, For Samuel Green, [1797].
HSP: [28] p. only (of 36; lacking last 4 leaves)

264. The New-England Almanac, and Gentlemen and Ladies Diary, Enlarged, For the Year ... 1800 .... By Nathan Daboll.
New London: Printed and Sold by Samuel Green, [1799].
HSP: [30] p. only (of 36; lacking 1st and last leaves and 1 leaf from sig. B). Sig. B is bound in a different order from that shown in the Early American Imprints microfilm of the MWA copy.

265. The New-England Almanack, For the Year ... 1810 .... By Nathan Daboll.
New London: Printed And Sold By Samuel Green, [1809].
HSP: [32] p. only (of 36; lacking 1st and last leaves)

266. The New-England Almanack, For the Year ... 1811 .... By Nathan Daboll.
New London: Printed And Sold By Samuel Green, [1810].
LCP: [28] p. only (of 36; lacking pp. 5-6, 17-20, 25-26); HSP [32] p. only (lacking 1st and last leaves)

267. The New-England Almanack, For the Year ... 1812 .... By Nathan Daboll.
New London: Printed And Sold By Samuel Green, [1811].
HSP: [30] p. only (of 32; lacking last leaf)

268. The New-England Almanack, For the Year ... 1813 .... The Astronomical Calculations Performed By Nathan Daboll.
New London: Printed And Sold By Samuel Green, [1812].
HSP: [20] p. only (of 32; lacking last 6 leaves)

269. The New-England Almanack, For the Year ... 1815 .... By Nathan Daboll.
New London: Printed And Sold By Samuel Green, [1814].
[32] p. LCP (lacking last 2 leaves); HSP

270. The New-England Almanack, For the Year ... 1816 .... The Astronomical calculations performed by Nathan Daboll.
New London: Printed And Sold By Samuel Green, [1815].
HSP: [28] p. only (of 32; lacking 1st 2 leaves)

271. The New-England Almanack, For the Year ... 1817 .... The Astronomical calculations performed by Nathan Daboll.
New London: Printed And Sold By Samuel Green, [1816].
[32] p. LCP

272. The New-England Almanack, For the Year ... 1818 .... The Astronomical calculations performed by Nathan Daboll.
New London: Printed And Sold By Samuel Green, [1817].
HSP: [30] p. only (of 32; lacking last leaf)

273. The New-England Almanack, And Farmer's Friend, For The Year ....
New London: Printed And Sold By Samuel Green, [1819-31].
HSP: severely imperfect copies, all lacking t.pp. and 2 to 4 additional leaves, for: 1820; 1822; 1823; 1827; 1831; 1832.

274. The New-England Almanack, And Farmer's Friend, For the Year ... 1825 .... The Astronomical calculations performed by Nathan Daboll.
New London: Printed And Sold By Samuel Green, [1824].
[32] p. LCP; HSP (lacking 1st 2 and last 2 leaves)

275. Green's New-England Almanack, And Farmer's Friend, For The Year ... 1826 .... The Astronomical calculations by Nathan Daboll.
New London: Printed And Sold By Samuel Green, [1825].
[32] p. LCP (2 copies: 1. on wove paper; 2. with 1st 10 leaves wove paper and remaining laid paper); HSP (lacking pp. 7-10 and last 2 leaves; on wove paper)

276. The New-England Almanack, And Farmer's Friend, For The Year ... 1830 .... The Astronomical calculations by Nathan Daboll.
New London: Printed And Sold By Samuel Green, [1829].
[32] p. HSP

277. The New-England Almanack And Farmer's Friend, For The Year ... 1833 .... The Astronomical Calculations Performed By Nathan Daboll.
New London: Printed And Sold By Samuel Green, [1832].
[28] p. LCP (lacking last 2 leaves); HSP

278. The New-England Almanack And Farmers' Friend, For The Year ... 1834 .... The Astronomical Calculations Performed By Nathan Daboll.
New London: Printed And Sold By Samuel Green, [1833].
32 [i.e. 31], [1] p. Copy with p. 31 misnumbered 32. HSP

279. The New-England Almanack And Farmers' Friend, For The Year ... 1834 ....
New London: Printed And Sold By Samuel Green, [1833].
31, [1] p. Copy with p. number corrected. LCP

280. The New-England Almanack, And Farmers Friend, For The Year ... 1835 .... The Astronomical Calculations Performed By Nathan Daboll.
New London: Printed And Sold By Samuel Green, [1834].
HSP: 28 p. only (of 32; lacking last 2 leaves)

281. The New-England Almanack, And Farmers' Friend, For The Year ... 1836 .... Tht Astronomical Calculations Performed By Nathan Daboll.
New London: Printed And Sold By Samuel Green, [1835].
31, [1] p. LCP; HSP

282. The New-England Almanack, And Farmers' Friend, For the Year ... 1837 .... The Astronomical Calculations Performed By Nathan Daboll.
New London: Printed And Sold By Samuel Green, [1836].
32 p. HSP

283. The New-England Almanack, And Farmers' Friend, For The Year ... 1838 .... The Astronomical Calculations Performed By Nathan Daboll.
New London: Published By Samuel Green; E. Williams, Printer, [1837].
32 p. LCP; HSP (1st 24 p. only)

284. The New-England Almanack, And Farmers' Friend, For The Year ... 1839 .... By Nathan Daboll, A.M.
New London: Printed And Published By E. Williams, And For Sale By W. & J. Bolles, [1838].
31, [1] p. HSP

285. The New-England Almanac, And Farmers' Friend, For The Year ... 1840 .... By Nathan Daboll, A.M.
[New London, 1839].
31, [1] p. HSP (with portion of t.p. damaged and illegible. See Drake 1144-46)

286. The New-England Almanac, And Farmers' Friend, For The Year ... 1841 .... By Nathan Daboll, A.M.
New London: Printed And Published By E. Williams, And For Sale By W. & J. Bolles, [1840].
31, [1] p. HSP

287. The New-England Almanac, And Farmers' Friend, For The Year ... 1842 .... By Nathan Daboll, A.M.
New London: Published By Bolles & Williams, [1841].
31, [1] p. HSP

288. The New-England Almanac, And Farmers' Friend, For The Year ... 1843 .... By Nathan Daboll, A.M.
New London: Published By Bolles & Williams; Norwich: For Sale By E. H. Starr, [1842].
31, [1] p. HSP

289. The New-England Almanac, And Farmers' Friend, For The Year ... 1846 .... By Nathan Daboll, A.M.
New London: Published By Bolles & Williams, [1845].
31, [1] p. HSP

290. The New-England Almanac, And Farmers' Friend, For The Year ... 1848 .... By Nathan Daboll.
New London: Published By Bolles & Williams, [1847].
31, [1] p. HSP

291. The New-England Almanac, And Farmers' Friend, For The Year ... 1849 .... By Nathan Daboll, A.M.
New London: Publihed [sic] By Bolles & Williams, [1848].
31, [1] p. HSP

292. The New-England Almanac, And Farmers' Friend, For The Year ... 1851 .... By Nathan Daboll, A.M.
New London: Published By Bolles & Co., [1850].
31, [1] p. HSP

293. The New-England Almanac, And Farmers' Friend, For The Year ... 1853 .... By Nathan Daboll, A.M. [wrapper title]
New London: Bolles & Co.; Norwich: L. & E. Edwards, 1852. [wrapper imprint]
48 p. HSP (wrappers)

294. The New-England Almanac, And Farmers' Friend. For The Year ... 1854 .... By Nathan Daboll, A.M. [wrapper title]
[New London, 1853].
32 p. HSP (wrappers; with portion of front wrapper damaged and illegible)

295. The New-England Almanac, And Farmer's Friend, For The Year ... 1855 .... By Nathan Daboll, A.M.
New London: Published By Bolles & Co.; Starr And Farnham, Printers, [1854].
31, [1] p. HSP

296. The New-England Almanac, And Farmer's Friend, For The Year ... 1856 .... By Nathan Daboll, A.M.
New London: Published By Starr & Co.; Norwich: Sold By Thomas Robinson; Starr And Farnham, Printers, [1855].
28, [2 (adv.)] p. HSP

297. The New-England Almanac, And Farmer's Friend, For The Year ... 1857 .... By Nathan Daboll, A.M.
New London: Published By Starr & Co.; Norwich: Sold By L. & E. Edwards; Starr And Farnham, Printers, [1856].
26, [18 (adv.)] p. HSP

298. The New-England Almanac, And Farmer's Friend, For The Year ... 1858 .... By Nathan Daboll, A.M.
New London: Published By Starr & Co.; Norwich: Sold By Thomas Robinson; Starr And Farnham, Printers, [1857].
HSP: pp. [1]-24, 35-39, [1] only.

299. The New-England Almanac, And Farmer's Friend, For The Year ... 1859 .... By Nathan Daboll, A.M.
New London: Published By Starr & Co.; Starr And Farnham, Printers, [1858].
29, [35 (adv.)] p. HSP

300. The New-England Almanac, And Farmer's Friend, For The Year ... 1860 .... By Nathan Daboll, A.M.
New London: Published By Starr & Co,; Norwich: Sold By L. & E. Edwards; Starr And Farnham, Printers, [1859].
29, [35 (adv.)] p. HSP

301. The New-England Almanac, And Farmer's Friend, For The Year ... 1861 .... By Nathan Daboll, A.M.
New London: Published and sold by Starr & Co.; Norwich: Sold by L. & E. Edwards; Starr & Farnham, Printers, [1860].
28, [20 (adv.)] p. HSP

__________. NEW ORLEANS.

302. Affleck's Southern Rural Almanac, And Plantation And Garden Calender, For 1858 ....
New Orleans: Published By B.M. Norman, [1857].
107 p. LCP (wrapper)

__________. NEW YORK.

303. Farmer's Almanac, For The Year ... 1832 .... By David Young, Philom.
New York: Printed and Sold by Christian Brown, [1831].
[36] p.; illus. HSP

304. Day's City And Country Almanac, For The Year ... 1834 .... By David Young, Philom.
New York: Printed And Sold By Mahlon Day, [1833].
[36] p. LCP

305. The Farmer's Almanac, For The Middle States, For The Year ... 1835 .... Calculated for the Meridian of New-York City .... By Thomas Spofford.
New York: Published By Charles Small; Boston: Printed By Samuel N. Dickinson, [1834].
[36] p. T.p. headed: Vol. 3. - No. 3.) An Astronomical Diary for 1835. (Whole No. 19. HSP

306. Day's City And Country Almanac, For The Year ... 1837 .... Calculated For The Horison [sic] and Meridian Of New-York .... By David Young, Philom.
New York: Mahlon Day, [1836].
[36] p. HSP

307. The Farmer's Almanac, For The Year ... 1839 ....
New York: Turner & Fisher, [1838].
35, [1] p. HSP

308. The Yankee Farmers' Almanac, For The Year ... 1842 .... Calculated for Boston .... By Thomas Spofford.
New York: Published by David Felt & Co., [1841].
[36] p. T.p. headed: Vol. 4. No. 2.) An Astronomical Diary For 1842. (Whole No. 26. HSP

309. The United States Farmers' Almanac, For The Year ... 1843 .... Fitted to Five different Meridians; and calculated for the whole United States .... By Thomas Spofford.
New York: Published By David Felt & Co., [1842].
[36] p. T.p. headed: Vol. 4. No. 3.) An Astronomical Diary For 1843. (Whole No. 27. HSP

310. The American Agriculturist's Almanac For 1844. By A.B. Allen, Editor of "The American Agriculturist."
New York: J. Winchester, New World Press, [1843].
62, [2 (adv.)] p. Wrapper imprint: J. Winchester, Office Of The "American Agriculturist;" Philadelphia: Zieber & Co.; Baltimore: W. Taylor, et al. LCP (wrappers)

311. The American Farmer's Almanac 1844.
New York and Philadelphia: Turner & Fisher, [1843].
[36] p.; illus. HSP

312. Farmer's Almanac 1844 Adapted To The Meridian Of Charleston, North Carolina, Tennessee, Georgia, Alabama, Mississippi, Louisiana, And Arkansas.
New York: Greeley & McElrath; Robinson, Pratt, & Co., [1843].
[32] p.; illus. HSP

313. The United States Farmers' Almanac, For The Year ... 1844 .... By Thomas Spofford.
New York: Published by David Felt & Co., [1843].
[36] p. T.p. headed: Vol. 4. No. 4.) An Astronomical Diary For 1844. (Whole No. 28. HSP

314. Spofford's United States Farmers' Almanac, For The Year ... 1845 .... By Thomas Spofford.
New York: Published By Charles Small; G.B. Maigne, Book and Job Printer, [1844].
[36] p. T.p. headed: Vol. 4. No. 4.) An Astronomical Diary, For 1845. (Whole No. 29. HSP

315. The American Farmer's Almanac 1846.
New York and Philadelphia: Published By Turner & Fisher, [1845].
HSP: [30] p. only (of 32; lacking last leaf); illus. Fuller title p. 3 ends: ... carefully calculated for the Middle And Western States. Original Calculations By Edward Hagerty, M.D. T.p. headed: For The Middle And Western States.

316. The United States Farmers' Almanac, For The Year ... 1846 ....
New York: Published by David Felt & Co., [1845].
[36] p. T.p. headed: Vol. 4 No. 6.) An Astronomical Diary For 1846. (Whole No. 30. HSP

317. The United States Farmers' Almanac, For The Year ... 1847 ....
New York: Published by David Felt & Co., [1846].
[36] p. T.p. headed: Vol. 5. No. 1.) An Astronomical Diary For 1847. (Whole No. 31. HSP

318. The American Farmer's Almanac 1848.
New York And Philadelphia: Turner & Fisher, [1847].
[36] p.; illus. HSP

319. The Farmer's Almanac, For The Year 1848.
New York: Collins And Brother, [1847].
LCP: 30, [2] p. At head of title is stamped: R.P. Johnson & Co.'s. Added at back is a leaf advertising the Johnson bookstore in Morristown, N.J.

320. Farmer's Almanac, And Ephemeris ... For The Year ... 1850 .... By David Young, Philom.
New York: Published And Sold ... By Samuel Raynor, [1849].
[36] p.; illus. T.p. headed: (Thirty-second Year of Publication.) HSP

321. The Northern Farmer's Almanac. Edited By Solon Robinson .... Calculated For The Meridian Of New York City ....
[New York:] Published By A.B. Allen & Co., New-York Agricultural Warehouse, [1850].
40 p.; illus. T.p. headed: 1851. HSP

322. Elton's Agricultural Almanac, For 1853 ....
New York: Published By Elton & Co., [1852].
48 p.; illus. HSP (wrappers)

323. The Farmers' Almanac For 1857 ....
New York: Francis Hart, Printer, [1856].
31, [1] p. Wrapper title continues: Published By The Lodi Manufacturing Co. 60 Cortlandt Street, New-York City. HSP (wrappers)

324. Stadt-und Land-Calender Auf das Jahr Christi 1857 ....
New York: Zu haben bei Wm. Radde; Philadelphia: Verlag von King und Baird, [1856].
HSP: 32, [2] p. only (of 36; lacking last leaf); illus.

325. The Farmers' Almanac For 1859.
New York: Published By Griffing, Brother & Co. Agents For The Lodi Manufacturing Co., [1858].
[48] p.; illus. HSP (wrappers)

__________. NORWICH, CT.

326. The New-England Almanac, And Farmers' Friend, For The Year ... 1841 .... By Nathan Daboll, A.M.
Norwich: Published And For Sale By Samuel C. Starr, [1840].
31, [1] p. LCP

327. The New-England Almanac, And Farmers' Friend, For The Year ... 1844 .... By Nathan Daboll, A.M.
Norwich: Published By L. & E. Edwards, [1843].
31, [1] p. HSP

328. The New-England Almanac, And Farmers' Friend, For The Year ... 1845 .... By Nathan Daboll, A.M.
Norwich: Published By L. & E. Edwards, [1844].
31, [1] p. HSP

329. The New-England Almanac, And Farmers' Friend, For The Year ... 1847 .... By Nathan Daboll.
Norwich: Published By L. & E. Edwards, [1846].
31, [1] p. HSP

330. The New-England Almanac, And Farmers' Friend, For The Year ... 1850 .... By Nathan Daboll, A.M.
Norwich: Published By T. Robinson, [1849].
31, [17 (adv.)] p. HSP

331. The New-England Almanac, And Farmers' Friend, For The Year ... 1851 .... By Nathan Daboll, A.M.
Norwich: Published By L. & E. Edwards, [1850].
31, [1] p. HSP

__________. PETERSBURG, VA.

332. The Virginia & North Carolina Almanack For the Year ... 1802 .... The Astronomical Part by Isaac Briggs. The Editorial Part by Americanus Urban.
Petersburg: Published By Ross & Douglas, Booksellers and Stationers, (For the Year 1802).
[48] p.; illus. T.p. headed: No. III. HSP

__________. PHILADELPHIA.

333. Poor Will's Almanack, For The Year ... 1787 .... Also, The Gardeners Kalender; Or, Useful Memorandums of Work necessary to be done, monthly, in the Garderns [sic] and Orchards of the middle States.
Philadelphia: Printed and Sold by Joseph Crukshank, [1786].
[44] p. HSP in LCP; LCP (lacking pp. 35-36 and last leaf)

334. Poulson's Town And Country Almanac, For The Year ... 1789 .... The Astronomical Calculations By William Waring ....
Philadelphia: Printed And Sold By Zachariah Poulson, Junior, [1788].
[40] p. Copy with long rule above the line "The Transit of Mercury" p. 4. HSP in LCP

335. Poulson's Town And Country Almanac ... 1789 ....
Philadelphia: Printed And Sold By Zachariah Poulson, Junior, [1788].
[40] p. Copy without long rule p. 4. HSP in LCP; LCP

336. Poulson's Town And Country Almanac, For The Year ... 1790 .... The Astronomical Calculations By William Waring.
Philadelphia: Printed And Sold By Zachariah Poulson, Junior, [1789].
[36] p. HSP in LCP; LCP

337. Poulson's Town And Country Almanac, For The Year ... 1791 .... The Astronomical Calculations by William Waring.
Philadelphia: Printed And Sold By Zachariah Poulson, Junior, [1790].
[36] p. HSP in LCP; LCP

338. Poulson's Town And Country Almanac, For The Year ... 1792 .... The Astronomical Calculations by William Waring.
Philadelphia: Printed And Sold By Zachariah Poulson, Junior, [1791].
[36] p. HSP in LCP; LCP

339. Poulson's Town and Country Almanac, For The Year ... 1793 .... [Calculations ... By William Waring (p. 2)].
Philadelphia: Printed And Sold By Zachariah Poulson, Junior, [1792].
[40] p. HSP in LCP

340. Poulson's Town and Country Almanac, For The Year ... 1793 .... [Calculations ... By William Waring (p. 2)].
Philadelphia: Printed And Sold By Zachariah Poulson, Junior, [1792].
[40] p. T.p. headed: The Second Edition. LCP

341. Poulson's Town and Country Almanac, For The Year ... 1794 .... [Calculations ... By William Waring (p. 2)].
Philadelphia: Printed And Sold By Zachariah Poulson, Junior, [1793].
[44] p. Drake 10360 has most copies with 24 leaves, but has two with 22; Evans 26421 has [44] p. HSP in LCP; LCP

342. Poulson's Town and Country Almanac, For The Year ... 1795 .... [Calculations ... By Abraham Shoemaker (p. 2)].
Philadelphia: Printed And Sold By Zachariah Poulson, Junior, [1794].
[44] p. HSP in LCP; LCP

343. Poulson's Town and Country Almanac, For The Year ... 1796 .... [Calculations ... By Abraham Shoemaker (p. 2)].
Philadelphia: Printed And Sold By Zachariah Poulson, Junior, [1795].
[48] p. HSP in LCP (lacking pp. 39-40 and last leaf); LCP

344. Poulson's Town and Country Almanac, For The Year ... 1797 .... [Calculations ... By Abraham Shoemaker (p. 2)].
Philadelphia: Printed And Sold By Zachariah Poulson, Junior, [1796].
[48] p. HSP in LCP; LCP

345. Poulson's Town and Country Almanac, For The Year ... 1798 .... [Calculations ... By Abraham Shoemaker (p. 2)].
Philadelphia: Printed And Sold By Zachariah Poulson, Junior, [1797].
[48] p. HSP in LCP; LCP

346. Poulson's Town and Country Almanac, For The Year ... 1799 .... [Calculations ... By Abraham Shoemaker (p. 2)].
Philadelphia: Printed And Sold By Zachariah Poulson, Junior, [1798].
[48] p. HSP in LCP; LCP

347. Poulson's Town and Country Almanac, For The Year ... 1800 ....
Philadelphia: Printed And Sold By Zachariah Poulson, Junior, [1799].
[48] p. HSP in LCP; LCP; HSP

348. Poulson's Town and Country Almanac, For The Year ... 1801 ....
Philadelphia: Printed And Sold By Zachariah Poulson, Junior, [1800].
[48] p. HSP in LCP; LCP; HSP

349. Citizen's and Farmer's Almanac, For the Year 1802 .... Astronomical Calculations, by Joshua Sharp ....
Philadelphia: Printed and sold by John M'Culloch, [1801].
[36] p. HSP

350. Poulson's Town and Country Almanac, For The Year ... 1802 ....
Philadelphia: Printed And Sold By John Bioren, [1801].
[48] p. HSP in LCP; LCP; HSP

351. Citizen's and Farmer's Almanac, For the Year 1803 .... Astronomical Calculations, by Joshua Sharp ....
Philadelphia: Printed and sold by John M'Culloch, [1802].
[36] p. HSP

352. Poulson's Town and Country Almanac, For The Year ... 1803 ....
Philadelphia: Printed And Sold By John Bioren, [1802].
[48] p. Copy with misprint "corrups" at end of B2r (i.e. p. 7). HSP in LCP; LCP

353. Poulson's Town and Country Almanac, For The Year ... 1803 ....
Philadelphia: Printed by John Bioren, for Benjamin Johnson; and Jacob Johnson, [1802].
[48] p. Copy with misprint on B2r corrected. HSP in LCP; LCP; HSP

354. Poulson's Town and Country Almanac, For The Year ... 1804 ....
Philadelphia: Printed And Sold By John Bioren, [1803].
48 p. HSP in LCP; LCP; HSP

355. Citizen's and Farmer's Almanac, For the Year 1805 .... Astronomical Calculations, by Joshua Sharp ....
Philadelphia: Printed for John M'Culloch, [1804].
[36] p. HSP

356. Poulson's Town and Country Almanac, For The Year ... 1805 ....
Philadelphia: Printed by John Bioren, [1804].
48 p. HSP in LCP (lacking last leaf); LCP; HSP

357. Citizen's and Farmer's Almanac, For the Year 1806 .... Astronomical Calculations, by Joshua Sharp ....
Philadelphia: Printed for John M'Culloch, [1805].
HSP: [30] p. only (of 36; lacking last 3 leaves)

358. Poulson's Town and Country Almanac, For The Year ... 1806 ....
Philadelphia: Printed by John Bioren, [1805].
48 p. Copy with Table of Interest p. 39. LCP; HSP in LCP; HSP

359. Poulson's Town and Country Almanac ... 1806 ....
Philadelphia: Printed by John Bioren, for Conrad & Co. and sold at their stores ..., [1805].
48 p. With article on Preserving p. 39. LCP

360. Poulson's Town and Country Almanac ... 1806 ....
Philadelphia: Printed by John Bioren, [1805].
48 p. Copy with Preserving p. 39 and with different arrangement of matter pp. 40-48. HSP in LCP

361. Poulson's Town And Country Almanack, For The Year ... 1807 ....
Philadelphia: Printed by John Bioren, [1806].
48 p. HSP in LCP; LCP; HSP

362. Bioren's Town And Country Almanack, (Formerly Published By Mr. Zachariah Poulson,) For The Year ... 1808 .... Calculated By Abraham Shoemaker ....
Philadelphia: Printed and Sold by John Bioren, [1807].
48 p. HSP

363. Washington Citizen and Farmer's Almanac, For the Year 1808 .... Astronomical Calculations by Joshua Sharp ....
Philadelphia: Printed for Ebenezer M'Culloch, [1807].
[48] p. HSP

364. Bioren's Town And Country Almanack ... For The Year ... 1809 .... Calculated By Abraham Shoemaker ....
Philadelphia: Printed and Sold by John Bioren, [1808].
48 p. LCP; HSP

365. Washington's Citizen's and Farmer's Almanac, For the Year 1809 .... Astronomical Calculations by Joshua Sharp ....
Philadelphia: Printed for Ebenezer M'Culloch, [1808].
[36] p. HSP

366. Bioren's Town And Country Almanack, For The Year ... 1810 .... Calculated By Abraham Shoemaker ....
Philadelphia: Printed and Sold by John Bioren, [1809].
48 p. LCP (lacking t.p. and last 7 leaves); HSP

367. Kite's Town & Country Almanac, For The Year 1810 .... The Astronomical Calculations By Joshua Sharp.
Philadelphia: Printed For Benjamin And Thomas Kite, [1809].
[48] p. HSP

368. Bioren's Town And Country Almanack, For The Year ... 1812 .... Calculated By William Collom ....
Philadelphia: Printed and Sold by John Bioren, [1811].
42 p. Drake 10920 has 24 leaves. HSP

369. Bioren's Town And Country Almanack, For The Year ... 1813 .... Calculated By William Collom ....
Philadelphia: Printed and Sold by John Bioren, [1812].
48 p. HSP in LCP; LCP; HSP (lacking last leaf)

370. Bioren's Town And Country Almanack, For The Year ... 1814 .... Calculated By William Collom ....
Philadelphia: Printed and Sold by John Bioren, [1813].
48 p. HSP

371. Kite's Town and Country Almanac, For The Year 1814 .... Calculated By William Collom.
Philadelphia: Printed For Benjamin And Thomas Kite; Lydia R. Bailey, Printer, [1813].
[48] p. HSP

372. Bioren's Town And Country Almanack, For The Year ... 1815 .... Calculated By William Collom ....
Philadelphia: Printed and Sold by John Bioren, [1814].
HSP: 1st 34 p. only (of 48)

373. Bioren's Town And Country Almanack, For The Year ... 1816 .... Calculated By William Collom.
Philadelphia: Printed and Sold by John Bioren, [1815].
36 p. LCP; HSP

374. Farmers' And Citizens' Almanac, For The Year ... 1816 ....
Philadelphia: Printed By David Dickinson, For George W. Mentz, [1815].
LCP: 36 p.; with first 22 p. of Almanac for 1818 bound at back.

375. Kite's Town and Country Almanac, For The Year 1816 .... Calculated By William Collom.
Philadelphia: Printed For Benjamin And Thomas Kite; Lydia R. Bailey, Printer, [1815].
[36] p. HSP

376. The Agricultural Almanack For The Year ... 1817. Patronized By The Philadelphia Society For Promoting Agriculture.
Philadelphia: Published By Solomon W. Conrad, [1816].
[40] p.; illus. LCP; APS; HSP

377. Bioren's Town And Country Almanack, For The Year ... 1817 .... Calculated By William Collom ....
Philadelphia: Printed and Sold by John Bioren, [1816].
48 p. LCP; HSP (lacking last 2 leaves)

378. Kite's Town and Country Almanac, For The Year 1817 .... Calculated By Joshua Sharp.
Philadelphia: Printed For Benjamin And Thomas Kite; Lydia R. Bailey, Printer, [1816].
HSP: [26] p. only (of 36; lacking last 5 leaves)

379. The Agricultural Almanack For The Year ... 1818. Patronized By The Philadelphia Society For Promoting Agriculture.
Philadelphia: Published By Solomon W. Conrad, [1817].
[48] p.; illus. HSP in LCP; LCP; HSP

380. Bioren's Town And Country Almanack, For The Year ... 1818 .... Calculated By William Collom ....
Philadelphia: Printed and Sold by John Bioren, [1817].
48 p. HSP

381. Kite's Town and Country Almanac, For The Year 1818 .... Calculated By Joshua Sharp.
Philadelphia: Printed For Benjamin And Thomas Kite; Lydia R. Bailey, Printer, [1817].
[36] p. HSP

382. The Agricultural Almanack For The Year ... 1819. Patronized By The Philadelphia Society For Promoting Agriculture.
Philadelphia: Published By Solomon W. Conrad, [1818].
[48] p.; illus. LCP; APS; HSP

383. Bioren's Town And Country Almanack, For The Year ... 1819 .... Calculated By William Collom ....
Philadelphia: Printed and Sold by John Bioren, [1818].
48 p. LCP; HSP

384. The Farmer's Almanac, For The Year ... 1819 .... Calculated for the Meridian of Philadelphia .... By Andrew Beers, Philom.
Philadelphia: Published By S. Potter & Co., [1818].
[36] p. HSP

385. Kite's Town and Country Almanac, For The Year 1819 .... Calculated By Joshua Sharp.
Philadelphia: Printed For Benjamin And Thomas Kite; Lydia R. Bailey, Printer, [1818].
[36] p. LCP; HSP

386. The Agricultural Almanack For The Year ... 1820 .... Patronized By The Philadelphia Society For Promoting Agriculture.
Philadelphia: Published By Solomon W. Conrad, [1819].
[48] p.; illus. HSP in LCP; LCP; APS; HSP

387. Bioren's Town And Country Almanack, For The Year ... 1820 .... Calculated By William Collom ....
Philadelphia: Printed and Sold by John Bioren, [1819].
48 p. LCP; HSP (1st 22 p. only)

388. Citizens & Farmers' Almanack, For The Year ... 1820 ....
Philadelphia: Published For The Booksellers, [1819].
[28] p. HSP

389. Kite's Town and Country Almanac, For The Year 1820 .... Calculated By William Collom.
Philadelphia: Printed For Benjamin And Thomas Kite, [1819].
[36] p. LCP; HSP (lacking pp. 7-8)

390. The Agricultural Almanack For The Year ... 1821. Patronized By The Philadelphia Society For Promoting Agriculture.
Philadelphia: Published By Solomon W. Conrad, [1820].
[48] p.; illus. LCP

391. The Agricultural Almanack For The Year ... 1821 ....
Philadelphia: Published By Solomon W. Conrad, [1820].
Copy with [52] p., last 4 p. pub. adv.; illus. HSP in LCP; LCP; APS

392. Citizens & Farmers' Almanack, For The Year .... 1821 .... Calculated By Joshua Sharp.
Philadelphia: Published For The Booksellers, By Griggs & Dickinson, [1820].
[36] p. HSP in LCP

393. The Farmers' Almanac, For The Year 1821 .... Carefully calculated for the Latitude and Meridian of Philadelphia, By John Ward.
Philadelphia: Published by M'Carty & Davis, [1820].
36 p. HSP

394. Kite's Town and Country Almanac, For The Year 1821 .... Calculated By William Collom.
Philadelphia: Printed For Benjamin And Thomas Kite, [1820].
[36] p. LCP (lacking last leaf); HSP

395. The Pennsylvania Farmer's Almanac, For The Year ... 1821 .... Calculated for the Meridian of Philadelphia .... By Andrew Beers, Philom.
Philadelphia: Published By S. Potter & Co. Booksellers and Stationers, [1820].
[36] p. HSP

396. The Agricultural Almanack For The Year ... 1822. Patronized By The Philadelphia Society For Promoting Agriculture.
Philadelphia: Published By Solomon W. Conrad, [1821].
[48] p.; illus. HSP in LCP (lacking first 2 leaves); LCP (lacking pp. 17-18); APS; HSP

397. Bioren's Town And Country Almanack, For The Year ... 1822 ....
Philadelphia: Printed and Sold by John Bioren, [1821].
36 p. HSP

398. Citizens & Farmers' Almanack, For The Year ... 1822 .... Calculated By Joshua Sharp.
Philadelphia: Published For The Booksellers, By Griggs & Dickinson, [1821].
[36] p. HSP

399. The Farmers' Almanac, For The Year 1822 .... By John Ward.
Philadelphia: Printed by M'Carty & Davis, [1821].
36 p. HSP

400. Kite's Town and Country Almanac, For The Year 1822 .... Calculated By William Collom.
Philadelphia: Printed For Benjamin And Thomas Kite, [1821].
[36] p. HSP in LCP (lacking pp. 21-24); HSP

401. The Pennsylvania Farmer's Almanac, For The Year ... 1822 .... By Andrew Beers, Philom.
Philadelphia: Published By S. Potter, & Co. Booksellers and Stationers, [1821].
[32] p. HSP

402. The Agricultural Almanack For The Year ... 1823. Patronized By The Philadelphia Society For Promoting Agriculture.
Philadelphia: Published By Solomon W. Conrad, [1822].
[48] p.; illus. LCP; HSP

403. Citizens & Farmers' Almanack, For The Year ... 1823 .... Calculated By Joshua Sharp.
Philadelphia: Published For The Booksellers, By Griggs & Dickinson, [1822].
[36] p.; illus. HSP

404. The Farmers' Almanac, For The Year 1823 .... By John Ward.
Philadelphia: Published by M'Carty & Davis, [1822].
36 p. HSP

405. Kite's Town and Country Almanac, For The Year 1823 .... Calculated By William Collom.
Philadelphia: Printed For Benjamin And Thomas Kite, [1822].
HSP: [26] p. only (of 36; lacking last 5 leaves)

406. The Agricultural Almanack For The Year ... 1824. Patronized By The Philadelphia Society For Promoting Agriculture.
Philadelphia: Published By Solomon W. Conrad, [1823].
[48] p.; illus. LSP; HSP

407. Bioren's Town And Country Almanack, For The Year ... 1824 .... Calculated By William Collom ....
Philadelphia: Printed and Sold by John Bioren, [1823].
36 p. LCP; HSP

408. Citizens & Farmers' Almanack, For The Year ... 1824 .... Calculated By Joshua Sharp ....
Philadelphia: Published For The Booksellers, By Griggs & Dickinson, [1823].
[36] p.; illus. HSP in LCP; LCP; HSP

409. The Farmers' Almanac, For The Year 1824 .... By John Ward.
Philadelphia: Published by M'Carty & Davis, [1823].
36 p. HSP

410. The Farmers & Mechanics' Almanac, For The Year ... 1824 .... By William Collom.
Philadelphia: Printed for Robert Desilver, [1823].
[34] p.; Drake 11404 has 18 leaves. HSP

411. Kite's Town and Country Almanac, For The Year 1824 .... Calculated By William Collom.
Philadelphia: Published By Benjamin And Thomas Kite, [1823].
[48] p. HSP

412. The Agricultural Almanack For The Year ... 1825. Patronized By The Philadelphia Society For Promoting Agriculture.
Philadelphia: Published And Sold By S.W. Conrad; Sold Also By Marot & Walter; Bennet & Walton; And Kimber & Sharpless, [1824].
[48] p.; illus. HSP

413. Citizens & Farmers' Almanack, For The Year ... 1825 .... Calculated By Joshua Sharp.
Philadelphia: Printed And Published By Griggs & Dickinson, [1824].
[36] p.; illus. HSP in LCP; HSP

414. Kite's Town and Country Almanac, For The Year 1825 .... Calculated By William Collom.
Philadelphia: Published By Benjamin & Thomas Kite, [1824].
[36] p. HSP

415. The Agricultural Almanack For The Year 1826. Patronized By The Philadelphia Society For Promoting Agriculture.
Philadelphia: Published And Sold By S.W. Conrad; Sold Also By Marot & Walter; Bennet & Walton; And Kimber & Sharpless, [1825].
[48] p.; illus. LCP; HSP

416. Citizens & Farmers' Almanack For The Year ... 1826 .... Calculated By Joshua Sharp.
Philadelphia: Printed And Published By Griggs & Dickinson, For Towar & Hogan, [1825].
[36] p.; illus. HSP

417. The Farmers' Almanac For The Year 1826 .... By John Ward.
Philadelphia: Published by M'Carty & Davis, 1826.
36 p. HSP in LCP (lacking last leaf); LCP; HSP (lacking last leaf)

418. Farmers' & Mechanicks' Almanack, For The Year ... 1826 .... Calculated By Joshua Sharp.
Philadelphia: Published By Marot & Walter, [1825].
[36] p. LCP; HSP

419. Kite's Town and Country Almanac, For The Year 1826 .... Calculated By William Collom.
Philadelphia: Published By Benjamin & Thomas Kite, [1825].
[34] p.; Drake 11498 has 18 leaves. HSP

420. The Agricultural Almanack For The Year ... 1827. Patronized By The Philadelphia Society For Promoting Agriculture.
Philadelphia: Published And Sold By S.W. Conrad; Sold Also By Marot & Walter; Bennet & Walton; And Kimber & Sharpless, [1826].
[48] p.; illus. LCP (with MS. notes of Deborah Logan); APS; HSP

421. Citizens & Farmers' Almamack For The Year ... 1827 .... Calculated By Joseph Cramer.
Philadelphia: Printed And Published By Griggs & Dickinson, For The Booksellers ..., [1826].
[36] p.; illus. LCP; HSP

422. The Farmers' Almanac, For The Year 1827 .... By John Ward.
Philadelphia: Published by M'Carty & Davis, 1827.
36 p. HSP

423. Farmers Almanac, For The Year ... 1827 .... Carefully Calculated For the Latitude and Meridian of Philadelphia, By Joseph Crammer [sic].
Philadelphia: Published By Isaac Pugh; Printed On The Vertical Press, By D. & S. Neal, [1826].
[36] p. HSP

424. Farmers & Mechanics Almanack For The Year ... 1827 .... Calculated By Joshua Sharp.
Philadelphia: Published By Marot & Walker, [1826].
[36] p.; illus. HSP

425. Kite's Town and Country Almanac, For The Year 1827 .... Calculated By William Collom.
Philadelphia: Published By Benjamin & Thomas Kite, [1826].
[36] p. LCP; HSP

426. Citizens & Farmers' Almanack For The Year ... 1828 .... Calculated By Joseph Cramer.
Philadelphia: Printed And Published By Griggs & Dickinson, For The Booksellers ..., [1827].
[36] p.; illus. HSP

427. The Farmers' Almanac, For The Year 1828 .... By John Ward.
Philadelphia: Published by M'Carty & Davis, 1828.
36 p. LCP; HSP

428. Kite's Town and Country Almanac, For The Year 1828 .... Calculated By William Collom.
Philadelphia: Printed For Benjamin And Thomas Kite, [1827].
[36] p. HSP

429. The Agricultural Almanack For The Year ... 1829. Patronized By The Philadelphia Society For Promoting Agriculture.
Philadelphia: Published And Sold By Thomas Desilver, [1828].
[36] p.; illus. HSP

430. The Farmers' Almanac, For The Year 1829 .... By John Ward.
Philadelphia: Published by M'Carty & Davis, 1829.
36 p. LCP (lacking pp. 11-16); HSP

431. Kite's Town and Country Almanac, For The Year 1829 .... Calculated By William Collom.
Philadelphia: Printed And Published By Thomas Kite, [1828].
[36] p. HSP

432. The Farmers' Almanac, For The Year 1830 .... By John Ward.
Philadelphia: Published by M'Carty & Davis, 1830.
36 p. HSP

433. The Farmers And Mechanics Almanack, For The Year ... 1830 .... Carefully Calculated For The Latitude And Meridian Of Philadelphia, By Charles Frederick Egelmann ....
Philadelphia: Published And Sold By George W. Mentz, [1829].
35, 6] p. incl. wrappers; illus. LCP (wrappers); HSP (wrappers)

434. Kite's Town and Country Almanac, For The Year 1830 .... Calculated By William Collom.
Philadelphia: Printed By Thomas Kite, [1829].
[36] p. HSP

435. Towar & Hogan's Farmer's And Citizen's Almanack, For The Year ... 1830 .... Carefully Calculated For The Latitude And Meridian Of Philadelphia. By Joseph Cramer.
Philadelphia: Published By Towar & Hogan; W. Pilkington & Co. Printers, [1829].
34, [2] p. HSP

436. Citizens & Farmers' Almanack, For The Year ... 1831 ....
Philadelphia: Printed And Published By Griggs & Dickinson, For John Grigg, [1830].
[32] p.; illus. Drake 11714 gives 18 leaves. HSP

437. The Farmers' Almanac, For The Year 1831 .... By John Ward.
Philadelphia: Published by M'Carty & Davis, 1831.
36 p. LCP; HSP

438. The Farmers And Mechanics Almanack, For The Year ... 1831 .... By Charles Frederick Egelmann ....
Philadelphia: Published And Sold By George W. Mentz & Son, [1830].
35, [1] p. incl. wrappers; illus. LCP (wrappers); HSP (wrappers; with added adv. leaf at back)

439. Kite's Town and Country Almanac, For The Year 1831 .... Calculated By William Collom.
Philadelphia: Printed And Published By Thomas Kite, [1830].
[36] p. HSP

440. The Farmers' Almanac, For The Year 1832 .... By John Ward.
Philadelphia: Published by M'Carty & Davis, 1832.
32 p. LCP; HSP

441. The Farmers And Mechanics Almanack, For The Year ... 1832 .... By Charles Frederick Egelman ....
Philadelphia: Published And Sold By George W. Mentz & Son, [1831].
35, [1] p. incl. wrappers; illus. HSP (wrappers)

442. The Farmers And Mechanics Almanack, For The Year ... 1832 .... By Charles Frederick Egelmann ....
Philadelphia: Published And Sold By George W. Mentz & Son; Easton, Pa.: Henry Hammam [sic], [1831].
35, [1] p. incl. wrappers; illus. LCP (wrappers)

443. The Farmers' Almanac, For The Year 1833 .... By John Ward.
Philadelphia: Published by M'Carty & Davis, 1833.
36 p. HSP

444. The Farmers And Mechanics Almanack, For The Year ... 1833 .... By Charles Frederick Egelmann ....
Philadelphia: Published And Sold By George W. Mentz & Son, [1832].
35, [1] p. incl. wrappers; illus. HSP (wrappers)

445. The Farmers And Mechanics Almanack, For The Year ... 1833 .... By Charles Frederick Egelman ....
Philadelphia: Published And Sold By George W. Mentz & Son; Easton, Pa.: And sold by Henry Hammann, [1832].
34, [2] p.; illus. LCP

446. Neuer Calender für die Bauern und Handwerker auf das Jahr ... 1833 .... Fur den Pennsylvanischen Horizont und die angrenzenden Staaten berechnet von Carl Egelmann.
Philadelphia: Herausgegeben von Georg W. Mentz und Sohn, [1832].
35, [1] p. incl. wrappers; illus. LCP (wrappers); HSP (wrappers)

447. The Farmers' Almanac, For The Year 1834 .... By John Ward.
Philadelphia: Published by M'Carty & Davis, 1834.
36 p. HSP

448. The Farmers And Mechanics Almanack, For The Year ... 1834 .... By Charles Frederick Egelmann ....
Philadelphia: Published By George W. Mentz & Son, [1833].
34, [2] p. incl. wrappers; illus. LCP (wrappers); HSP (wrappers)

449. Neuer Calender für die Bauern und Handwerker auf das Jahr ... 1834 .... berechnet von Carl Egelmann.
Philadelphia: Herausgegeben von Georg W. Mentz und Sohn, [1833].
35, [1] p. incl. wrappers; illus. HSP in LCP (wrappers)

450. The Farmers And Mechanics Almanack, For The Year ... 1835 .... By Charles Frederick Egelmann ....
Philadelphia: Published By George W. Mentz, [1834].
HSP: 34 p. only (of 36 incl. wrappers; lacking back wrapper); illus.

451. The Farmer's Calendar, For The Year ... 1835 .... By Charles F. Egelmann.
Philadelphia: Published By Grigg & Elliot, [1834].
[32] p.; illus. HSP

452. Neuer Calender für Bauern und Handwerker auf das Jahr ... 1835 .... berechnet, von Carl F. Egelmann.
Philadelphia: Herausgegeben von Georg W. Mentz und Sohn, [1834].
35, [1] p. incl. wrappers; illus. HSP in LCP (wrappers); LCP (wrappers)

453. The Farmers' Almanac, For The Year 1836 .... By John Ward.
Philadelphia: Published by M'Carty & Davis, 1836.
36 p. LCP; HSP

454. The Farmers And Mechanics Almanack, For The Year ... 1836 .... By Charles Frederick Egelmann ....
Philadelphia: Published By George W. Mentz & Son, [1835].
34, [2] p. incl. wrappers; illus. Wrapper imprint: Easton, Pa: Published By H. Hammann. LCP (wrappers)

455. Neuer Calender für Bauern und Handwerker auf das Jahr ... 1836 .... berechnet, von Carl F. Egelmann.
Philadelphia: Herausgegeben von Georg W. Mentz und Sohn, [1835].
35, [1] p. incl. wrappers; illus. HSP in LCP (wrappers); LCP (wrappers)

456. The Farmers' Almanac, For The Year 1837 .... By John Ward.
Philadelphia: Published by M'Carty & Davis, 1837.
36 p. HSP

457. The Farmers And Mechanics Almanack, For The Year ... 1837 .... By Charles Frederick Egelmann ....
Philadelphia: Published By George W. Mentz & Son, [1836].
34, [2] p. incl. wrappers; illus. LCP (wrappers); HSP (front wrapper only)

458. Neuer Calender für Bauern und Handwerker auf das Jahr ... 1837 .... berechnet, von Carl F. Egelmann.
Philadelphia: Herausgegeben von Georg W. Mentz und Sohn, [1836].
35, [1] p. incl. wrappers; illus. HSP in LCP (wrappers)

459. The Farmers' Almanac, For The Year 1838 .... By John Ward.
Philadelphia: Published by M'Carty & Davis, 1838.
36 p. LCP; HSP

460. The Farmers And Mechanics Almanack, For The Year ... 1838 .... By Charles Frederick Egelmann ....
Philadelphia: Published By George W. Mentz & Son, [1837].
HSP: 32 p. only (of 36 incl. wrappers; lacking first and last leaf, i.e. wrappers); illus. HSP

461. Neuer Calender für Bauern und Handwerker, auf das Jahr ... 1838 .... berechnet, von Carl F. Egelmann.
Philadelphia: Herausgegeben von Georg W. Mentz und Sohn, [1837].
35, [1] p. incl. wrappers; illus. HSP (wrappers)

462. The Farmers' Almanac, For The Year 1839 .... By John Ward.
Philadelphia: Published by M'Carty & Davis, 1839.
HSP: 32 p. only (of 36; lacking last 2 leaves)

463. The Farmer's Almanac For The Year ... 1839 ....
Philadelphia: Turner & Fisher, [1838].
35, [1] p. HSP in LCP

464. The Farmers And Mechanics Almanack, For The Year ... 1839 .... By Charles Frederick Egelmann.
Philadelphia: Published By George W. Mentz & son, [1838].
34, [2] p. incl. wrappers; illus. LCP (lacking back wrapper); HSP (wrappers)

465. Keystone Agricultural Almanac, For 1839 ....
Philadelphia: Published and sold by William W. Walker, [1838].
35, [1] p.; illus. LCP; HSP

466. Neuer Calender für Bauern und Handwerker auf das Jahr ... 1839 .... berechnet, von Carl F. Egelmann.
Philadelphia: Herausgegeben von Georg W. Mentz und Sohn, [1838].
35, [1] p. incl. wrappers; illus. HSP in LCP (wrappers); LCP (wrappers)

467. The Farmers' Almanac, For The Year 1840 .... By John Ward.
Philadelphia: Published by M'Carty & Davis, 1840.
36 p. HSP

468. The Farmers And Mechanics Almanack, For The Year ... 1840 .... By Charles Frederick Egelmann.
Philadelphia: Published By George W. Mentz & Son, [1839].
LCP: 32 p. only (of 36 incl. wrappers; lacking first and last leaves, i.e. wrappers); illus.

469. Fithian's Silk Growers Almanac 1840.
Philadelphia: M. Fithian, Publisher, [1839].
31, [1] p.; illus. HSP

470. Keystone Agricultural Almanac, For 1840 ....
Philadelphia: Published and sold by William W. Walker, [1839].
35, [1] p.; illus. HSP

471. Neuer Calender für Bauern und Handwerker auf das Jahr ... 1840 .... berechnet, von Carl F. Egelmann.
Philadelphia: Herausgegeben von Georg W. Mentz und Sohn, [1839].
35, [1] p. incl. wrappers; illus. HSP in LCP (wrappers)

472. The Farmers' Almanac, For The Year 1841 .... By John Ward.
Philadelphia: Printed by M'Carty & Davis, 1841.
36 p. HSP

473. The Farmers And Mechanics Almanack, For The Year ... 1841 .... By Charles Frederick Egelmann ....
Philadelphia: Published By George W. Mentz & Son, [1840].
34, [2] p. incl. wrappers; illus. HSP (wrappers)

474. Keystone Agricultural Almanac, For 1841 ....
Philadelphia: Published and sold by William W. Walker, [1840.]
35, [1] p.; illus. LCP; HSP (lacking last leaf)

475. Neuer Calender für Bauern und Handwerker auf das Jahr ... 1841 .... berechnet, von Carl F. Egelmann.
Philadelphia: Herausgegeben von Georg W. Mentz und Sohn, [1840].
35, [1] p. incl. wrappers; illus. HSP in LCP (wrappers)

476. Agricultural Almanac, Designed for all parts of the Union ... For The Year ... 1842 .... To Which Is Added, A General Catalogue Of Agricultural Machines ... Garden, Field, and Flower Seeds. For sale at the Seed and Implement Warehouse .... Published And For Sale By D.O. Prouty ..., Philadelphia.
[Philadelphia:] Gihon, Fairchild & Co., printers, [1841].
69, [3] p.; illus. Incl. (pp. 33-45): Catalogue Of Vegetable Seeds; (pp. 45-66) Catalogue of Implements. LCP

477. The Farmers' Almanac, For The Year 1842 .... By John Ward.
Philadelphia: Published by M'Carty & Davis, 1842.
36 p. LCP (lacking last leaf); HSP

478. The Farmers And Mechanics Almanack, For The Year ... 1842 .... By Charles Frederick Egelmann ....
Philadelphia: Published By George W. Mentz, [1841].
34, [2] p. incl. wrappers; illus. HSP (wrappers)

479. Neuer Calender für Bauern und Handwerker auf das Jahr ... 1842 .... berechnet, von Carl F. Egelmann.
Philadelphia: Herausgegeben von Georg W. Mentz und Sohn, [1841].
35, [1] p. incl. wrappers; illus. HSP in LCP (wrapppers)

480. The Farmers' Almanac, For The Year 1843 .... By John Ward.
Philadelphia: Published by M'Carty & Davis, 1843.
36 p. LCP; HSP

481. The Farmers And Mechanics Almanack, For The Year ... 1843 .... By Charles Frederick Egelmann ....
Philadelphia: Published By Mentz & Rovoudt, (Successors To Geo. W. Mentz & Son), [1842].
34, [2] p. incl. wrappers; illus. LCP (wrappers); HSP (wrappers)

482. Neuer Calender für Bauern und Handwerker auf das Jahr ... 1843 .... berechnet, von Carl F. Egelmann.
Philadelphia: Herausgegeben von Mentz und Rovoudt, [1842].
35, [1] p. incl. wrappers; illus. HSP in LCP (wrappers)

483. The Citizens' And Farmers' Almanac, For The Year ... 1844 .... Original Calculations By Henry Frost ....
Philadelphia: Published by Desilver & Muir, Booksellers and Stationers, [1843].
[32] p. LCP

484. The Farmers' Almanac, For The Year 1844 .... By John Ward.
Philadelphia: Published by M'Carty & Davis, 1844.
36 p. HSP

485. The Farmers And Mechanics Almanack For The Year ... 1844 .... By Charles Frederick Egelmann ....
Philadelphia: Published By Mentz & Rovoudt, (Successors To Geo. W. Mentz & Son), [1843].
34, [2] p. incl. wrappers; illus. LCP (wrappers); HSP (wrappers)

486. Keystone Agricultural Almanac, For 1844 ....
Philadelphia: Published & Sold By William W. Walker, [1843].
35, [1] p.; illus. LCP; HSP

487. Neuer Calender für Bauern und Handwerker auf das Jahr ... 1844 .... berechnet, von Carl F. Egelmann.
Philadelphia: Herausgegeben von Mentz und Rovoudt, [1843].
35, [1] p. incl. wrappers; illus. HSP in LCP (wrappers)

488. The American Farmer's Almanac 1845.
Philadelphia and New York: Published By Turner & Fisher; Baltimore: J.B. Keller, [1844].
[32] p.; Drake 12368, with Philadelphia only listed in imprint, gives 18 leaves. HSP

489. The Delaware & Maryland Farmers' Almanac, For The Year 1845 ....
Philadelphia: Published By J. M'Dowell, [1844].
[36] p. HSP

490. The Farmers' Almanac, For The Year 1845 .... By John Ward.
Philadelphia: Published by Thomas Davis, 1845.
36 p. HSP

491. The Farmers And Mechanics Almanack, For The Year ... 1845 .... By Charles Frederick Egelmann ....
Philadelphia: Published By Mentz & Rovoudt, (Successors To Geo. W. Mentz & Son), [1844].
34, [2] p. incl. wrappers; illus. HSP (wrappers)

492. The Farmers' Almanac, For The Year 1846 .... By John Ward.
Philadelphia: Published by Thomas Davis, 1846.
36 p. HSP

493. The Farmer's And Farrier's Alamnac [sic] For 1846.
Philadelphia: John B. Perry; New York: Nafis & Cornish, [1845].
34, [2] p.; illus. LCP

494. The Farmer's And Farrier's Almanac For 1846.
Philadelphia: John B. Perry; New York: Nafis & Cornish, [1845].
34, [2] p.; illus. Copy with t.p. misprint corrected. HSP

495. The Farmers And Mechanics Almanack, For The Year ... 1846 .... By Charles Frederick Egelmann ....
Philadelphia: Published By Mentz & Rovoudt, (Successors To Geo. W. Mentz & Son), [1845].
34, [2] p. incl. wrappers; illus. HSP (wrappers)

496. Neuer Calender für Bauern und Handwerker auf das Jahr ... 1846 .... berechnet, von Carl F. Egelmann.
Philadelphia: Herausgegeben von Mentz und Rovoudt, [1845].
35, [1] p. incl. wrappers; illus. HSP in LCP (wrappers)

497. The Farmers' Almanac, For The Year 1847 .... By John Ward.
Philadelphia: Published by Thomas Davis, 1847.
36 p. HSP

498. The Farmer's And Farrier's Almanac For 1847.
Philadelphia: John B. Perry; New York: Nafis & Cornish, [1846].
35, [1] p.; illus. LCP; HSP

499. The Farmers And Mechanics Almanack, For The Year ... 1847 .... By Charles Frederick Egelmann ....
Philadelphia: Published By Mentz & Rovoudt, (Successors To Geo. W. Mentz & Son), [1846].
34, [2] p. incl. wrappers; illus. LCP (lacking back wrapper); HSP (wrappers)

500. Neuer Calender für Bauern und Handwerker auf das Jahr ... 1847 .... berechnet, von Carl F. Egelmann.
Philadelphia: Herausgegeben von Mentz und Rovoudt, [1846].
35, [1] p. incl. wrappers; illus. HSP in LCP (wrappers)

501. The Farmer's And Farrier's Almanac ... For 1848.
Philadelphia: John B. Perry; New York: Nafis & Cornish, [1847].
35, [1] p.; illus. HSP

502. The Farmers And Mechanics Almanack, For The Year ... 1848 .... By Charles Frederick Egelmann ....
Philadelphia: Published By Mentz & Rovoudt, (Successors To Geo. W. Mentz & Son), [1847].
34, [2] p. incl. wrappers; illus. Wrapper imprint: Philadelphia: Isaac M. Moss. LCP; HSP (imprint portion of wrapper mutilated)

503. Americanischer Stadt-und Land-Calender Auf das 1849ste Jahr ....
Philadelphia: Gedruckt und zu haben bey Conrad Zentler, [1848].
27, [3] p.; illus. HSP

504. The Farmers And Mechanics Almanack, For The Year ... 1849 .... By Charles Frederick Egelmann ....
Philadelphia: Published By Mentz & Rovoudt, [1848].
34, [2] p. incl. wrappers; illus. Wrapper imprint: Philadelphia: Isaac M. Moss And Brother. LCP (wrappers); HSP (wrappers)

505. Americanischer Stadt-und Land-Calender Auf das 1850ste Jahr ....
Philadelphia: Gedruckt und zu haben bei Ch. G. Sauer; Ehedem von Conrad Zentler herausgegeben, [1849].
35, [1] p.; illus. HSP

506. The Farmers And Mechanics Almanack, For The Year ... 1850 .... By Charles F. Eagelmann [sic] ....
Philadelphia: Published By William G. Mentz, [1849].
34, [2] p. incl. wrappers; illus. HSP (wrappers)

507. Neuer Kalender für Bauern und Handwerker auf das Jahr ... 1850 .... Von Carl F. Egelmann.
Philadelphia: Herausgegeben von Wilhelm G. Mentz, [1849].
HSP: 34 p. only (of 36 incl. wrappers; lacking back wrapper); illus.

508. Stadt-und Land-Calender Auf das Jahr Christi 1850 ....
Philadelphia: Zu haben bei John H. Simon, [1849].
35, [1] p.; illus. HSP

509. Der Bauern Calender. 1851.
Philadelphia: Herausgegeben von M. Brunner u. Co., [1850].
35, [1] p.; illus. HSP

510. Farmer's Almanac, And Housekeeper's Receipt Book. For 1851.
Philadelphia: Published By King & Baird; King & Baird, Printers, [1850].
35, [1] p.; illus. LCP; HSP

511. The Farmers And Mechanics Almanack, For The Year ... 1851 .... By Charles F. Eagelmann [sic] ....
Philadelphia: Published By William G. Mentz, [1850].
HSP: 34 p. only (of 36 incl. wrappers; lacking back wrapper); illus. HSP

512. The American Farmer's Almanac 1852.
Philadelphia; New York; Boston; Baltimore: Fisher & Brother, Publishers, [1851].
[36] p.; illus. HSP in LCP; HSP

513. Farmer's Almanac 1852.
Philadelphia: Published by John C. Davis, [1851].
34 p.; illus. HSP

514. Farmer's Almanac, And Housekeeper's Receipt Book, For 1852.
Philadelphia: Published By King & Baird; King & Baird, Printers, [1851].
[36] p.; illus. HSP

515. The Poultry Breeder's Almanac, For 1852.
Philadelphia: Published By King & Baird; King & Baird, Printers, [1851].
35 [i.e. 34], [2] p.; illus. P. 34 misnumbered 35. LCP; HSP (lacking t.p. and last leaf)

516. The United States Farmers Almanac 1852.
Philadelphia: Published by Sower & Barnes, [1851].
34, [2] p.; illus. HSP

517. The American Farmer's Almanac 1853.
Philadelphia; New York; Boston; Baltimore: Fisher & Brother, [1852].
[36] p.; illus. Fuller title p. 3 begins: Fisher & Brother's American Farmer's Almanac .... HSP

518. The Farmers And Mechanics Almanack, For The Year ... 1853 .... By Charles F. Eagelmann [sic] ....
Philadelphia: Published By William G. Mentz, [1852].
34, [2] p. incl. wrappers; illus. HSP (wrappers)

519. Neuer Kalender für Bauern und Handwerker auf das Jahr ... 1853 .... Von Carl F. Egelmann.
Philadelphia: Herausgegeben von Wilhelm G. Mentz, [1852].
HSP: Pp. [3]-34 only (of 36; lacking wrappers, i.e. pp. 1-2 and 35-36); illus.

520. Der Bauern Calender. 1854.
Philadelphia: Herausgegeben von John C. Davis, [1853].
HSP: 34 p. only (of 36; lacking last leaf); illus.

521. The Farmer's And Family Almanac. For The Year 1854.
Philadelphia: King & Baird; Weik & Wieck, [1853].
34, [2] p.; illus. LCP; HSP

522. The Farmers And Mechanics Almanack, For The Year ... 1854 .... By Charles F. Eagelmann ....
Philadelphia: Published By William G. Mentz, [1853].
34, [2] p. incl. wrappers; illus. HSP (wrappers)

523. Neuer Kalender für Bauern und Handwerker auf das Jahr ... 1854 .... Von Carl F. Egelmann.
Philadelphia: Herausgegeben von Wilhelm G. Mentz, [1853].
34, [2] p. incl. wrappers; illus. Wrapper imprint: Philadelphia: Isaac M. Moss und Bruder. HSP in LCP (wrappers)

524. Neuer Kalender für Bauern und Handwerker ... 1854 ....
Philadelphia: Herausgegeben von Wilhelm G. Mentz, [1853].
34, [2] p. incl. wrappers; illus. Wrapper imprint: Philadelphia: Peck & Bliss. HSP in LCP (wrappers)

525. The American Farmer's Almanac 1855.
[Philadelphia:] Published By J.W. Clothier & Co., Stationers, [1854].
HSP: [32] p. only (of 36; lacking last 2 leaves); illus. Fuller title p. 3 begins: Fisher & Brother's American Farmer's Almanac ....; at bottom of p. 3: Published annually, by Fisher & Brother, Philadelphia, New York and Baltimore. HSP also holds a 2nd copy without t.p. but with text to p. 34.

526. The Farmer's And Family Almanac. For The Year 185[5].
Philadelphia: Published By King & Baird, [1854].
34, [2] p.; illus. HSP (with last digit of date supplied in MS.)

527. The Farmers And Mechanics Almanack, For The Year ... 1855 .... By Charles F. Eagelmann ....
Philadelphia: Published By William G. Mentz, [1854].
34, [2] p. incl. wrappers; illus. HSP (wrappers)

528. Neuer Kalender für Bauern und Handwerk [sic] auf das Jahr ... 1855 .... Von Carl F. Egelmann ....
Philadelphia: Herausgegeben von Wilhelm G. Mentz, [1854].
32, [4] p. incl. wrappers; illus. HSP in LCP (wrappers)

529. The Poultry Breeder's Almanac. For The Year 1855.
Philadelphia: Published By King & Baird, [1854].
34, [2] p.; illus. HSP

530. The American Farmer's Almanac 1856.
Philadelphia; New York; Boston; Baltimore: Fisher & Brother, [1855].
[36] p.; illus. Fuller title p. 3 begins: Fisher & Brother's American Farmer's Almanac .... HSP

531. The Farmer's And Family Almanac. For The Year 1856.
Philadelphia: Published By Bast & Miller, (Successors to John H. Simon,) [1855].
34, [2] p.; illus. HSP

532. Landreths' Rural Register And Almanac For 1856. For Gratuitous Distribution.
[Philadelphia:] Stavely & McCalla Printers, [1855].
84 p.; illus. HSP in LCP; LCP

533. Neuer Kalender für Bauern und Handwerker auf das Jahr ... 1856 .... Von Carl F. Egelmann ....
Philadelphia: Herausgegeben von Wilhelm G. Mentz, [1855].
34, [2] p. incl. wrappers; illus. HSP in LCP (wrappers)

534. The Poultry Breeder's Almanac. For The Year 1856.
Philadelphia: Published By King & Baird, [1855].
HSP: 34 p. only (of 36; lacking last leaf); illus.

535. The American Farmer's Almanac 1857.
Philadelphia; New York; Boston; Baltimore: Fisher & Brother, [1856].
35, [1] p.; illus. Fuller title p. 3 begins: Fisher & Brother's American Farmer's Almanac .... HSP

536. Der Amerikanische Stadt-und Land-Kalender auf das Jahr 1857 .... Für den Pennsylvanischen Horizont und die angrenzenden Staaten berechnet.
Philadelphia: Herausgegeben von Ch. G. Sauer, [1856].
34 p.; illus. HSP

537. The Farmer's And Family Almanac, For The Year 1857.
Philadelphia: Published By King & Baird, [1856].
34, [2] p.; illus. HSP

538. The Farmers And Mechanics Almanack, For The Year ... 1857 .... By Charles F. Eagelmann ....
Philadelphia: Published By William G. Mentz, [1856].
34, [2] p. incl. wrappers; illus. HSP (wrappers)

539. Landreths' Rural Register And Almanac For 1857 ....
[Philadelphia:] Stavely & McCalla Printers, [1856].
55, [1] p.; illus. HSP

540. The Poultry Breeder's Almanac: For The Year 1857.
Philadelphia: Published By King & Baird, [1856].
34, [14] p.; illus. Last 7 leaves chiefly adv. HSP

541. The American Farmer's Almanac 1858.
Philadelphia; New York; Boston; Baltimore: Fisher & Brother, [1857].
31, [1] p.; illus. Fuller title p. 3 begins: Fisher & Brother's American Farmer's Almanac .... HSP

542. Amerikanischer Stadt-und Land-Calender Auf das 1858ste Jahr ....
Philadelphia: Gedruckt und zu haben bei Sauer & Barnes, [1857].
34, [2] p.; illus. HSP

543. The Farmer's And Family Almanac, For The Year 1858.
Philadelphia: Published By King & Baird, [1857].
33, [3] p.; illus. HSP

544. The Farmer's And Family Almanac, For The Year 1858.
Philadelphia: Published By S.A. Roberts, [1857].
33, [3] p.; illus. HSP

545. Landreths' Rural Register And Almanac For 1858 ....
[Philadelphia:] J.S. McCalla Printer, [1857].
60 p. HSP

546. The Poultry Breeder's Almanac: For The Year 1858.
Philadelphia: Published By S.A. Roberts [Printed by King & Baird (colop.)], [1857].
33, [3] p.; illus. HSP

547. Stadt-und Land-Calender Auf das Jahr Christi 1858 ....
Philadelphia: Verlag von King und Baird, [1857].
34, [2] p.; illus. HSP

548. The United States Farmers Almanac 1858.
Philadelphia: Published by Sower & Barnes, [1857].
33, [3] p.; illus. HSP

549. The American Farmer's Almanac 1859.
Philadelphia; New York; Boston; Baltimore: Fisher & Brother, [1858].
35, [1] p.; illus. Fuller title p. 3 begins: Fisher & Brother's American Farmer's Almanac .... HSP

550. Amerikanischer Stadt-und Land-Calender Auf das 1859ste Jahr ....
Philadelphia: Gedruckt und zu haben bei Sauer & Barnes, [1858].
34, [2] p.; illus. HSP

551. Cottage Garden Almanac For The Year 1859.
Philadelphia: Published by King & Baird, [1858].
33, [3] p.; illus. LCP.

552. Farmer's Almanac, For The Year 1859.
Philadelphia: Published By King & Baird, [1858].
33, [3] p.; illus. HSP

553. Landreths' Rural Register And Almanac For 1859 ....
[Philadelphia: 1858].
60 p. LCP (lacking last leaf [adv.]); HSP

554. Stadt-und Land-Calender Auf das Jahr Christi 1859 ....
Philadelphia: Herausgegeben von King und Baird, [1858].
33, [3] p.; illus. HSP

555. The United States Farmers Almanac 1859.
Philadelphia: Published by Sower & Barnes, [1858].
33, [3] p.; illus. HSP

556. The American Farmer's Almanac 1860.
Philadelphia; New York; Boston; Baltimore: Fisher & Brother, [1859].
35, [1] p.; illus. Fuller title p. 3 begins: Fisher & Brother's American Farmer's Almanac .... HSP

557. Amerikanischer Stadt-und Land-Calender Auf das 1860ste Jahr ....
Philadelphia: Gedruckt und zu haben bei Sauer & Barnes, [1859].
34, [2] p.; illus. HSP

558. Farmer's Almanac, For The Year 1860.
Philadelphia: Published By King & Baird, [1859].
33, [3] p.; illus. HSP

559. Landreths' Rural Register And Almanac 1860 ....
[Philadelphia:] J.S. McCalla, Printer, [1859].
60 p. LCP (lacking last leaf [adv.]); HSP

560. Stadt-und Land-Calender Auf das Jahr Christi 1860 ....
Philadelphia: Verlag von Hoffmann und Morwitz, [1859].
34, [2] p.; illus. HSP

561. The United States Farmers Almanac 1860.
Philadelphia: Published by Sower & Barnes, [1859].
33, [3] p.; illus. HSP

562. The American Farmer's Almanac 1861.
Philadelphia: J. & E. Newbert, [1860].
35, [1] p.; illus. Fuller title p. 3 begins: Fisher & Brother's American Farmer's Almanac ...; at bottom of p. 3: Published Annually, by Fisher & Brother, Philadelphia, New York, Boston and Baltimore. HSP

563. Farmer's Almanac, For The Year 1861.
Philadelphia: Published By King & Baird, [1860].
33, [3] p.; illus. HSP

564. The Farmer's And Gardener's Almanac For 1861 ....
Philadelphia: Published & Sold by McDowell & Halloway, [1860].
35, [1] p.; illus. HSP

565. The Farmer's And Gardener's Almanac For 1861 ....
Philadelphia: Almanacs printed and for Sale, By J. Van Court, [1860].
35, [1] p.; illus. HSP

566. Fruit Grower's Almanac, For The Year 1861.
Philadelphia: Published By Hollowbush & Carey [King & Baird, Printers (colop.)], [1860].
33, [3] p.; illus. HSP

567. Stadt-und Land-Calender Auf das Jahr Christi 1861 ....
Philadelphia: Zu haben bei Milton Benner; Gedruckt bei King und Baird, [1860].
32, [4] p.; illus. HSP

568. Stadt-und Land-Calender Auf das Jahr Christi 1861 ....
Philadelphia: Herausgegeben von King und Baird, [1860].
33, [3] p.; illus. HSP

569. The United States Farmers Almanac 1861.
Philadelphia: Published by Sower & Barnes, [1860].
33, [3] p.; illus. HSP

__________. PITTSBURGH.

570. The Western Farmer's Almanac, For The Year ... 1828 .... Calculated By The Rev. John Taylor ....
Pittsburgh: Published by H. Holdship; Printed By D. & M. Maclean, [1827].
[36] p. T.p. headed: Number II. LCP

__________. PORTLAND, MAINE.

571. The Farmers' Almanac, For The Year ... 1823 .... Calculated For The Meridian Of Portland .... Portland: Printed And Published At The Gazette And Argus Offices, And ... at William Hyde's and Isaac Adams' Bookstores ..., [1822].
[48] p. HSP

572. The Farmer's Almanack ... For The Year ... 1838 .... Calculated For The State Of Maine .... By Robert B. Thomas.
Portland: Published By William Hyde, [1837].
[42] p.; illus. T.p. headed: Number XLVI. Drake 2061 has 22 leaves. HSP

573. The Farmer's Almanack ... For The Year ... 1844 .... By Robert B. Thomas.
Portland, Maine: Sanborn & Carter, [1843].
46 p.; illus. T.p. headed: Maine Edition. Number LII. Drake 2112 has 24 leaves. HSP

574. The Farmer's Almanack ... For The Year ... 1846 .... By Robert B. Thomas.
Portland, Me.: Hyde, Lord & Duren, [1845].
46 p.; illus. T.p. headed: Maine Edition. Number LIV. Drake 2119 has 24 leaves. HSP

__________. POUGHKEEPSIE, N.Y.

575. The Farmer's Almanac, For The Year ... 1817 .... Calculated for the Meridian of Poughkeepsie .... By Andrew Beers, Philom.
Poughkeepsie: Printed By P. & S. Potter; Sold ... by P. Potter and by Stockholm & Brownejohn; Newburgh: P.B. Pratt, _et al_., [1816].
[36] p. HSP

__________. PROVIDENCE.
See The (Old) Farmer's Almanack ... 1851; Boston. Wrapper has Providence imprint.

__________. READING, PA.

576. Der Neue, Americanische Landwirthschafts-Calender Auf das Jahr ... 1811 .... Zum Sechstenmal herausgegeben.
Reading: gedruckt und zu haben bey Johann Ritter und Comp., [1810].
[42] p. incl. wrapper. HSP in LCP (wrapper)

577. Der Neue, Americanische Landwirthschafts-Calender, Auf das Jahr ... 1812 .... Zum Siebentenmal herausgegeben.
Reading: gedruckt und zu haben bey Johann Ritter und Comp., [1811].
[40] p. incl. wrapper; illus. HSP in LCP (wrapper); LCP (lacking wrapper and last 3 leaves)

578. Der Neue, Americanische Landwirthschafts-Calender, Auf das Jahr ... 1813 .... Zum Achtenmal herausgegeben.
Reading: gedruckt und zu haben bey Johann Ritter und Comp., [1812].
[42] p. incl. wrapper; illus. HSP in LCP (wrapper)

579. Der Neue, Americanische Landwirthschafts-Calender Auf das Jahr ... 1814 .... Zum Neuntenmal herausgegeben.
Reading: gedruckt und zu haben bey Johann Ritter und Comp., [1813].
[40] p. incl. wrapper; illus. Drake 11018 has most copies with 21 leaves. HSP in LCP (wrapper); LCP (1st 28 p. only)

580. Der Neue, Americanische Landwirthschafts-Calender, Auf das Jahr ... 1815 .... Zum Zehntenmal herausgegeben.
Reading: gedruckt und zu haben bey Johann Ritter und Comp., [1814].
[40] p. incl. wrapper; illus. Drake 11055 has most copies with 21 leaves. HSP in LCP (wrapper)

581. Der Neue, Americanische Landwirthschafts-Calender, Auf das Jahr ... 1816 .... Zum Elftenmal herausgegeben.
Reading: gedruckt und zu haben bey Johann Ritter und Comp., [1815].
[42] p. incl. wrapper; illus. HSP in LCP (wrapper)

582. Der Neue, Americanische Landwirthschafts-Calender, Auf das Jahr ... 1817 .... Zum Zwolftenmal herausgegeben.
Reading: gedruckt und zu haben bey Johann Ritter und Comp., [1816].
[36] p. incl. wrapper; illus. HSP in LCP (wrapper)

583. Der Neue, Americanische Landwirthschafts-Calender, Auf das Jahr ... 1818 .... Zum Dreyzehntenmal herausgegeben.
Reading: gedruckt und zu haben bey Joseph Ritter und Comp., [1817].
[36] p. incl. wrappers; illus. HSP in LCP (wrappers); LCP (wrappers)

584. Der Neue, Americanische Landwirthschafts-Calender, Auf das Jahr ... 1819 .... Zum Vierzehntenmal herausgegeben.
Reading: gedruckt und zu haben bey Johann Ritter und Comp., [1818].
[36] p. incl. wrappers; illus. HSP in LCP (wrappers); LCP (wrappers)

585. Der Neue, Americanische Landwirthschafts-Calender, Auf das Jahr ... 1820 .... Zum Funfzehntenmal herausgegeben.
Reading, (Penn.): gedruckt und zu haben bey Johann Ritter und Comp., [1819].
[36] p. incl. wrappers; illus. HSP in LCP (wrappers); LCP (lacking front wrapper)

586. Der Neue, Americanische Landwirthschafts-Calender, Auf das Jahr ... 1821 .... Zum Sechszehntenmal herausgegeben.
Reading, (Penn.): gedruckt und zu haben bey Johann Ritter und Comp., [1820].
[36] p. incl. wrappers; illus. HSP in LCP (lacking front wrapper and last 2 leaves); LCP (wrappers)

587. Der Neue, Americanische Landwirthschafts-Calender, Auf das Jahr ... 1822 .... Zum Siebzehntenmal herausgegeben.
Reading, (Penn.): gedruckt und zu haben bey Johann Ritter und Comp., [1821].
[36] p. incl. wrappers; illus. HSP in LCP (wrappers); LCP (wrappers)

588. Der Neue, Americanische Landwirthschafts-Calender, Auf das Jahr ... 1823 .... Zum Achtzehntenmal herausgegeben.
Reading, (Penn.): gedruckt und zu haben bey Johann Ritter und Comp., [1822].
[36] p. incl. wrappers; illus. HSP in LCP (wrappers); LCP (wrappers)

589. Der Neue, Americanische Landwirthschafts-Calender, Auf das Jahr ... 1824 .... Von Carl Friedrich Egelmann .... Zum Neunzehntenmal herausgegeben.
Reading, (Penn.): gedruckt und zu haben bey Johann Ritter und Comp., [1823].
[36] p. incl. wrappers; illus. HSP in LCP (wrappers); LCP (wrappers)

590. Der Neue, Americanische Landwirthschafts-Calender, Auf das Jahr ... 1825 .... Von Carl Friedrich Egelmann .... Zum Zwanzigstenmal herausgegeben.
Reading, (Penn.): gedruckt und zu haben bey Johann Ritter, [1824].
36 p. incl. wrappers. HSP in LCP (wrappers)

591. Der Neue, Americanische Landwirthschafts-Calender, Auf das Jahr ... 1825 ....
Reading, (Penn.): gedruckt und zu haben bey Johann Ritter, [1824].
36 p. incl. back wrapper. Copy without front wrapper; with t.p. as 1st leaf; new matter on A2; and remaining pp. as in above copy. LCP

592. Der Neue, Americanische Landwirthschafts-Calender, Auf das Jahr ... 1826 .... Von Carl Friedrich Egelmann .... Zum Einundzwanzigstenmal herausgegeben.
Reading, (Penn.): gedruckt und zu haben bey Johann Ritter, [1825].
36 p. incl. wrappers. HSP in LCP (wrappers); LCP (wrappers)

593. Der Neue, Americanische Landwirthschafts-Calender, Auf das Jahr ... 1827 .... Von Carl Friedrich Egelmann .... Zum Zweyundzwanzigstenmal herausgegeben.
Reading, (Penn.): gedruckt und zu haben bey Johann Ritter, [1826].
36 p. incl. wrappers. HSP in LCP (wrappers)

594. Der Neue, Americanische Landwirthschafts-Calender, Auf das Jahr ... 1828 .... Von Carl Friedrich Egelmann .... Zum Dreyundzwanzigstenmal herausgegeben.
Reading, (Penn.): gedruckt und zu haben bey Johann Ritter, [1827].
36 p. incl. wrappers. HSP in LCP (wrappers; lacking pp. 33-34); LCP (wrappers)

595. Der Neue, Americanische Landwirthschafts-Calender, Auf das Jahr ... 1829 .... Von Carl Friedrich Egelmann .... Zum Vierundzwanzigstenmal herausgegeben.
Reading, (Penn.): gedruckt und zu haben bey Johann Ritter u. Comp., [1828].
36 p. incl. wrappers. HSP in LCP (wrappers); LCP (wrappers)

596. Der Neue, Americanische Landwirthschafts-Calender, Auf das Jahr ... 1830 .... Von Carl Friedrich Egelmann .... Zum Fünfundzwanzigstenmal herausgegeben.
Reading, (Penn.): gedruckt und zu haben bey Johnann Ritter u. Comp., [1829].
36 p. incl. wrappers. HSP in LCP (wrappers); LCP (wrappers)

597. Der Neue, Americanische Landwirthschafts-Calender, Auf das Jahr ... 1831 .... Von Carl Friedrich Egelmann .... Zum Sechsundzwanzigstenmal herausgegeben.
Reading, (Penn.): gedruckt und zu haben bey Johann Ritter u. Comp., [1830].
36 p. incl. wrappers; illus. LCP (wrappers)

598. Der Neue, Americanische Landwirthschafts-Calender, Auf das Jahr ... 1832 .... Von Carl Friedrich Egelmann .... Zum Siebenundzwanzigstenmal herausgegeben.
Reading, (Penn.): gedruckt und zu haben bey Johann Ritter u. Comp., [1831].
[32] p. incl. wrappers; illus. LCP (wrappers)

599. Der Neue, Americanische Landwirthschafts-Calender, Auf das Jahr ... 1833 .... Von Carl Friedrich Egelman .... Zum Achtundzwanzigstenmal herausgegeben.
Reading, (Penn.): gedruckt und zu haben bey Johann Ritter u. Comp., [1832].
[32] p. incl. wrappers; illus. HSP in LCP (wrappers); LCP (wrappers)

600. Der Neue, Americanische Landwirthschafts-Calender, Auf das Jahr ... 1834 .... Von Carl Friedrich Egelmann .... Zum Neunundzwanzigstenmal herausgegeben.
Reading, (Penn.): gedruckt und zu haben bey Johann Ritteru. [sic] Comp., [1833].
[32] p. incl. wrappers. HSP in LCP (wrappers); LCP (wrappers)

601. Der Neue, Americanische Landwirthschafts-Calender, Auf das Jahr ... 1835 .... Von Carl Friedrich Egelmann .... Zum Dreyssigstenmal herausgegeben.
Reading, (Penn.): gedruckt und zu haben bey Johann Ritter u. Comp., [1834].
[32] p. incl. wrappers. HSP in LCP (wrappers); LCP (wrappers)

602. Der Neue, Americanische Landwirthschafts-Calender, Auf das Jahr ... 1836 .... Von Carl Friedrich Egelmann .... Zum ein und dreyssigstenmal herausgegeben.
Reading, (Penn.): gedruckt und zu haben bey Johann Ritter u. Comp., [1835].
[32] p. incl. wrappers; illus. HSP in LCP (wrappers); LCP (wrappers)

603. Der Neue, Americanische Landwirthschafts-Calender, Auf das Jahr ... 1837 .... Von Carl Friedrich Egelmann .... Zum Zwey und dreyssigstenmal herausgegeben.
Reading, (Penn.): gedruckt und zu haben bey Johann Ritter u. Comp., [1836].
[32] p. incl. wrappers; illus. HSP in LCP (wrappers); LCP (wrappers)

604. Der Neue, Americanische Landwirthschafts-Calender, Auf das Jahr ... 1838 .... Von Carl Friedrich Egelmann .... Zum drey und dreyssigstenmal herausgegeben.
Reading, (Penn.): gedruckt und zu haben bey Johann Ritter u. Comp., [1837].
[32] p. incl. wrappers; illus. HSP in LCP (wrappers); LCP (lacking back wrapper)

605. Der Neue, Americanische Landwirthschafts-Calender, Auf das Jahr ... 1839 .... Von Carl Friedrich Egelmann .... Zum vier und dreyssigstenmal herausgegeben.
Reading, (Penn.): gedruckt und zu haben bey Johann Ritter u. Comp., [1838].
[32] p. incl. wrappers; illus. HSP in LCP (wrappers); LCP (wrappers)

606. Der Neue, Amerikanische Landwirthschafts-Calender, Auf das Jahr ... 1840 .... Von Carl Friederich Egelmann .... Zum fünf und dreyssigstenmal herausgegeben.
Reading, (Penn.): gedruckt und zu haben bey John Ritter und Comp., [1839].
35, [1] p. incl. wrappers; illus. HSP in LCP (wrappers); LCP (wrappers)

607. Der Neue, Amerikanische Landwirthschafts-Calender, Auf das Jahr ... 1841 .... Von Carl Friederich Egelmann .... Zum sechs und dreyssigstenmal herausgegeben.
Reading, (Penn.): gedruckt und zu haben bey John Ritter und Comp., [1840].
35, [1] p. incl. wrappers; illus. LCP (wrappers)

608. Der Neue, Amerikanische Landwirthschafts-Calender, Auf das Jahr ... 1842 .... Von Carl Friederich Egelmann .... Zum sieben und dreyssigstenmal herausgegeben.
Reading, (Penn.): gedruckt und zu haben bey John Ritter und Comp., [1841].
35, [1] p. incl. wrappers; illus. HSP in LCP (wrappers)

609. Der Neue, Amerikanische Landwirthschafts-Calender, Auf das Jahr ... 1843 .... Von Carl Friederich Egelmann .... Zum acht und dreyssigstenmal herausgegeben.
Reading, (Penn.): gedruckt und zu haben bey John Ritter und Comp., [1842].
31, [1] p. incl. wrappers; illus. HSP in LCP (wrappers)

610. Der Neue, Amerikanische Landwirthschafts-Calender, Auf das Jahr ... 1844 .... Von Carl Friederich Egelmann .... Zum neun und dreyssigstenmal herausgegeben.
Reading, (Penn.): gedruckt und zu haben bey John Ritter und Comp., [1843].
31, [1] p. incl. wrappers; illus. HSP in LCP (wrappers); LCP (wrappers)

611. Der Neue, Amerikanische Landwirthschafts-Calender, Auf das Jahr ... 1845 .... Von Carl Friederich Egelmann .... Zum vierzigstenmal herausgegeben.
Reading, (Penn.): gedruckt und zu haben bey John Ritter und Comp., [1844].
31, [1] p. incl. wrappers; illus. HSP in LCP (wrappers)

612. Der Neue, Amerikanische Landwirthschafts-Calender, Auf das Jahr ... 1846 .... Von Carl Friederich Egelmann .... Zum ein und vierzigstenmal herausgegeben.
Reading, (Penn.): gedruckt und zu haben bey John Ritter und Comp., [1845].
31, [1] p. incl. wrappers; illus. HSP in LCP (wrappers)

613. Der Neue, Amerikansiche Landwirthschafts-Calender, Auf das Jahr ... 1847 .... Von Carl Friederich Egelmann .... Zum zwey und vierzigstenmal herausgegeben.
Reading, (Penn.): gedruckt und zu haben bey John Ritter und Comp., [1846].
31, [1] p. incl. wrappers; illus. HSP in LCP (wrappers)

614. Der Neue, Amerikanische Landwirthschafts-Calender, Auf das Jahr ... 1848 .... Von Carl Friederich Egelmann .... Zum drey und vierzigstenmal herausgegeben.
Reading, (Penn.): gedruckt und zu haben bey John Ritter und Comp., [1847].
31, [1] p. incl. wrappers; illus. HSP in LCP (wrappers)

615. Der Neue, Amerikanische Landwirthschafts-Calender, Auf das Jahr ... 1849 .... Von Carl Friederich Egelmann .... Zum vier und vierzigstenmal herausgegeben.
Reading, (Penn.): gedruckt und zu haben bey John Ritter und Comp., [1848].
30, [2] p. incl. wrappers; illus. HSP in LCP (wrappers)

616. Der Neue, Amerikanische Landwirthschafts-Calender, Auf das Jahr ... 1850 .... Von Carl Friederich Egelmann .... Zum fünf und vierzigstenmal herausgegeben.
Reading, (Penn.): gedruckt und zu haben bey John Ritter und Comp., [1849].
31, [1] p. incl. wrappers; illus. HSP in LCP (wrappers)

617. Der Neue, Amerikanische Landwirthschafts-Calender, Auf das Jahr ... 1851 .... Von Carl Friederich Egelmann .... Zum sechs und vierzigstenmal herausgegeben.
Reading, (Penn.): gedruckt und zu haben bey John Ritter und Comp., [1850].
35, [1] p. incl. wrappers; illus. HSP in LCP (wrappers)

618. Der Neue, Amerikanische Landwirthschafts-Calender, Auf das Jahr ... 1852 .... Von Carl Friederich Egelmann .... Zum sieben und vierzigstenmal herausgegeben.
Reading, (Penn.): gedruckt und zu haben bey John Ritter und Comp., [1851].
31, [1] p. incl. wrappers; illus. HSP in LCP (wrappers)

619. Der Neue, Amerikanische Landwirthschafts-Calender, Auf das Jahr ... 1854 .... Von Carl Friederich Egelmann .... Zum neun und vierzigstenmal herausgegeben.
Reading, (Penn.): gedruckt und zu haben bey Ritter und Co., [1853].
31, [1] p. incl. wrappers; illus. HSP in LCP (wrappers)

620. Der Neue, Amerikanische Landwirthschafts-Calender, Auf das Jahr ... 1856 .... Von Carl Friederich Egelmann .... Zum einundfünfzigstenmal herausgegeben.
Reading, (Penn.): gedruckt und zu haben bey Ritter u. Co, [1855].
39, [1] p. incl. wrappers; illus. HSP in LCP (wrappers)

__________. SAINT LOUIS, MO.

621. Stadt-und Land-Calender Auf das Jahr Christi 1856 ....
St. Louis, Mo.: Zu haben bei C. Witters, [1855].
32, [4] p.; illus. HSP

__________. SANDY HILL, N.Y.

622. Howland, Harvey & Co's Farmers Almanac, For 1849. Calculations Being For The Meridian Of Albany.
Sandy Hill, N.Y.: Published By Howland, Harvey & Co., [1848].
[24] p. HSP

__________. STILLWATER, N.Y.

623. The Farmers' Almanac For 1846: .... Calculated for the Meridian of Stillwater.
Stillwater, Saratoga County, N.Y.: Published For The General Store Of Samuel G. Eddy; Albany: J. Munsell, Printer, [1845].
[36] p.; illus. T.p. headed: No. 2. HSP

__________. TRENTON.

624. Wilson's Farmers' Almanac, For The Year ... 1817.... Calculated for the Latitude and Meridian of Philadelphia By Joshua Sharp ....
Trenton: Published By James J. Wilson, [1816].
[36] p. HSP

625. Farmers, [sic] Almanac, For The Year ... 1852 .... Calculated For The Horizon And Meridian Of New Jersey .... By David Young, Philom.
Trenton, N.J.: Published And Sold ... By Charles Scott, [1851].
35, [1] p.; illus. HSP

626. Farmer's Almanac, For The Year ... 1856 .... By Samuel Hart Wright ....
Trenton, N.J.: Published And Sold ... By Charles Scott & Co., [1855].
35, [1] p.; illus. T.p. headed: No. 31. HSP

627. Farmer's Almanac, For The Year ... 1857 .... By Samuel Hart Wright ....
Trenton, N.J.: C. Scott & Co., [1856].
35, [1] p.; illus. T.p. headed: No. 32. HSP

__________. TROY, N.Y.

628. The Troy Almanac And Farmers' Calendar For 1855 .... Calculated for the Latitude and Longitude of Albany ....
Troy, N.Y.: William H. Young, Wholesale And Retail Bookseller And Stationer, [1854].
[36] p.; illus. HSP

__________. WETHERSFIELD, CONN.

629. The Gardener's Almanac 1850, Calculated For Boston, New York, Baltimore And Charleston .... Containing General Directions For The Management Of The Vegetable Garden; A Description Of The Finest Culinary Vegetables; And A Select List Of Flowers. (Fourth Edition.) [wrapper title]
Wethersfield, Conn: Published By Comstock, Ferre & Co., Proprietors Of The Wethersfield Seed Garden, [1849]. [wrapper imprint]
[12], 48 p.; illus. The Almanac takes up the 1st 12 p.; sep. t.p. and pagination for: Descriptive Catalogue Of Garden Seeds Cultivated And Sold At The .... LCP (wrappers)

__________. WHEELING, [W.] VA.

630. The Farmer's Almanac, For The Year ... 1820 ... Calculated for the Horizon and meridian of Philadelphia .... By Anbrew [sic] Beers, Philom.
Wheeling, [W.] Virginia: Published By S. Potter, & Co., Booksellers and Stationers; Philadelphia: Sold also at their Book and Stationary Store, [1819].
[36] p.; Drake 14147 has 16 leaves. HSP

631. Fisher's Farmers' And Mechanics' Almanac, For The Year ... 1839. Calculated By John Armstrong ....
Wheeling: Published By Robt. Fisher, Bookseller; John M. M'Creary, Printer, [1838].
60 p.; illus. Incl. (pp. 37-60): Magazine Almanac, For 1839. HSP

632. Fisher's Farmers' And Mechanics' Almanac, For The Year ... 1840. Calculated By John Armstrong ....
Wheeling: Published By A. & R. Fisher, Booksellers; John M. M'Creary, Printer, [1839].
58 p.; illus. Incl. (pp. 37-58): Magazine Almanac. For 1840. HSP

633. Farmers' And Mechanics' Almanac, For The Year ... 1841. Calculated By John Armstrong ....
Wheeling: Published By Stephenson & Garwood, Booksellers; John M. M'Creary, Printer, [1840].
60 p.; illus. Incl. (pp. 37-60): Magazine Almanac. 1841. HSP

634. Farmers' And Mechanics' Almanac, For The Year ... 1842. Calculated By John Armstrong ....
Wheeling: Published By Robb & Stephenson, Booksellers, &c.; John M. M'Creary, [1841].
HSP: 58 p. only (of 60; lacking last leaf); illus. Incl. (pp. 37-58): Magazine Almanac For 1842. HSP

635. Farmers' And Mechanics' Almanac, For The Year ... 1843. Calculated By John Armstrong ....
Wheeling: Published By Robb & Stephenson Booksellers, &c.; John M. M'Creary, [1842].
60 p.; illus. Incl. (pp. 37-60): Magazine Almanac. 1843. HSP

__________. WILMINGTON, DEL.

636. Bonsal and Niles' Town And Country Almanac, For The Year ... 1801 .... Calculated to the Meridian of the Middle States ....
Wilmington: Printed and sold by Bonsal And Niles, [1800].
[48] p. HSP

637. Bonsal & Niles' Citizen And Farmer's Almanac, For The Year ... 1803 .... The astronomical calculations by Isaac Briggs.
Wilmington: Printed and sold by Bonsal & Niles, [1802].
HSP: [44] p. only (of 48; lacking last 2 leaves)

638. The Annual Visiter: And Citizen And Farmers Almanac, For The Year 1808 .... The Astronomical Calculations, suited to the Meridian of the middle states, by Joshua Sharp ....
Wilmington, De.: Printed and sold by James Wilson; & M.R. Lockerman, [1807].
[48] p. LCP; HSP

639. The Annual Visiter: And Citizen And Farmers Almanac, For The Year 1809 .... Calculations ... by Joshua Sharp ....
Wilmington, De.: Printed and Sold by James Wilson; & M.R. Lockerman, [1808].
LCP: [48] p. on 30 leaves; HSP: [48] p.

640. The Annual Visiter And Citizen and Farmers Almanac, For The Year 1810 .... Calculations ... by Joshua Sharp ....
Wilmington, Del.: Printed and sold by James Wilson, [1809].
[48] p. LCP

641. The Annual Visiter And Citizen and Farmers Almanac, For The Year 1811. .... Calculations ... by Joshua Sharp ....
Wilmington, Del.: Printed and sold by James Wilson, [1810.]
[50] p.; Drake 1431 has 24 leaves. LCP

642. The Annual Visiter And Citizen And Farmer's Almanac For the Year 1812 .... Calculations ... by Joshua Sharp ....
Wilmington, Del: Printed and sold by James Wilson, [1811].
[48] p.; LCP

643. The Citizen & Farmer's Almanac, For, The Year ... 1812 ....
Wilmington, Del.: Printed And Sold By Robert Porter, 1812.
[40] p.; Drake 1434 has most copies with 24 leaves. HSP

644. The Annual Visiter And Citizen And Farmer's Almanac, For The Year 1813 .... Calculations ... by Joshua Sharp ....
Wilmington, Del: Printed and sold by James Wilson, [1812.]
[48] p. LCP.

645. The Citizen & Farmers Almanac, For The Year ... 1813 ....
Wilmington, Del.: Printed And Sold By Robert Porter, 1813.
[36] p.; Drake 1439 has 24 leaves. HSP

646. The Annual Visiter And Citizen And Farmer's Almanac, For The Year 1814 .... Calculations ... by Joshua Sharp ....
Wilmington, Del: Printed and sold by James Wilson, [1813.]
[48] p. LCP.

647. The Annual Visiter And Citizen And Farmer's Almanac, For The Year 1815 .... Calculations ... by Joshua Sharp ....
Wilmington, Del.: Printed and sold by James Wilson, [1814.]
[48] p. LCP.

648. The Annual Visiter And Citizen and Farmer's Almanac, For The Year 1818 .... Calculations ... by Joshua Sharp ....
Wilmington, Del.: Printed and sold by James Wilson, [1817].
[38] p.; Drake 1451 has 18 leaves. LCP

649. The Annual Visiter And Citizen And Farmer's Almanac, For the Year 1819 .... Calculated for the Meridian of the States of Pennsylvania, New Jersey, Delaware And Maryland. By Nathan Bassett.
Wilmington, Del.: Printed And Sold By J. Wilson, [1818].
[36] p. LCP; HSP

650. The Annual Visiter And Citizen And Farmer's Almanac For the Year 1820 .... Calculated ... By Joshua Sharp.
Wilmington, Del.: Printed And Sold By J. Wilson, [1819.]
[36] p. LCP; HSP

651. The Annual Visiter And Citizen And Farmer's Almanac, For the Year 1821 ....
Wilmington, Del.: Printed And Sold By J. Wilson, [1820.]
[36] p. LCP

652. The Annual Visiter And Citizen And Farmer's Almanac, For the Year ... 1824 ....
Wilmington, Del.: Printed And Sold By J. Wilson, [1823.]
[36] p. LCP

__________. WINDSOR, VT.

653. The New-England Farmer's Almanac, For The Year ... 1834 .... Fitted to the Latitude of Windsor, Vt. ... By Truman Abell.
Windsor: Published and sold by Ide & Goddard; Lempster: sold also ... by the Author; Alstead: Newton & Tufts, et al., [1833].
[48] p.; illus. T.p. headed: No. XX. LCP; HSP

__________. WORCESTER, MASS.
See also The Farmer's Almanack ... 1842; 1853; Boston. Wrappers have Worcester imprint.

654. Isaiah Thomas, Junior's Town & Country Almanack, Or Complete Farmer's Calendar, For The Year ... 1812 ....
Worcester: Printed By Isaiah Thomas, Jun., [1811].
[48] p.; illus. LCP

__________. YORK, PA.

655. Stadt-und Land-Calender Auf das Jahr Christi 1852 ....
York, Pa.: Herausgegeben von A.J. Warfield, [1851].
34, [2] p.; illus. HSP

END OF ALMANACS

656. Alta California: Embracing Notices Of The Climate, Soil, And Agricultural Products Of Northern Mexico And The Pacific Seaboard; Also, A History Of The Military and Naval Operations of the United States directed against the Territories of Northern Mexico, In The Year 1846-'47 .... By A Captain Of The Volunteers.
Philadelphia: H. Packer & Co., Publishers; Sold By The Booksellers Generally, 1847.
Pp. [1-4], [7]-64. Cowan p. 4 and Howes C-37 both call for 64 p. LCP (wrappers)

AMERICAN [pseud.]
See CHRISTY, DAVID, b. 1802.

657. The American Agriculturist. [With varying subtitles]
New York.
LCP: Vol. 1, nos. 1, 3-12 (Apr., June 1842-Mar. 1843); vol. 7, nos. 8, 12 (Aug., Dec. 1848); vol. 8, no. 12 (Dec. 1849); vol. 9, nos. 1-2, 4, 9-10, 12 (Jan.-Feb., Apr., Sept.-Oct., Dec. 1850); vol. 10, nos. 1, 4 (Jan., Apr. 1851); and scattered issues between 1863 and 1875.
HSP: Vols. 1-2, 5-7 (April 1842-Dec. 1843; Jan. 1846-Nov. 1848); 20-28 (Jan. 1861-Dec. 1869); and scattered issues.

658. AMERICAN EMIGRATION LAND COMPANY, Philadelphia.
American Emigration Land Company. 50,000 Acres Of Land In Farms Of 50 And 100 Acres. Office, No. 59 Seventh St., Above Arch, Philadelphia. [wrapper title]
Philadelphia: J.H. Jones & Co., Printers, 1857. [wrapper imprint]
16 p. Wrapper headed: Homes For The People. The company's lands lay in Bath Co., Va. LCP (wrappers); HSP (wrappers. 2 copies, one with company prospectus in German pasted inside back wrapper)

659. The American Farmer .... John S. Skinner, Editor.
Baltimore.
LCP: Vols. 1-2 (April 2, 1819-March 23, 1821; Third Edition, pub. 1821); vols. 3-10 (March 30, 1821-March 13, 1829); [s4] vol. 10, no. 7; vol. 13, nos. 1-3, 6 (Jan. 1855; July-Sept., Dec. 1857)
HSP in LCP: Vols. 1-13 (April 2, 1819-March 9, 1832)
APS: [s4] vol. 12, no. 2 (Aug. 1856)
HSP: Vols. 1-5 (April 2, 1819-March 19, 1824 + 3rd edn. vol. 1); vol. 10 (March 21, 1828-March 13, 1829); [s4] vols. 13-14 (July 1857-June 1859)

660. The American Farmer's Guide: Or A New And Excellent Treatise On Agriculture. Wherein The Planter And Farmer Will See Such A Judicious Method For The Management Of Stock - Rasiing [sic] Manure - Enriching And Improving His Lands That They Shall Produce The Most Plentiful Crops, And In A Few Years Become As Valuable As Those Of England Or France.
Philadelphia: Reprinted for the Rev. Mason L. Weems, by Jacob Johnson & Co., [1795?].
83 p. See Ford/Skeel's Mason Locke Weems no. 234 for discussion of attribution and date. The Early American Imprints Microfilm of the MWA copy shows an unnumbered leaf of Contents at end. APS (without Contents leaf); HSP (with Contents leaf after p. 12)

661. American Farmer's Magazine.
Philadelphia; New York.
LCP: Vols. 1-3 (July 1848-June 1851); vol. 4, nos. 7-12 (Jan.-June 1852); vol. 5, nos. 2-3 (Feb.-Mar. 1853); vol. 7, nos. 1, 5 (July, Nov. 1854); vol. 8, no. 5 (Nov. 1855 [fragment]); vol. 12, no. 1 (Jan. 1859)
APS: Vols. 1-7 (July 1848-June 1855)
HSP: Vols. 1-5 (July 1848-June 1853)
Also (1848-1857) known as: The Plough, The Loom And The Anvil.

American Geographical And Statistical Society, New York. Agricultural Section.
See JAY, JOHN, 1817-1894. A Statistical View of American Agriculture .... An Address Delivered ... on the Organization of the Agricultural Section. 1859.

662. American Guano Company, New York.
Report Of Experiments With American Guano.
New York: W.H. Arthur & Co., Printers And Stationers, 1859.
27 p. APS (wrappers)

663. American Husbandry. Containing an Account of the Soil, Climate, Production and Agriculture, Of The British Colonies In North-America and the West- Indies; with observations on the Advantages and Disadvantages of settling in them, compared with Great Britain and Ireland. By An American. In Two Volumes. Vol. I [-II].
London: Printed for J. Bew, 1775.
2 vols. Vol. I: [4], 472 p. II: [4], 319, [17] p. Possibly by either John Mitchell (d. 1768) or Arthur Young (1741-1820); see H.J. Carman's edn. of the work (NY 1939). LCP; HSP

AMERICAN INSTITUTE OF THE CITY OF NEW YORK.
See NEW YORK CITY. AMERICAN INSTITUTE.

664. American Journal of Agriculture and Science. Conducted by Dr. E. Emmons and Dr. A.J. Prime.
Albany.
LCP: Vol. 6, [no. 12] (Dec. 1847); vol. 7, no. 7 (July 1848)
APS: Vol. 1, no. 1 (Jan.-March 1845)
Also (vols. 1-4) known as: American Quarterly Journal of Agriculture and Science.

665. American Medical And Philosophical Register; Or, Annals Of Medicine, Natural History, Agriculture, And The Arts.
New York.
LCP: Vols. 1-4 (July 1810-Apr. 1814). All pub.

666. [American Medical and Philosophical Register]. Prospectus Of A New Periodical Work, To Be Entitled The American Medical And Philosophical Register; Or, Annals Of Medicine, Natural History, Agriculture, And The Arts. Conducted By A Society Of Gentlemen.
[New York: 1810].
Brs. Dated: New York, June 1, 1810. LCP

667. AMERICAN PHILOSOPHICAL SOCIETY, Held in Philadelphia, for promoting useful knowledge.
Premiums. The American Philosophical Soceity, Held At Philadelphia, For Promoting Useful Knowledge ... have agreed to appropriate, annually, a part of their funds to be disposed of in Premiums to the Authors of the best performances, inventions, or improvements ....
[Philadelphia: 1796].
Brs. Incl. notice of premiums for preventing decay in peach trees and for vegetable dyes. APS

__________.
See also
BOISSIER DE SAUVAGES DE LA CROIX, PIERRE AUGUSTIN, 1706-1767, and SAMUEL PULLEIN, fl. 1734-1760. Directions For The Breeding And Management Of Silk-Worms .... 1770. Translated and arranged for the Society.
MATLACK, TIMOTHY, 1736-1829. An Oration, Delivered ... Before .... 1780.
RUSH, BENJAMIN, 1745-1829. An Account Of The Sugar Maple-Tree .... Read In .... 1792.

668. AMERICAN POMOLOGICAL SOCIETY.
Proceedings.
Philadelphia; Boston; etc.
LCP: Vols. 2-3 (1852-54); 6-8 (1856-60); 10 (1864)
APS: Vols. 2-3; 6-8; 11 (1868)
HSP: Vols. 2-3, 6.

__________.
See also WILDER, MARSHALL PINCKNEY, 1798-1886. Address Delivered At .... 1856.

669. The American Pomologist: Containing Finely Colored Drawings, Accompanied By Letter Press Descriptions Of Fruits Of American Origin. Edited by W.D. Brinckle.
Philadelphia.
LCP: Vol. 1 [no. 1] ([July] 1851). All pub.

American Quarterly Journal of Agriculture and Science.
See American Journal of Agriculture and Science.

670. The American Rose Culturist; Being A Practical Treatise On The Propagation, Cultivation, And Management Of The Rose In All Seasons; With A List Of Choice And Approved Varieties, Adapted To The Climate Of The United States; To Which Are Added Full Directions For The Treatment Of The Dahlia. Illustrated by Engravings.
New York: C.M. Saxton, Agricultural Book Publisher, [1852].
96 p.; illus. T.p. headed: Saxton's Cottage And Farm Library. LCP

671. The American Silk Grower And Farmer's Manual. Edited By Ward Cheney & Brothers.
Philadelphia.
LCP: Vol 1, nos. 1, 3-4, 6-7, 10, 12 (July, Sept.-Oct., Dec. 1838; Jan., April, June 1839)
APS: Vol. 1, nos. 3, 7, 10-12 (Sept. 1838; Jan., April-June 1839)
HSP: Vol. 1 (July 1838-June 1839); Vol. 2, nos. 1-6 (July-December 1839). All pub.

672. AMERICAN SILK SOCIETY, Baltimore.
Journal Of The American Silk Society, And Rural Economist. Gideon B. Smith, Editor.
Baltimore.
APS: Vols. 1-2 (Jan. 1839-Dec. 1840)
HSP: Vol. 1; Vol. 2, nos. 1-8 (Jan. 1839-Aug. 1840)

The American Traveller .... 1770.
See CLUNY, ALEXANDER.

673. Americanisches Noth-und Hülfs-Büchlein; Enthaltend Hausmittel und Recepte für allerley Krankheiten und Zufälle an Menschen und Vieh. Hauptsächlich aus den hinterlassenen Manuscripten eines zu seiner Zeit berühmt gewesenen Artzes in Virginia gezogen.
Osnaburg, Stark County, Ohio: Gedruckt und zu haben bey Heinrich Kurtz, 1837.
[2], 57, [3] p. HSP in LCP

Der Amerikanische Pferdearzt .... 1832.
See TOWAR, ALEXANDER.

ANDERSON, JAMES.
See
MARSHALL, CHARLES d. 1818. An Introduction .... To Which Is Added, An Essay On Quick-Lime ... By .... 1799.
WASHINGTON, GEORGE, 1732-1799. Selections From The Correspondence Of George Washington And James Anderson .... 1800.

674. Antidote To The Merino-Mania Now Progressing Through The United States; Or, The Value Of The Merino Breed, Placed By Observation And Experience, Upon A Proper Basis.
Philadelphia: Printed And Sold By J. & A.Y. Humphreys, 1810.
52 p. Consists of extracts from Communications to the [British] Board of Agriculture, the Bath Society's Papers, and Dickson's Agricultural Magazine. LCP

AQUIDNECK [R.I.]. AGRICULTURAL SOCIETY.
See
HALL, JONATHAN PRESCOTT, 1796-1862. An Address Delivered .... 1854.
HAZARD, THOMAS ROBINSON, 1797-1886. Address Delivered Before .... 1853.

675. ARCHBALD, WILLIAM A.
Prospectus D'un Mode Perfectionné de Fabrication pour le Sucre Brut, d'apres les Brevets de William A. Archbald. [cap. title]
[New Orleans: 1830].
[2] p. Dated at end: Nouvelle-Orleans, 5 Janvier 1830. LCP

676. Archives Of Useful Knowledge, A Work Devoted To Commerce, Manufactures, Rural And Domestic Economy, Agriculture .... By James Mease.
Philadelphia.
LCP: Vols. 1-3 (July 1810-Apr. 1813). All pub.
APS: Vols. 1-2 (July 1810-April 1812)
HSP: Vols. 2-3 (July 1811-April 1813)

677. [ARMSTRONG, JOHN, 1758-1843].
A Treatise On Agriculture; Comprising A Concise History Of Its Origin And Progress, The Present Condition Of The Art, Abroad And At Home, And The Theory and Practice of Husbandry, Which Have Arisen Out Of The Present State Of Philosophical Attainments In Europe. By A Practical Farmer.
Albany: Printed By J. Buel, And for sale at the Bookstores; New York: Wiley & Co.; Utica: W. Williams; Geneva: J. Bogert; Canandaigua: J.D. Bemis & Co.; Auburn: T.M. Skinner; Johnstown: A. Childs, et al., [1819].
168 p. HSP in LCP

678. ________.
A Treatise On Agriculture; Comprising A Concise History Of Its Origin And Progress .... To Which Is Added, A Dissertation On The Kitchen And Fruit Garden. With Notes By J. Buel.
New York: Harper & Brothers, 1840.
Pp. [1-2], [5]-282. LCP

The Art Of Making Common Salt .... 1776.
See PAINE, ROBERT TREAT, 1731-1814.

679. [ARTHUR, BURNHAM & CO., can mfrs., Philadelphia].
Fresh Fruits And Vegetables, All The Year, At Summer Prices; And How To Obtain Them.
Philadelphia: T.B. Peterson [Stereotyped By George Charles (t.p. verso)], 1855.
36 p.; illus. On preserving with "Arthur's Self-Sealing, Air-Tight Cans." Company name on verso of front wrapper. LCP; HSP (wrappers)

680. ATKINS, JEARUM.
Atkins' Automaton, Or Self-Raking Reaper & Mower. Invented By Mr. Jearum Atkins, Of Chicago. J.S. Wright, Proprietor and Manufacturer, "Prairie Farmer" Warehouse, Chicago, Building 700; Rufus Dutton, Dayton, Ohio, Building Five Hundred; Probably 500 To Be Built In New York and Philadelphia ....
Chicago: Democratic Press Steam Print, 1855.
32 p.; errata slip tipped on to t.p. LCP

AUTHOR OF CONVERSATIONS ON CHEMISTRY.
Conversations On Vegetable Physiology .... 1830 etc.
See MARCET, JANE HALDIMAND, 1769-1858.

681. [AYSCOUGH, SAMUEL, 1745-1804].
Remarks On The Letters From An American Farmer; Or A Detection of the Errors Of Mr. Hector St. John; Pointing out the pernicious Tendency of these Letters to Great Britain.
London: Printed for J. Fielding, 1783.
[2], 26 p. HSP in LCP (2 pub. adv. clippings pasted t.p. verso)

B. or
B., J.B.
See BORDLEY, JOHN BEALE, 1727-1804.

682. BACHE, ALEXANDER DALLAS, 1806-1867.
Anniversary Address Before The American Institute, Of the City of New-York, At The Tabernacle, October 28th, 1856, During The Twenty-Eighth Annual Fair.
New York: Pudney & Russell, Printers, 1857.
60 p. APS (wrapper); HSP

683. Back Lands And Improvements. The Outlines Of A Plan For settling and improving a select Body of New Lands in the State of Pennsylvania, so as to make them productive to the Owners, within the shortest possible Period.
[Philadelphia?: 1800?].
Brs. HSP

684. BACON, LEONARD, 1802-1881.
An Address Delivered At The Annual Fair Of The New Haven County Horticultural Society, September 27, 1848. With Transactions Of The Society.
New Haven: Printed By B.L. Hamlen, 1848.
31 p. LCP

685. [BADCOCK, JOHN].
Farriery, Taught On A New And Easy Plan; Being A Treatise On The Diseases And Accidents Of The Horse; With Instructions To The Shoeing-Smith, Farrier, And Groom .... By John Hinds [pseud.] .... With Considerable Additions And Improvements, Particularly Adapted To This Country. By Thomas M. Smith .... With A Supplement: Comprising An Essay On Domestic Animals, Especially The Horse .... By J.S. Skinner ....
Philadelphia: J.B. Lippincott & Co., 1857.
LCP: xiv, 24 [i.e. 224], 101 p.; front.; 1 pl. only (of 2; lacking 1). P. 224 misnumbered 24. T.p. headed: Hinds' Farriery and Stud Book-New Edition. Sep. t.p. and pagination for: Supplement To Mason And Hind's [sic] Popular System Of Farriery .... By J.S. Skinner ....

686. [________].
The Groom's Oracle, And Pocket Stable Directory: In Which The Management Of Horses Generally, As To Health, Dieting, And Exercise, Are Considered, In A Series Of Familiar Dialogues, Between Two Grooms Engaged In Training Horses to their Work. By John Hinds [pseud.] .... From The Second London Edition. With Notes And An Appendix, Including Extracts From The Receipt Book Of The Author.
Philadelphia: E.L. Carey And A. Hart; Sold By J. Grigg; New York: G. & C. & H. Carvill And Collins & Hannay; Boston: Carter, Hendee & Babcock; Baltimore: W. & J. Neal; Washington: And Thompson & Homans, 1831.
228, [2 (pub. adv.)] p. HSP in LCP (lacking front.)

687. [__________].
The Veterinary Surgeon; Or Farriery Taught On A New And Easy Plan; Being A Treatise On All The Diseases And Accidents To Which The Horse Is Liable ... With Instructions To The Shoeing-Smith, Farrier, And Groom, How To Acquire Knowledge In The Art Of Farriery .... By John Hinds [pseud.] .... With Considerable Additions and Improvements, particularly Adapted to this Country. By Thomas M. Smith ....
Philadelphia: John Grigg [Mifflin & Parry, Printers (t.p. verso)],
xiv, 1830.
284 p.; fold. front.; 3 fold. pls. LCP; HSP (with last pl. present in duplicate)

688. [__________].
The Veterinary Surgeon: Or, Farriery Taught In A New And Easy Plan .... By John Hinds [pseud.] .... With Considerable Additions and Improvements .... By Thomas Smith ....
Philadelphia: John Grigg [Stereotyped by J. Crissy & G. Goodman (t.p. verso)], 1833.
xiv, 224, [2 (pub. adv.)] p.; fold. front.; 3 fold. pls.; illus. LCP

689. [__________].
The Veterinary Surgeon: Or, Farriery Taught In A New And Easy Plan .... By John Hinds [pseud.] .... With Considerable Additions and Improvements .... By Thomas Smith ....
Philadelphia: John Grigg [Printed By T.K. And P.G. Collins (t.p. verso)], 1838.
xiv, 224, [2 (pub. adv.)] p.; fold. front.; 3 fold. pls.; illus. HSP

690. BAKEWELL, ROBERT, 1768-1843.
Observations On The Influence Of Soil And Climate Upon Wool; From Which Is Deduced, A Certain And Easy Method Of Improving The Quality Of English Clothing Wools, And Preserving The Health Of Sheep; With Hints For The Management Of Sheep After Shearing .... With Occasional Notes And Remarks, By The Right Hon. Lord Somerville.
Philadelphia: Published By Kimber And Conrad; William Fry, Printer, 1814.
120 p. HSP in LCP; LCP

691. BALDWIN, HENRY, 1780-1844.
Anniversary Address Delivered Before The American Institute Of The City Of New-York, At The Chatham Street Chapel, October 9, 1834, During The Seventh Annual Fair.
New York: Printed And Published At The Office Of The Mechanics' Magazine, 1834.
24 p. HSP (wrappers)

692. BALL, LEVI CHANDLER, 1809-1875.
Address Delivered At The Annual Fair And Cattle Show, Of The Bennington Co. Agricultural Society, Held At North Bennington, Vt., September 22 & 23, 1852.
Troy, N.Y.: From The Power Press Of The Daily Whig, 1852.
28 p.; 1 pl. Incl. (pp. 23-28): description of, and testimonial letters for, Manny's Patent Adjustable Reaping And Mowing Machines, manufactured by Ball and J. Russell Parsons in Hoosick Falls, Rensselaer Co., N.Y. The pl. illustrates the machine. LCP

693. BALMER, DANIEL.
Eine Sammlung von Neuen Recepten und Bewahrten Curen, fur Menschen und Vieh. Zusammengetragen und zum Druck befordert von Daniel Ballmer, bey Chambersburg, Anno 1826.
Shellsburg [Pa.]: Gedruckt bey Friedrich Goeb, 1827.
Pp. [1-6], [11]-40. LCP (wrappers)

694. [BARBOUR, I. RICHMOND, comp.].
The Silk Culture In The United States: Embracing Complete Accounts Of The Latest And Most Approved Modes Of Hatching, Rearing And Feeding The Silk-Worm, Managing A Cocoonery, Reeling, Spinning .... Compiled From The Most Approved And Reliable Works .... To Which Is Added Blydenburgh's Manual Of The Silk Culture ....
New York: Greeley & McElrath, Tribune Office, 1844.
Pp. [1-10], [5]-80; illus. "It has been prepared ... with the aid of Mr. I. Barbour ...." (Advertisement). LCP; APS

BARCLAY, CAPTAIN, pseud.
See ALLARDICE, ROBERT BARCLAY, 1779-1854.

695. BARD, SAMUEL, 1742-1821.
A Guide For Young Shepherds; Or Facts And Observations On The Character And Value Of Merino Sheep: With Rules And Precepts For Their Management, And The Treatment Of Their Diseases, As well as of Sheep In General. Collected By The Latest And Best Writers On These Subjects, And Confirmed By The Experience Of The Author And His Friends.
New York: Printed And Sold by Collins & Co., 1811.
112 p.; illus. LCP

696. BARNARD, DANIEL DEWEY, 1797-1861.
Anniversary Address, Delivered Before The American Institute, At The Tabernacle, In New-York, On The 20th October, 1843. Published by request of the Institute.
New York: James Van Norden & Co., Printers, 1843.
28 p. HSP (wrapper)

BARNSTABLE CO. [MASS.]. AGRICULTURAL SOCIETY.
See DODGE, ALLEN WASHINGTON, 1804-1878. Address Before .... 1846.

697. BARNUM, H.L.
The American Farrier; Containing A Minute Account Of The Formation Of Every Part Of The Horse, From The Extremity Of The Head To The Hoof. With A Description Of All The Diseases To Which Each Part Is Liable .... Accompanied with A Copious Alphabetical List Of Medicines .... Illustrated By Engravings, And Arranged On A New Plan.
Philadelphia: Uriah Hunt; Cincinnati: H.L. Barnum, At The Farmer's Reporter Office; And booksellers generally throughout the United States, 1832.
286 p.; illus. LCP

698. __________.
The Farmer's Own Book; Or Family Receipts For The Husbandman And Housewife; Being A Compilation Of The Very Best Receipts On Agriculture, Gardening And Cookery; With Rules For Keeping Farmers' Accounts. Stereotype Edition.
Boston: Published By Carter And Hendee, 1832.
165, [1] p.; illus. LCP

BARREDA, F., & BROTHER, Agents, Baltimore, Md.
See ROBINSON, SOLON, 1803-1880. Guano .... Prepared And Published ... For .... 1852.

699. BARRY, PATRICK, 1816-1890.
The Fruit Garden; A Treatise Intended To Explain And Illustrate The Physiology Of Fruit Trees, The Theory And Practice Of All Operations Connected With The Propagation, Transplanting, Pruning And Training Of Orchard And Garden Trees, As Standards, Dwarfs, Pyramids, Espaliers, Etc., The Laying Out And Arranging Different Kinds Of Orchards And Gardens, The Selection Of Suitable Varieties For Different Purposes And Localities ... Illustrated With Upwards Of 150 Figures ....
New York: Charles Scribner [Stereotyped by C.W. Benedict (t.p. verso)], 1851.
xiv, [2], 398, [1 (pub. adv.)] p.; front.; illus. LCP

700. __________.
The Fruit Garden; A Treatise Intended To Explain And Illustrate The Physiology Of Fruit Trees ....
New York: Charles Scribner, 1852.
xiv, [2], 398, [1 (adv.)] p.; front.; illus. LCP

__________, jt. proprietor.
See ELLWANGER & BARRY, Nurserymen, Rochester, N.Y.

701. BARTLET, JOHN, 1716?-1772.
The Gentleman Farrier's Repository, Of Elegant and approved Remedies For The Diseases Of Horses; In Two Books. Containing, I. The Surgical; II. The Medical Part Of Practical Farriery; Also, Directions for the proper Treatment of Post Chaise and other Horses .... The Third Edition.
Philadelphia: Printed and sold by Joseph Crukshank, 1775.
xii, 293, [3 (pub. adv.)] p. HSP in LCP; LCP

702. [BARTLETT, EDWIN, 1796-1867].
Guano, Its Origin, Properties And Uses, Showing Its Importance To The Farmers Of The United States As Cheap And Valuable Manure: With Directions For Using It.
New York: Wiley And Putnam, 1845.
80 p. Author's name p. 2. LCP

703. [__________].
Guano, Its Origin, Properties And Uses .... Second Edition, With An Appendix.
New York: Wiley And Putnam, 1845.
92 p. LCP

BARTRAM, JOHN, 1699-1777.
See
[STORK, WILLIAM]. An Account of East-Florida, With A Journal Kept By John Bartram .... [1766].
[__________]. A Description Of East-Florida, With A Journal .... 1769.

BARTRAM'S BOTANIC GARDEN, Kingsessing, Pa.
Established by John Bartram (1699-1777) and taken over upon his death by John the younger (1743-1812) in partnership with his brother William (1738-1823). John the younger's daughter Ann and her husband Robert Carr took possession after John's death. See DAB.

704. __________.
Catalogue D'Arbres D'Arbustes Et De Plantes Qui croissent en Amerique, & produisent des Graines en Maturite dans le Jardin de John Bartram, pres de Philadelphie ....
[Paris: Pierres, ca. 1783].
Brs. APS

705. __________.
Catalogue of American Trees, Shrubs and Herbaceous Plants, most of which are now growing, and produce ripe Seed in John Bartram's Garden, near Philadelphia. The Seed and growing Plants of which are disposed of on the most reasonable Terms.
[Philadelphia: ca. 1783].
Brs. APS; HSP

706. __________.
A Catalogue Of Trees, Shrubs, And Herbaceous Plants, Indigenous To The United States Of America; Cultivated And Disposed Of By John Bartram & Son, At their Botanical Garden, Kingsess, near Philadelphia. To Which Is Added A Catalogue Of Foreign Plants, Collected From Various Parts Of The Globe.
Philadelphia: Printed By Bartram And Reynolds, 1807.
33 p. LCP; HSP

707. __________.
Periodical Catalogue Of Fruit And Ornamental Trees And Shrubs, Green House Plants, &c. Cultivated And For Sale At Bartram's Botanic Garden, Kingsessing, Near Gray's Ferry - Four miles from Philadelphia. Robert Carr, Proprietor.
Philadelphia: Russell And Martien, Printers, 1828.
47, [1] p. HSP

708. __________.
Periodical Catalogue Of Fruit And Ornamental Trees And Shrubs, Green House Plants, &c. ....
Philadelphia: Russell And Martien, Printers, 1828.
47, [1], [4] p. First 48 p. as above; last 4 p.: Additions. (March, 1830). LCP

709. __________.
Periodical Catalogue Of American Trees, Shrubs, Plants, And Seeds, Cultivated And For Sale At The Bartram Botanic Garden ....
Philadelphia: Printed by Russell & Martien, 1831.
Pp. [1-2], [63]-84. LCP

710. ________.
Periodical Catalogue Of Fruit And Ornamental Trees And Shrubs, Green-House Plants, &c. Cultivated And For Sale ....
Philadelphia: William S. Martien, Printer, 1836.
LCP: 56 p. only, with pp. 49-56 defective. Presumably lacks the last 7 leaves; a p. 69 is listed in the index.

711. BATES, ISAAC CHAPMAN, 1780-1845.
An Address To The Hampshire, Franklin and Hampden Agricultural Society, Northampton, Mass. October 23, 1823.
Northampton: Printed By Thomas W. Shepard, 1823.
35 p. LCP

712. BATSTO FARM AND AGRICULTURAL COMPANY, Atlantic Co., N.J.
An Investment Full of Interest to Every Man A Home For All! Batsto Farm and Agricultural Company. [cap. title]
[Philadelphia?: 1855?]
7, [1] p. LCP

713. BAUDRY DES LOZIÈRES, LOUIS NARCISSE, 1761-1841.
Second Voyage A La Louisiane, Faisant Suite Au Premier De L'Auteur De 1794 A 1798. Contenant la vie militaire du général Grondel ... un Détail sur les productions les plus avantageuses, les plus extraordinaires de cette belle Colonie, et sur ses quartiers les plus fertiles .... Tome Premier [- Second].
Paris: Chez Charles, Imprimeur, An XI.-1803 (Mars).
2 vols. Vol. I: [2], xvi, 414, [1] p.; fold. tbl. II: [4], 410, [1] p. Title of Vol. II continues after 1st sentence: Contenant un Mémoire sur la découverte du Coton Animal: un Manuel Botanique à l'usage des jeunes Colons ... Observations sur la Botanique Médicinale, d'une grande utilité pour la santé et l'instruction des Planteurs américains: réflexions sur les chambres d'Agriculture .... BM gives author's forenames as Louis Nicolas. APS (lacking ½ t.p. and t.p. to Vol. I; with fold. tbl. in Vol. II); HSP

714. BEARDSLEY, HENRY.
The Domestic Instructor, And Manufacturer's Guide, Containing The necessary instructions in the art of Woollen Manufacturing, and Dying. With Some Remarks on Wool Growing.
Utica: Printed By William Williams, 1830.
46, [1] p. LCP

715. BEATSON, ALEXANDER, 1759-1833.
A New System Of Cultivation, Without Lime, Or Dung, Or Summer Fallows: As Practised At Knowle-Farm, In The County Of Sussex. Illustrated By Three Engravings.
Philadelphia: M. Carey & Son, 1821.
Pp. [i]-iv, [3]-5, [1], [1]-77; 3 pls. LCP (wrappers); APS; HSP (lacking pls.)

716. BEATTY, ADAM, 1757-1858.
Essays On Practical Agriculture, Including His Prize Essays, Carefully Revised ....
Maysville, Ky.: Collins & Brown, 1844.
298 p. Pp. 295-8 are Testimonials. LCP

717. BECK, LEWIS CALEB, 1798-1853.
Mineralogy Of New-York; Comprising Detailed Descriptions Of The Minerals Hitherto Found In The State Of New-York, And Notices Of Their Uses In The Arts And Agriculture.
Albany: Printed By W. & A. White & J. Visscher, 1842.
xxiv, [2], 536, [3] p.; eng. general t.p.; 8 pls.; illus. This is Division III of the Natural History Of New-York. LCP; APS

BECK, THEODRIC ROMEYN, 1791-1855, jt. auth.
See EATON, AMOS, 1776-1842, and ________.

BECQUEREL, ANTOINE CÉSAR, 1788-1878. The Action Of Salt .... 1848.
See DUMAS, JEAN BAPTISTE ANDRE, 1800-1884. Manures .... 1848. Incl. Becquerel's work.

718. BEECHER, HENRY WARD, 1813-1887.
Plain And Pleasant Talk About Fruits, Flowers and Farming.
New York: Derby & Jackson [W.H. Tinson, Stereotyper; Geo. Russell & Co., Printers; Somerville & Brother, Binders (t.p. verso)], 1859.
420 p. LCP

719. BEMAN, DAVID.
The Mysteries Of Trade, Or The Great Source Of Wealth: Containing Receipts and Patents In Chemistry And Manufacturing; With Practical Observations On The Useful Arts. Original And Compiled.
Boston: Printed For The Author, By Wm. Bellamy, 1825.
152 p. Chiefly on brewing and distilling. LCP

720. BEMENT, CALEB N., 1791?-1868.
The American Poulterer's Companion: A Practical Treatise On The Breeding, Rearing, Fattening, And General Management Of The Various Species Of Domestic Poultry, With Illustrations, And Portraits Of Fowls Taken From Life.
New York: Saxton And Miles [S.W. Benedict & Co., Ster. And Print. (t.p. verso)], 1845.
[2], 379 p.; front.; illus. LCP

721. ________.
The American Poulterer's Companion. A Practical Treatise On The Breeding, Rearing, And General Management Of Various Species Of Domestic Poultry. Illustrated With Portraits Of Fowls, Mostly Taken From Life; Poultry-Houses, Coops, Nests, Feeding-Hoppers, &c., &c., &c. A New Edition, Enlarged And Improved. With 120 Illustrations.
New York: Harper & Brothers, Publishers, [1856].
304 p. incl. 2 fronts.; illus. LCP

722. BENNETT, JOHN C.
The Poultry Book: A Treatise On Breeding And General Management Of Domestic Fowls; With Numerous Original Descriptions, and Portraits from Life.
Boston: Phillips, Sampson And Company [Stereotyped by Hobart And Robbins; New England Type and Stereotype Foundery (t.p. verso)], 1856.
310 p. incl. front.; illus. LCP

BENNINGTON CO. [VT.]. AGRICULTURAL SOCIETY.
See BALL, LEVI CHANDLER, 1809-1875. Address Delivered At The Annual Fair .... 1852.

723. BENTWRIGHT, JEREMIAH.
The American Horse Tamer, Showing How To Cure The Wildest & Most Vicious Horse In The World Of Kicking, Balking, and other Bad Habits; Also, Directions To Purchasers, Hints On Stable Economy, And Remedies For All Diseases To Which Horses Are Liable Such As Spavin, Distemper, Ringbone, &c. &c.
New York: A.O. Moore, Agricultural Book Publisher (Late C.M. Saxton & Co.), 1858.
Pp. [1-2], [7]-86, 1-6 (pub. adv.); front; 2 pls. LCP

724. BENTZ, LOUIS, and A. J. CHRETIEN DE ROVILLE.
Elements Of Agriculture For The Use Of Primary And Secondary Schools. Approved And Recommended By The Governmental Council Of The Department Of The Meurthe. Translated And Adapted To The Use Of The Rural Primary Schools Of The United States Of America. By F.G. Skinner.
New York: C.M. Saxton, Agricultural Book Publisher, [c1848].
4 (pub. adv.), [2], 91 p. T.p. headed: A Book For Every Boy In The Country. LCP

BERKSHIRE [MASS.]. AGRICULTURAL SOCIETY.
See
GOLD, THOMAS, 1759-1827. Address Of ... Delivered Before .... [1816].
SEDGWICK, THEODORE, 1780-1839. An Address, Delivered Before .... 1830.

725. [BETTON, SAMUEL].
Address Of The President And Fellows Of Rittenhouse College.
[Philadelphia: 1837].
7 p. Author's name at end. The college emphasized instruction in animal husbandry. LCP (wrappers); HSP (wrappers)

726. BEVAN, EDWARD, 1770-1860.
The Honey Bee; Its Natural History, Physiology, And Management. With Thirty-Five Engravings On Wood, By R.S. Gilbert.
Philadelphia: Carey And Hart, 1843.
128 p.; illus. Wrapper headed: Bevan on the Bee - Cheap Edition. LCP (wrappers); HSP (wrappers)

BICKELL, CHARLES, jt. auth.
See
HIGGINS, JAMES. The Sixth Agriculural Report .... 1858. Incl. Appendix by Bickell.
HIGGINS, JAMES, and __________. Report ... on an examination ... of Sea Island Cotton Fibre .... 1854.

727. BIDDLE, CRAIG, 1823-1910.
Address: Delivered Before The Philadelphia Society for Promoting Agriculture, On Its Seventy-Fifth Anniversary, February 11th, 1860. Printed By Order Of The Society.
Philadelphia: King & Baird, Printers, 1860.
28 p. LCP (wrapper); HSP (wrappers)

728. BIDDLE, JOHN BARCLAY, 1815-1879.
An Address Delivered Before The Chester County Agricultural Society, At Westchester, Pa. On the 4th of October, 1856. (Published By Order Of The Society.)
Philadelphia: Merrihew & Thompson, Printers, 1856.
12 p. HSP (wrappers)

729. BIDDLE, NICHOLAS, 1786-1844.
Address Delivered Before The Philadelphia Society for promoting Agriculture, At Its Annual Meeting, On The Fifteenth Of January, 1822. Printed by Order Of The Society.
Philadelphia: Clark & Raser, Printers, 1822.
39 p. LCP; APS; HSP

730. BIGELOW, JOHN PRESCOTT, 1797-1872.
Statistical Tables: Exhibiting The Condition And Products Of Certain Branches Of Industry In Massachusetts, For The Year Ending April 1, 1837. Prepared From The Returns Of The Assessors.
Boston: Dutton And Wentworth, State Printers, 1838.
viii, 209, [3] p. Incl. data on sheep, manufacture of agricultural machinery, etc. LCP; APS; HSP (wrappers)

731. BIGELOW, LEWIS, 1785-1838.
Address, Delivered Before The Worcester [Mass.] Agricultural Society, October 12, 1820, Being Their Anniversary Cattle Show and Exhibition of Manufactures. Published By Order Of The Trustees.
Worcester: Printed By Manning & Trumbull, Feb. 1821.
32, [2] p. LCP

BIGLER, WILLIAM, 1814-1880.
See PHILADELPHIA. SOCIETY FOR PROMOTING AGRICULTURE. Proceedings Of ..., And The Address Of Ex-Governor Bigler. 1857.

732. BINNS, JOHN ALEXANDER, ca. 1761-1813.
A Treatise On Practical Farming; Embracing Particularly The Following Subjects, Viz. The Use of Plaister of Paris, with Directions for Using it; and General Observations on the Use of Other Manures. On Deep Ploughing; Thick Sowing of Grain; Method of Preventing Fruit Trees from Decaying, and Farming in General.
Frederick, Maryland: Printed by John B. Colvin, 1803.
[4], 72 p. LCP; APS (with appended fragment, paged 13-22, beginning: Having received information that there are some persons who doubt what I have said ... in my Treatise on Farming ... I have thought proper to subjoin a few certificates ....)

733. BIRKBECK, MORRIS, 1764-1825.
Bemerkungen auf einer Reise in America, von der Küste von Virginien bis zum Gebiet von Illinois. Nebst einer Vollstandigen geographisch-topographisch-statistisch-politischen Ubersicht der vereinigten Staaten von Nordamerica, nach ihrem neuesten Zustande. Aus dem Englischen ....
Jena: bei August Schmidt und Compagnie, 1818.
iv, 276 p. LCP

734. ________.
Extracts From A Supplementary Letter From The Illinois, Dated January 31st, 1819. Address To British Emigrants Arriving In The Eastern Ports. July 13th, 1819. Reply To William Cobbett, Esq. July 31st, 1819.
New York: Published By C. Wiley And Co.; J. Seymour, Printer, 1819.
29 p. The 3rd essay, "Reply to William Cobbett," refers to 2 letters criticizing Birkbeck and pub. in Part III of Cobbett's Year's Residence In The United States .... LCP; APS

735. ________.
Letters From Illinois. Illustrated By A Map Of The United States, Shewing Mr. Birkbeck's Journey From Norfolk To Illinois And A Map Of English Prairie And The Adjacent Country, By John Melish.
Philadelphia: Published By M. Carey And Son, 1818.
154 p. incl. pub. adv. pp. 3-4; 2 fold. maps. Encourages emigration and describes opportunities for farmers. LCP (lacking pub. adv. p.); APS

736. ________.
Lettres Sur Les Nouveaux Établissemens Qui Se Forment Dans Les Parties Occidentales Des États-Unis D'Amérique. Traduit Sur L' Édition Originale De Philadelphie.
Paris: Chez L'Huillier; Delaunay, Libraire [De L'Imp. De Me Jeunehomme-Crémière (t.p. verso)], 1819.
[2], xvi, 156 p.; fold. map. APS

737. ________.
Notes On A Journey In America, From The Coast Of Virginia To The Territory Of Illinois. With Proposals For The Establishment Of A Colony Of English.
Philadelphia: Published By Caleb Richardson; J.R.A. Skerrett, Printer, 1817.
189 p. HSP

738. ________.
Notes On A Journey In America, From The Coast Of Virginia To The Territory Of Illinois. The Fourth Edition.
London: Printed by Severn and Co.; For James Ridgway; And To Be Had Of All Other Booksellers, 1818.
156, 16 (pub. adv.) p.; fold. front. map. APS

________.
See also
COURTAULD, GEORGE. Address To Those Who May Be Disposed To Remove To The United States .... Including Remarks On Mr. Birkbeck's Opinions .... 1820.
FAUX, WILLIAM. Memorable Days In America .... Including Accounts Of Mr. Birkbeck's Settlement In The Illinois .... 1823.
FEARON, HENRY BRADSHAW, b. ca. 1770. Sketches of America .... With Remarks On Mr. Birkbeck's "Notes" And "Letters" .... 1818 etc.

739. BLACK, JEREMIAH SULLIVAN, 1810-1883.
Address, Delivered Before The Agricultural Society Of Somerset County [Pa.]. At Its Annual Exhibition. October 6, 1854.
Pittsburgh: Published by the Society; Printed By Keenan & Co., [1854].
16 p. LCP; HSP (wrappers)

BLAIR CO. [PA.]. AGRICULTURAL SOCIETY.
See TAYLOR, GEORGE. Annual Address Before The Agricultural Societies Of Blair and Cambria Counties .... 1859.

740. BLAKE, JOHN LAURIS, 1788-1857.
The Farm And The Fireside, Or The Romance Of Agriculture. Being Half Hour Sketches Of Life In the Country.
Auburn: Alden, Beardsley & Co; Rochester: Wanzer, Beardsley & Co. [J.P. Jones & Co., Stereotypers (t.p. verso)], 1853.
480 p. incl. extra eng. t.p.; illus. T.p. headed: To Improve The Soil And The Mind. LCP

741. ________.
The Farmer's Every-Day Book; Or, Sketches Of Social Life In The Country: With The Popular Elements Of Practical And Theoretical Agriculture.
New York: Miller, Orton & Co. [Auburn: Miller, Orton, & Mulligan, Stereotypers and Printers (t.p. verso)], 1857.
654 p.; front.; 7 pls. LCP

742. ________.
The Modern Farmer; Or, Home In The Country: Designed For Instruction And Amusement On Rainy Days And Winter Evenings.
Auburn: Derby And Miller; Buffalo: Derby, Orton And Mulligan; Cincinnati: Henry W. Derby [New York: J.P. Jones & Co., Stereotyper (t.p. verso)], 1854.
[4 (pub. adv.)], 456, [4], [3]-8 (pub. adv.) p.; front; 21 pls. T.p. headed: Third Thousand. LCP

________, ed.
See [MARCET, JANE HALDIMAND, 1769-1858]. Conversations On Vegetable Physiology .... 1830 etc.

743. Des Blinden Rosshandlers Arzneymittel.
[Frederick, Md.:] Gedruckt bey M. Martgis, [ca. 1800].
Brs. LCP

744. BLOCKLEY AND MERION [PA.]. AGRICULTURAL SOCIETY.
Rules & Regulations Of The Blockley And Merion Society For promoting Agriculture and Rural Economy. [cap title]
[Philadelphia: 1808].
Pp. [1]-4, [1-4], 5-7. LCP (wrappers)

________.
See also BRECK, SAMUEL, 1771-1862. Address Delivered Before .... 1828.

745. BLODGET, LORIN, 1823-1901.
Climate Of The Summer Of 1853 In Its Relation To The Agricultural Interests of the United States. (A Report To The Secretary Of The Smithsonian Institution.)
Washington: Henry Polkinhorn, Printer, November, 1853.
26 p. T.p. headed: Smithsonian Institution. LCP

746. __________.
Climatology Of The United States, And Of The Temperate Latitudes Of The North American Continent. Embracing A Full Comparison Of These With The Climatology Of The Temperate Latitudes Of Europe And Asia. And Especially In Regard To Agriculture, Sanitary Investigations, And Engineering. With Isothermal and Rain Charts For Each Season ....
Philadelphia: J.B. Lippincott And Co.; London: Trübner And Co., 1857.
536 p.; 12 fold. maps; 1 chart. LCP; HSP

BLYDENBURGH, S.
See
[BARBOUR, I. RICHMOND, comp.] The Silk Culture ... To Which Is Added Blydenburgh's Manual .... 1844.
The Silkworm [pdcl.]

747. BOAS & SPANGLER, Mfrs. of agricultural machinery, Reading, Pa.
Ketchum's Improved Combined Harvester, Manufactured By Boas & Spangler, Reading, Berks County, Pa. All Orders Addressed To The Manufacturers, Or To Boas, Spangler & Co., Sole Agents For Philadelphia, Will Receive Prompt Attention.
Philadelphia: Printed By A.C. Bryson, 1857.
24 p.; illus. APS

748. BOISSIER DE SAUVAGES DE LA CROIX, PIERRE AUGUSTIN, 1706-1767, and SAMUEL PULLEIN, fl. 1734-1760.
Directions For The Breeding And Management Of Silk-Worms. Extracted from the Treatises of The Abbé Boissier de Sauvages, and Pullein. With A Preface, Giving Some Account Of the Rise and Progress Of The Scheme For encouraging the Culture Of Silk, In Pennsylvania, and the adjacent Colonies.
Philadelphia: Printed by Joseph Crukshank, and Isaac Collins, 1770.
xv, [1], 32 p. Extracted and translated by Jonathan Odell (1737-1818) for the American Philosophical Society. HSP in LCP; LCP; APS

749. BOLMER, PAUL.
A Collection of New Receipts, and Approved Cures for Man & Beast.
Germany [i.e. Pennsylvania?]: Printed for the Purchaser, 1831.
34 p. LCP (upper edges of all leaves damaged)

750. __________.
Eine Sammlung von Neuen Rezepten und Erprobten Kuren für Menschen und Thiere.
Deutschland [i.e. Pennsylvania?]: Gedruckt für den Käufer, 1831.
Pp. [1-2], [7]-35. HSP in LCP (lacking pp. 26-34; wrappers); LCP (lacking last leaf; wrappers)

BOLTON FARM, Bucks Co., Pa.
See FELLENBERG INSTITUTION, Bolton Farm, Bucks Co., Pa.

751. BOMMER, GEORGE.
New Method, Which Teaches How To Make Vegetable Manure, By A Course Of High Fermentation In Fifteen Days, Without Cattle, As Good And More Durable Than Farm Manure: To Appropriate It To The Nature Of Soils And Families Of Plants, And With Great Economy ....
New York: Herald Building, 1843.
86, [1] p. Wrapper title: New Edition, Improved And Greatly Enlarged, Of The Bommer Method Of Making Manure .... LCP (wrappers)

752. __________.
New Method, Which Teaches How To Make Vegetable Manure .... Second Edition, Revised And Corrected.
New York: Stereotyped By Redfield & Savage, 1845.
90 p.; illus. LCP

753. BONOEIL, JOHN.
His Maiesties Graciovs Letter To The Earle Of Sovth-Hampton, Treasurer, and to the Councell and Company of Virginia heere: commanding the present setting up of Silkeworks, and planting of Vines in Virginia .... Also a Treatise of the Art of making Silke: Or, Directions for the making of lodgings, and the breeding, nourishing, and ordering of Silkewormes, and for the planting of Mulbery trees .... Together with instructions how to plant and dresse Vines, and to make Wine, and how to dry Raisins, Figs, and other fruits, and to set Oliues, Oranges, Lemons, Pomegranates, Almonds .... Published by Authority.
London: Printed by Felix Kyngston, 1622.
[10], 88 p.; illus. Incl. brief prefatory letters of James I and Southampton. HSP in LCP

754. BONYNGE, FRANCIS.
The Future Wealth Of America. Being A Glance At The Resources Of The United States And The Commercial And Agricultural Advantages Of Cultivating Tea, Coffee, And Indigo, The Date, Mango, Jack, Leechee, Guava, And Orange Trees, Etc. With A Review Of The China Trade.
New York: Published By The Author, 1852.
242 p.; errata slip tipped in at back. LCP

755. The Book Of Commerce By Sea And Land, Exhibiting Its Connection With Agriculture, The Arts, And Manufactures. To Which Are Added, A History Of Commerce, And A Chronological Table; Designed For The Use Of Schools. Illustrated By A Map And Numerous Engravings.
Boston: Allen And Ticknor, 1834.
Pp. [1]-6, [9]-185; front.; illus. LCP (lacking map)

756. The Book of Commerce By Sea And Land .... Illustrated By A Map And Numerous Engravings.
Philadelphia: Uriah Hunt; And For Sale By The Principal Booksellers Throughout The United States, 1837.
Pp. [1]-6, [9]-185; front.; 1 col. fold. map; illus. LCP

757. The Book Of Commerce By Sea And Land .... To Which Are Added A History Of Commerce, And A Chronological Table.
Philadelphia: Uriah Hunt & Son; Cincinnati: Applegate & Co., 1854.
Pp. [1]-6, [9]-185; illus. HSP

758. BOOTH, JAMES CURTIS, 1810-1888.
Memoir Of The Geological Survey Of The State Of Delaware: Including The Application Of The Geological Observations To Agriculture.
Dover: Printed By S. Kimmey, 1841.
Pp. [i]-xi, [1], [9]-188. LCP; APS; HSP

759. ________.
Practical Value Of The Analysis Of Soils. (Read before the Philadelphia Society for the Promotion of Agriculture). [cap. title]
[Philadelphia: 1853].
8 p. LCP

760. [BORDLEY, JOHN BEALE, 1727-1804].
Cattle pastured and soiled in Summer: kept and fattened in Winter: with a comparative Estimate between pasturing and tying up, in houses, the year through. [cap. title]
[Philadelphia: Charles Cist, 1798].
12 p. Signed at end: B. March, 1798. LCP; APS; HSP

761. [________].
Country Habitations. [cap. title]
[Philadelphia: Charles Cist, 1798].
13 p.; front. LCP (with numerous MS. notations by Bordley; signed in MS. at end: B. July 1798); HSP in LCP (signed in MS. at end: B. July 1798)

762. ________.
Essays And Notes On Husbandry And Rural Affairs.
Philadelphia: Printed By Budd And Bartram, For Thomas Dobson, 1799.
viii, 591, [1], 7, [8] p.; 4 pls., 2 fold. LCP (with MS. notes by Bordley throughout)

763. ________.
[Essays And Notes On Husbandry And Rural Affairs.] Additions And Amendments. [Cap. title; text begins:] A few of the Errors in the preceding Essays and Notes on Husbandry and Rural Affairs, are corrected: Additions and Amendments follow ....
[Philadelphia: 1800?]
Pp. 593-630, [1]; paginated continuously with the 1st edn. of Essays And Notes. Advertisement at back notes that the work was "produced during a painful state of the Author's health, and the injurious effects of a long course of opium: ... [but] the impulse on him to finish it was irresistible." APS

764. ________.
Essays And Notes On Husbandry And Rural Affairs. The Second Edition With Additions.
Philadelphia: Printed By Budd And Bartram, For Thomas Dobson, 1801.
viii, 536, 8, [22] p.; front.; 5 pls., 3 fold. HSP in LCP; APS

765. [________].
Hemp. [cap. title]
[Philadelphia: Charles Cist, 1799].
7 p. Signed at end: B. January, 1799. LCP (with extensive MS. corrections and annotations by Bordley)

766. [________].
Hemp. [cap. title]
[Philadelphia: Charles Cist, 1799?].
Pp. [1]-7, [1], 8-9. P. 8 headed: Supplement; the 1st 7 p. are from the same setting of type as is the above edn. The Supplement incorporates some of the MS. corrections made in the above LCP copy. HSP

767. [________].
[No title; text begins:] Intending to retire from the bustle of a town life ....
[Philadelphia: 1797?].
7 p.; fold. tbl. Text headed: "---Commerce feeds the passions;/Agriculture calms them." Tal. First edn.; not in Evans or Bristol. APS

768. [________].
[No title; text begins:] Intending to retire ....
[Philadelphia: 1797].
8 p.; fold. tbl. paginated 9-10. Text headed as above. 2nd edn., with added footnotes and minor changes in text. Signed at end: B. July, 1797. Bristol B9869. LCP

769. [________].
Intimations; On Agriculture And Manufactures; And on new Sources of Trade, which will interfere with the Produce of America in Foreign Markets. [cap. title]
[Philadelphia: Charles Cist, 1793?].
8 p. First edn., not in Evans or Bristol. HSP

770. [________].
Intimations; On Manufactures;--on the Fruits of Agriculture;--And On new Sources of Trade, interfering with Products of the United States of America in Foreign Markets. [cap. title]
[Philadelphia: Charles Cist, 1794].
8 p. 2nd edn., with minor additions; Evans 26681. Signed at end: B. January, 1794. HSP

771. [________].
Necessaries; Best Product Of Land; Best Staple Of Commerce.
Philadelphia: Printed By James Humphreys, Junr., 1776.
17, [2] p. Signed at end: Cresinus [pseud.]. LCP; APS

772. [________].
Outlines Of A Plan, For Establishing A State Society Of Agriculture In Pennsylvania.
Philadelphia: Printed By Charles Cist, 1794.
[2], 11 p. The Committee to prepare the outline was headed by Bordley. HSP in LCP; HSP

773. [________].
Purport of a Letter on Sheep. Written in Maryland, March the 30th, 1789. [cap. title]
[Philadelphia: Daniel Humphreys, 1789].
6 p. Signed at end: B. HSP in LCP; LCP.

774. [________].
Queries selected from a paper of the Board of Agriculture in London, on the Nature and Principles of Vegetation: with Answers and Observations by J.B.B. [cap. title]
[Philadelphia: Charles Cist, 1797].
19 p. Dated at end: June, 1797. LCP

775. [_________].
Sketches On Rotations Of Crops.
Philadelphia: Printed By Charles Cist, 1792.
[2], 47 p. Signed at end: B. HSP in LCP; LCP (with occasional MS. annotations by Bordley); APS

776. [_________].
Sketches / On / Rotations / Of / Crops, / And Other / Rural Matters. / To Which Are Annexed / Intimations On Manufactures; On The / Fruits Of Agriculture; And On New / Sources Of Trade, Interfering With / Products Of The United States Of / America In Foreign Markets.
Philadelphia: Printed By Charles Cist, 1796.
[2], 76 p.; illus. Signed at end of Sketches (p. 66): J.B.B.; signed at end of Intimations: B. January, 1794. Title-page A, with line endings as indicated by slashes. LCP (with imprint date changed in ink to read 1797)

777. [_________].
Sketches / On / Rotations / Of / Crops .... / To Which Are Annexed, / Intimations / On Manufactures; On The Fruits Of Agri- / culture; And On New Sources Of Trade, / Interfering With Products Of The United / States Of America in Foreign Markets.
Philadelphia: Printed By Charles Cist, 1796.
Identical to the above copy in all respects except for the last six lines of the title (Title-page B); line endings as indicated. HSP in LCP (with imprint date changed in ink to read 1797); LCP (2 copies, both unaltered; one copy has presentation inscription from the publisher dated 16 Mar. 1797)

_________. Sketches On Rotations Of Crops.
See also PARKINSON, RICHARD, 1748-1815. The Experienced Farmer's Tour .... 1805. Incl. extracts from Sketches, with commentary.

778. [_________].
A Summary View Of The Courses Of Crops, In The Husbandry Of England & Maryland; With A Comparison Of Their Products; And A System Of Improved Courses, Proposed For Farms In America.
Philadelphia: Printed By Charles Cist, 1784.
22 p. HSP in LCP; LCP; APS; HSP

[_________], ed.
See
FORSYTH, WILLIAM, 1737-1804. Epitome Of Mr. Forsyth's Treatise On ... Fruit-Trees .... By An American Farmer [i.e. Bordley].
Gleanings From The Most Celebrated Books On Husbandry .... Interspersed with Remarks and Observations by a Gentleman of Philadelphia [i.e. Bordley].

779. BOSSON, CHARLES PALFRAY.
Observations On The Potatoe, And A Remedy For The Potatoe Plague. In Two Parts. Containing A History Of The Potatoe, Its Cultivation, And Uses, Also A Treatise On The Potatoe Malady, Its Origin And Appearances ... And An Inquiry Into The Causes ....
Boston: Published By E. L. Pratt [Printed by S. N. Dickinson & Co. (t.p. verso)], 1846.
[6], 118, 4 (adv.) p. LCP (wrappers)

780. BOSTON AND ROXBURY MILL CORPORATION, Boston.
The Boston And Roxbury Mill Corporation. [Cap. title; text begins:] The subscriber and his associates now offer to the public the project of their great improvement, viz. the contemplated mill power and toll road ....
[Boston: 1818].
24 p. Incl. discussion of the advantages of a flour mill. LCP

781. BOSWELL, PETER.
Bees, Pigeons, Rabbits, And The Canary Bird, Familiarly Described: Their Habits, Propensities, And Dispositions Explained; Mode Of Treatment ... With An Appendix, Containing Directions For The Care Of Several American Singing-Birds.
New York: Wiley And Putnam [J.P. Wright, Printer (t.p. verso)], 1842.
Pp. [1-2], [7]-164; illus. LCP

782. _________.
The Poultry-Yard: A Practical View Of The Best Method Of Selecting, Rearing, And Breeding The Various Species Of Domestic Fowl. First American Edition.
New York: Wiley And Putnam [J.P. Wright, Printer (t.p. verso)], 1841.
Pp. [1-2], [7]-242, [1 (pub. adv.)]; illus. LCP

Both Sides Of The Grape Question .... 1860.
See SPANGLER, ANDREW M., pub.

783. BOUTWELL, GEORGE SEWALL, 1818-1905.
Address Delivered At Concord, September 18, 1850, Before The Middlesex Society Of Husbandmen And Manufacturers.
Boston: Printed By Abner Forbes, 1850.
28 p. LCP (wrappers)

784. [BOWEN, Eli, b. 1824].
Off-Hand Sketches; A Companion For The Tourist And Traveller Over The Philadelphia, Pottsville, And Reading Railroad. Describing The Scenery, Improvements, Mineral And Agricultural Resources, Historical Incidents, And Other Subjects Of Interest In The Vicinity Of The Route. With Numerous Engravings.
Philadelphia: Published By J.W. Moore [Stereotyped by Slote & Mooney; C. Sherman, Printer (t.p. verso)], 1854.
Pp. [1]-10, 15-194, [2 (pub. adv.)] incl. front.; illus. Consists of selections from Pictorial Sketch-Book (see below). HSP (2 copies: 1. in cloth; 2. wrappers, with wrapper title: Bromwells New Railway Guides)

785. __________.
The Pictorial Sketch-Book Of Pennsylvania. Or Its Scenery, Internal Improvements, Resources, And Agriculture, Popularly Described. Illustrated with over Two Hundred Engravings, And A Colored Map.
Philadelphia: Willis P. Hazard [Stereotyped by Slote & Mooney; C. Sherman, Printer (t.p. verso)], 1852.
Pp. [1]-268, [1-4], 13-192, [1-4 (adv.)] incl. front.; col. fold. map; illus. First leaf: timetable of the Philadelphia and Reading Railroad. Second section with sep. ½ t.p.: Locomotive Sketches, With Pen And Pencil. From Philadelphia To Pittsburgh. LCP (with adv. slip tipped in at front); HSP

786. __________.
The Pictorial Sketch-Book Of Pennsylvania. Or Its Scenery .... Illustrated ..., And A Colored Map.
Philadelphia: William Bromwell [Stereotyped by Slote & Mooney; C. Sherman, Printer (t.p. verso)], 1853.
Pp. [1]-309, [1-5], 13-207, [1-3]; front.; illus. First leaf and sep. ½ t.p. as above. LCP (in red gilt-and blind-stamped cloth); HSP (2 copies: 1. in brown gilt-and blind-stamped cloth; 2. in blue gilt-and blind-stamped cloth). None of the 3 copies incl. the map called for on t.p.

787. __________.
The Pictorial Sketch-Book Of Pennsylvania. Or Its Scenery .... illustrated ..., and a Colored Map. Eighth Edition, Revised And Greatly Enlarged.
Philadelphia: W. White Smith, Publisher [Stereotyped by Slote & Mooney; C. Sherman, Printer (t.p. verso)], 1854.
Pp. [1]-309, [1-5], 13-207, [1-3]; front.; illus. First leaf and sep. ½ t.p. as above. T.p. headed: Eighth Thousand. HSP

788. BOWEN, JOHN S.
Address, Delivered At The Delaware County Institute [of Science, Pa.], September 9, 1848. Published By Request Of The Institute.
Chester: Printed By Y.S. Walter, 1848.
8 p. On improving agriculture. HSP

789. BOWIE, WALTER WILLIAM WEEMS, 1814-1891.
Address Delivered By W.W.W. Bowie, Of Maryland, Before The Valley Agricultural Society Of Virginia, At Charlestown, on Friday, October 22, 1852.
Baltimore: From The Press Of Sands & Mills, Office Of The American Farmer, 1853.
20 p. LCP

790. BOWLER, METCALF, d. 1789.
A Treatise On Agriculture And Practical Husbandry, Designed for the Information of Land-Owners and Farmers. With a Brief Account of the Advantages arising from the new Method of Culture practiced in Europe.
Providence: Printed By Bennett Wheeler, 1786.
88 p. LCP

791. BRACKEN, HENRY, 1697-1764.
Farriery Improved; Or, A Complete Treatise On The Art Of Farriery. Wherein Are Fully Explained, The Nature and Structure of that useful Creature, a Horse; with the Diseases and Accidents he is liable to; and the Methods of Cure. Exemplified By Ten Elegant Cuts .... Likewise, Rules for Breeding and Training of Colts .... To Which Is Prefixed, Ten Minutes Advice to the Purchasers of Horses.
Philadelphia: Printed For Mathew Carey; By Lang And Ustick, March 17, 1796.
144 p.; front.; 9 pls., 8 fold. T.p. headed: A New Edition. LCP

792. __________.
Farriery Improved; Or, A Complete Treatise On The Art Of Farriery .... To Which Is Prefixed, Ten Minutes Advice ....
Philadelphia: Printed For Mathew Carey; By Stephen C. Ustick, Jan. 11, 1798.
LCP: 144 p.; front.; 8 fold. pls. only (of 9 pls.; lacking 1). T.p. headed: Second American Edition. LCP

__________. Farriery Improved.
See also BUCHAN, WILLIAM, 1729-1805. Every Man His Own Doctor .... 1816. Incl. unattributed excerpts.

793. __________.
Taplin Improved; Or A Complete Treatise On The Art Of Farriery, Wherein Are Fully Explained The Nature and Structure of that useful Creature, a Horse .... To Which Is Prefixed, Ten Minutes Advice ....
Troy: Printed And Sold By Francis Adancourt, And Also Sold By Whiting & Power; Albany: By H.C. Southwick; Hudson: W.E. Norman; N. York: Samuel Wood, 1815.
204 p.; front.; 9 pls. LCP

794. BRADFORD CO. [PA.]. AGRICULTURAL SOCIETY.
First Annual Report Of The Transactions Of The Bradford County Agricultural Society, Made February 6, 1854.
Towanda, Pa.: Printed By Elijah A. Parsons, 1854.
47, [1] p. LCP

795. BRECK, CHARLES H.B., Nurseryman, Boston.
New Seed Store. Garden Seeds, Of The Growth Of 184-, Raised For, And Put Up By Charles H.B. Breck, At his Seed Establishment, No. 45 North Market Street ....
[Boston: ca. 1845].
Brs. LCP

796. BRECK, JOSEPH, 1794-1873.
The Flower-Garden; Or, Breck's Book Of Flowers; In Which Are Described All The Various Hardy Herbaceous Perennials, Annuals, Shrubby Plants, And Evergreen Trees, Desirable For Ornamental Purposes, With Directions For Their Cultivation.
Boston: Published By John P. Jewett & Company [Stereotyped by Hobart & Robbins, New England Type And Stereotype Foundery (t.p. verso)], 1851.
336 p.; front. LCP

797. BRECK, SAMUEL, 1771-1862.
Address Delivered Before The Blockley And Merion Agricultural Society, On Saturday, September 20th, 1828. On The Death On [sic] Their Late President, The Hon. Richard Peters.
Philadelphia: Printed By Lydia R. Bailey, 1828.
27 p. Wrapper title: A Short Notice Of The Life Of The Late Hon. Richard Peters. LCP (wrappers); APS

798. BRIDGEMAN, THOMAS, d. 1850.
The Florist's Guide: Containing Practical Directions For The Cultivation Of Annual, Biennial, And Perennial Flowering Plants, Of Different Classes, Herbaceous and Shrubby, Bulbous, Fibrous and Tuberous-rooted; ... Second Edition, Improved.
New York: Printed And Sold By Mitchell & Turner; Sold by the Author; G.C. Thorburn; Alexander Smith; J. Stanley and Co; Bartlett and Raynor, et al., 1836.
128 p. LCP

799. _________.
The Florist's Guide; Containing Practical Directions For The Cultivation Of Annual, Biennial, And Perennial Flowering Plants .... Third Edition, Enlarged And Improved.
New York: For sale by T. Bridgeman; G.C. Thorburn; Alexander Smith; Flushing, L.I.: G.R. Garretson; Albany: Wm. Thorburn; Rochester: Reynolds, and Bateham; Boston: Hovey & Co.; Ellis and Bosson; Joseph Breck and Co.; Philadelphia: Hirst and Dreer; D. Landreth and Co.; et al, 1840.
180 p. LCP

800. _________.
The Kitchen Gardener's Instructor; Containing A Catalogue Of Garden And Herb Seeds, With Practical Directions Under Each Head, For The Cultivation Of Culinary Vegetables & Herbs: A New And Improved Edition, Of Five Thousand Copies, With A Calendar, Showing the Work necessary to be done in a Kitchen Garden, Every Month Throughout The Season; Also, Directions For Forcing Or Forwarding Vegetables Out Of The Ordinary Season.
New York: For sale by T. Bridgeman; G.C. Thorburn; Alexander Smith; Flushing, L.I.: G.R. Garretson; Albany: Wm. Thorburn, et al., 1840.
144, 4 p. LCP

801. _________.
The Kitchen Gardener's Instructor, Containing A Catalogue Of Garden And Herb Seed .... A New And Improved Edition.
New York: For Sale By The Author; And Edward Walker; J.M. Thorburn & Co.; Alex. Smith; Thomas Dunlap; Albany: Wm. Thorburn; Boston: Hovey & Co., Joseph Breck & Co., J.L.L.F. Warren, Walker & Co.; Philadelphia: David Landreth, H.A. Dreer, Robert Buist, et al., [1847].
Pp. [1-2], [5]-181, [1 (pub. adv.)]. LCP

802. _________.
The Young Gardener's Assistant: Containing A Catalogue Of Garden And Flower Seeds, With Practical Directions Under Each Head, For The Cultivation Of Culinary Vegetables And Flowers, With Directions For Cultivating Fruit Trees, The Grape Vine, &c. &c. &c. Sixth Edition.
New York: Printed And Sold By W. Mitchell; Sold by the Author; G.C. Thorburn; Alexander Smith; and other Seedsmen and Florists; by J. Stanley and Co.; J.G. Shaw; and the Booksellers in general, 1835.
viii, 349, [3 (pub. adv.)] p. LCP

803. _________.
The Young Gardener's Assistant, In Three Parts: Containing Catalogues Of Garden And Flower Seeds ... Also, Directions For Cultivating Fruit Trees, The Grape Vine, &c. To Which Is Added, A Calendar To Each Part .... Tenth Edition, Improved.
New York: For Sale By The Author; George C. Thorburn; Alex'r Smith; Albany: William Thorburn; Rochester: C.F. Crosman; Boston: Hovey & Co.; Joseph Breck & Co.; Philadelphia: D. Landreth & Munns; H.A. Dreer, et al.; Printed by A. Hanford [New York: Stereotyped By Vincent L. Dill (t.p. verso)], 1844.
Pp. [1]-164, [1-4], 9-164, [1-2], [1]-175, [1 (pub. adv.)]; front. Sep. ½ t.p. and pagination for each part. LCP

A Brief Narrative Of The Invention Of Reaping Machines .... 1854.
See STABLER, EDWARD, 1794-1883.

804. BRIGGS, CHARLES, 1791-1873.
A Discourse Delivered At Concord, October The Fifth, 1825.
Concord: Printed For The Society By John C. Allen, 1825.
34 p. Wrapper title: Mr. Briggs's Discourse Before The Society Of Middlesex [Mass.] Husbandmen and Manufacturers .... LCP (wrappers)

805. BRIGGS, ISAAC.
Statements And Remarks, Addressed To Thomas Newton, Chairman Of The Committee Of Commerce and Manufactures, On The Subject Of Agriculture, Maufactures [sic], and Commerce.
Washington: Printed By William A. Davis, 1816.
18 p. HSP in LCP; LCP

806. BRINCKLÉ, WILLIAM DRAPER, 1798-1862.
Remarks On Entomology, Chiefly In Reference To An Agricultural Benefit.
Lancaster, Pa.: W.B. Wiley, Printer, 1852.
8 p. Written for the Pennsylvania State Agricultural Convention held in Harrisburg, Jan. 1851. LCP

807. _________.
Remarks On Entomology, Chiefly In Reference To An Agricultural Benefit.
Philadelphia: Printed By T.K. And P.G. Collins, 1852.
24 p. LCP

_________, ed.
See The American Pomologist .... [pdcl.].

808. BRISSOT DE WARVILLE, JACQUES PIERRE, 1754-1793, and ÉTIENNE CLAVIÈRE, 1735-1793.
The Commerce Of America With Europe; Particularly With France And Great Britain; Comparatively Stated, And Explained. Shewing The Importance Of The American Revolution To The Interests Of France, And Pointing Out The Actual Situation Of The United States Of North America, In Regard To Trade, Manufactures, And Population. Translated From The Last French Edition, Revised By Brissot, and called the Second Volume of his View of America. With The Life Of Brissot, And An Appendix, By The Translator.
London: Printed For J.S. Jordan, 1794.
[2], lxiv, 348 p.; front. With general t.p.: New Travels In The United States Of America: Including The Commerce Of America .... Vol. II. This work, 1st published in English as Considerations on the Relative Situation .... 1788 (q.v. below), was also issued as the final vol. to some edns. of Brissot's New Travels, which is of slight agricultural interest. LCP

809. __________.
The Commerce Of America With Europe .... Translated from the last French Edition, Revised by Brissot, and called the Second Volume of his View of America. With the Life of Brissot, and an Appendix, By the Translator.
New York: Printed and sold by T. and J. Swords, 1795.
xxxv, [1], 228 p.; front. LCP; APS

810. __________.
Considerations On The Relative Situation Of France And The United States Of America: Shewing The Importance Of The American Revolution To The Welfare Of France: Giving Also An Account of their Productions ... And Finally, Pointing Out The Actual Situation Of The United States. Translated From The French Of ....
London: Printed At The Logographic Press, And Sold By Robson And Clarke; T. Longman; And W. Richardson, 1788.
[2], xvi, xxxiv, 326, [2 (pub. adv.)] p. Incl. remarks on American production of silk, grapevines, tobacco, corn, etc. APS

811. __________.
De La France Et Des États-Unis. Ou De L'Importance De La Révolution De L'Amérique Pour Le Bonheur De La France; Des rapports de ce Royaume et des États-Unis, des avantages réciproques qu'ils peuvent retirer de leurs liaisons de commerce, et enfin de la situation actuelle des États-Unis. Nouvelle Edition.
Paris: Chez Buisson, Imprimeur et Libraire, April 1791. [general t.p. imprint]
[4], xxiii, [1], 448 p. With general t.p.: Nouveau Voyage Dans Les États-Unis .... Tome Troisième. LCP

812. BRISTED, JOHN, 1778-1855.
The Resources Of The United States Of America; Or, A View Of The Agricultural, Commercial, Manufacturing, Financial, Political, Literary, Moral And Religious Capacity And Character Of The American People.
New York: Published By James Eastburn & Co.; Abraham Paul, printer, 1818.
xvi, 505, [1] p. LCP; APS

813. BRISTOL CO. [MASS.]. AGRICULTURAL SOCIETY.
Transactions.
Boston.
LCP: 1854.

814. BRODIN DE LA JUTAIS, PIERRE, d. 1765.
L'Abondance, Ou La Veritable Pierre Philosophale, Consistant En Un Moyen De Multiplier Abondamment Les Grains, Les Fruits, Les Fleurs, Et Tous Les Végétaux Généralement; Ouvrage publié à Paris en 1752 ... par le Chevalier Brodin de la Jutais ... Corrigé et augmenté de Principes physiques, par son gendre Faming De La Jutais ... Dedié a Son Excellence John [sic] Jefferson ....
Philadelphia: 1805.
47 p. APS

815. BROOM, JAMES MADISON, 1778-1850.
An Address Delivered Before The Philadelphia Society for Promoting Agriculture: At Its Anniversary Meeting, January 16, 1821.
Philadelphia: Printed for Littell & Henry; By Clark & Raser, 1821.
24 p. APS; HSP

816. __________.
An Address delivered before the Philadelphia Society for promoting Agriculture; at its anniversary meeting, January 16, 1821. Published By Order Of The Society. [cap. title]
[Philadelphia: 1821].
16 p. Title headed: Supplement To The National Recorder. HSP

BROOMALL, JAMES.
See DELAWARE CO. [PA.]. INSTITUTE OF SCIENCE. Report Of The Committee ... Also, The Address Of .... 1850.

817. BROWNE, DANIEL JAY, b. 1804.
The American Bird Fancier; Considered With Reference To The Breeding, Rearing, Feeding, Management, And Peculiarities Of Cage And House Birds; With Remarks On Their Diseases And Remedies; Drawn From Authentic Sources And Personal Observation. Illustrated with Engravings.
New York: Published By C.M. Saxton [Turney & Bro.'s Stereotype, c1850 (t.p. verso)].
107, [1] p.; illus. LCP

818. __________.
The American Bird Fancier ....
New York: Published By C.M. Saxton [Turney & Bro.'s Stereotype (t.p. verso)], 1851.
107, [1], [12 (pub. adv.)] p.; front.; illus. LCP

819. __________.
The American Poultry Yard; Comprising The Origin, History, And Description Of The Different Breeds Of Domestic Poultry; With Complete Directions For Their Breeding, Crossing, Rearing, Fattening .... With An Appendix, Embracing The Comparative Merits Of Different Breeds Of Fowls, By Samuel Allen.
New York: Published By C.M. Saxton, 1850.
322, [2], [12 (pub. adv.)] p.; front.; illus. LCP

820. BROWNE, PETER ARRELL, 1782-1860.
An Essay On Indian Corn. Read Before The Chester County Cabinet Of Natural Science, April 22, 1837, And Published By Order Of The Cabinet.
Philadelphia: Printed By J. Thompson, 1837.
32 p.; tinted front. LCP (wrappers); HSP

821. _________.
An Essay On The Veterinary Art; Setting Forth Its Great Usefulness, Giving An Account Of The Veterinary Colleges In France And England, And Exhibiting The Facility And Utility Of Instituting Similar Schools In The United States. To which is added, A Few Hints Upon The Propriety Of Connecting Therewith An Insurance Upon The Lives Of Horses.
Philadelphia: Printed By John Thompson, 1837.
22 p. LCP; HSP (wrappers)

822. _________.
Trichologia Mammalium, Or, A Treatise On The Organization, Properties And Uses Of Hair And Wool; Together With An Essay Upon The Raising And Breeding Of Sheep. Published Under The Patronage Of The Commonwealth Of Pennsylvania.
Philadelphia: J.H. Jones, Printer, 1853.
Pp. [1-6], [v]-viii, [7]-179, [1-2]; front.; 9 pls.; illus. LCP; APS (lacking pp. 127-179, [1-2], and last pl.); HSP

823. BROWNELL, THOMAS CHURCH, 1779-1865.
Annual Address, On The Theory Of Agriculture: Delivered By Appointment, Before The Society For The Promotion Of Useful Arts, At The Capitol, In The City Of Albany, On The 14th February, 1815.
Albany: Printed By Websters And Skinners, For The Society For The Promotion Of Useful Arts, 1815.
LCP: 1st 16 p. only (of 43; lacking last 14 leaves)

BROWNRIGG, WILLIAM, 1711-1800. Treatise On The Art Of Making Bay-Salt ....
See [PAINE, ROBERT TREAT, 1731-1814.] The Art Of Making Common Salt .... With An Extract from .... 1776.

BRUN CHAPPUIS, -----. Traité sur les Avantages de la Vigne .... 1824.
See FISHER, S.I. Observations On The Character And Culture Of The European Vine .... To Which Is Added, The Manual Of The Swiss Vigneron ... By .... 1834.

824. BRUSH, JOHN C.
A Small Tract Entitled A Candid And Impartial Exposition Of The Various Opinions on the subject of the Comparative quality of the Wheat And Flour In The Northern And Southern Sections Of The United States, With A View To Develope The True Cause Of The Difference .... In A Letter ... To Samuel Latham Mitchill, L.L.D. Of New-York.
Washington City: Printed By Jacob Gideon, Junior, 1820.
52 p. LCP

825. BUCHAN, WILLIAM, 1729-1805.
Every Man His Own Doctor; Or, A Treatise On The Prevention And Cure Of Diseases, By Regimen And Simple Medicines. To Which Is Added, A Treatise On The Materia Medica .... With An Appendix, Containing a Complete Treatise on the Art of Farriery; with Directions to the Purchasers of Horses ....
New Haven: Published By Nathan Whiting, 1816.
464, 144 p. Farriery treatise has sep. t.p. and pagination, and consists of excerpts from Henry Bracken's Farriery Improved (q.v.) without attribution to Bracken. LCP

826. BUCHANAN, ROBERT, 1797-1879.
The Culture Of The Grape, And Wine-Making. With An Appendix Containing Directions For The Cultivation Of The Strawberry, By N. Longworth.
Cincinnati: Moore & Anderson [C.A. Morgan & Co., Stereotypers (t.p. verso)], 1852.
Pp. [i]-x, 9-142. LCP

827. _________.
The Culture Of The Grape, And Wine-Making. With An Appendix Containing Directions For The Cultivation Of The Strawberry, By N. Longworth. Fifth Edition.
Cincinnati: Moore, Wilstach, Keys & Co. [C.A. Morgan & Co., Stereotypers (t.p. verso)], 1856.
Pp. [i]-x, 9-142. LCP

828. BUCKMINSTER, WILLIAM.
The Practical Farmer, Or Spirit of the Boston Cultivator, Containing A Collection Of Valuable Essays, On Practical Agriculture, &c.
Boston: David H. Williams, 1840.
300 p.; 1st leaf pub. adv. LCP

BUCKS CO. [PA.]. AGRICULTURAL SOCIETY.
See MORRIS, JAMES PEMBERTON. An Address, Delivered Before .... 1823 [2 edns.].

829. BUDD, THOMAS, d. 1698.
Good Order Established In Pennsilvania & New-Jersey In America, Being a true Account of the Country; With its Produce and Commodities there made. And the great Improvements that may be made by means of Publick Store-houses for Hemp, Flax and Linnen-Cloth .... With the advantages of publick Granaries. Likewise, several other things needful to be understood by those that are or do intend to be concerned in planting in the said Countries ....
[Philadelphia:] Printed [by William Bradford] in the Year 1685.
39, [1] p. See Hildeburn 2 for discussion of printer. HSP in LCP

830. BUEL, JESSE, 1778-1839.
Address of The Hon. Judge Buel, Delivered Before The Agricultural And Horticultural Societies Of New Haven County, Sept. 25, 1839: With The Reports Of The Managers Of The Societies.
New Haven: B. L. Hamlen, 1839.
35 p. LCP (wrappers)

831. __________.
The Farmer's Companion; Or, Essays On The Principles And Practice Of American Husbandry. With The Address Prepared To Be Delivered Before The Agricultural And Horticultural Societies Of New-Haven County, Connecticut, And An Appendix, Containing Tables, And Other Matter Useful To The Farmer. To Which Is Prefixed, A Eulogy On The Life And Character Of Judge Buel By Amos Dean, Esq.
Boston: Thomas H. Webb, Agent [Education Press (t.p. verso)], 1842.
[2], xxiv, 336 p.; extra eng. t.p.; illus. Half t.p.: The School Library. Published Under The Sanction Of The Board Of Education Of The State Of Massachusetts. Vol. XVI. LCP

832. __________.
The Farmers' Instructer [sic]; Consisting Of Essays, Practical Directions, And Hints For The Management Of The Farm And The Garden. Originally Published In The Cultivator; Selected And Revised For The School District Library. In Two Volumes. Vol. I [- II].
New York: Harper & Brothers, Publishers, 1854.
2 vols. Vol. I: pp. [1-2], [7]-276; illus. II: pp. [1-6], [3]-247; illus. LCP

833. BUIST, ROBERT, 1805-1880.
The American Flower Garden Directory: Containing Practical Directions For The Culture Of Plants In The Flower Garden, Hot-House .... Second Edition, With Numerous Additions.
Philadelphia: E.L. Carey & A. Hart, 1839.
[8 (pub. adv.)], 379, [1], 4 (pub. adv.) p. For the 1st edn., see: HIBBERT AND BUIST, Nurserymen, Philadelphia. See also later edns., below. LCP

834. __________.
The American Flower Garden Directory: Containing Practical Directions For The Culture Of Plants .... Third Edition, With Numerous Additions.
Philadelphia: Carey And Hart [T.K. & P.G. Collins, Printers (t.p. verso)], 1845.
344, [2], [1 (adv.)] p. LCP

835. __________.
American Flower-Garden Directory: Containing Practical Directions For The Culture Of Plants .... Fifth Edition, with Numerous Additions.
Philadelphia: A. Hart, Late Carey & Hart [C. Sherman, Printer (t.p. verso)], 1852.
Pp. [i]-xvi, [13]-339, [1-5 (pub. adv.)]. HSP

836. __________.
American Flower-Garden Directory: Containing Practical Directions For The Culture Of Plants .... Sixth Edition, with Numerous Additions.
New York: C.M. Saxton, Agricultural Book Publisher, 1854.
342, [2], 4 (pub. adv.) p. LCP

837. __________.
The Family Kitchen Gardener; Containing Plain And Accurate Descriptions Of All The Different Species And Varieties Of Culinary Vegetables; With Their Botanical, English, French, And German Names, Alphabetically Arranged, And The Best Mode of Cultivating them, in the Garden or under Glass; With A Description Of Implements and Medicinal Herbs in General Use .... Illustrated with Twenty-five Engravings.
New York: Published By J.C. Riker [Stereotyped By C. Davison & Co. (t.p. verso)], 1847.
[1 (adv.)], 216 p.; illus. LCP

838. __________.
The Family Kitchen Gardener; Containing Plain And Accurate Descriptions ....
New York: Published By C.M. Saxton; J.C. Riker [Stereotyped By C. Davison & Co. (t.p. verso)], 1850.
216, [2], [8 (pub. adv.)] p.; illus. LCP

839. __________.
The Rose Manual; Containing Accurate Descriptions Of All The Finest Varieties Of Roses, Properly Classed In Their Respective Families, Their Character And Mode Of Culture, With Directions For Their Propagation, And The Destruction Of Insects. With Engravings.
Philadelphia: For The Author And Messrs. Carey And Hart, 1844.
Pp. [1-2], [vii]-xii, [13]-182, [1-2 (pub. adv.)]; illus. LCP; HSP

840. __________.
Twelfth Edition Of R. Buist's Select Catalogue Of Rare And Popular Flowering Green-House, Hot-House, And Hardy Plants, Including New Species And Varieties Lately Introduced ... [wrapper title]
[Philadelphia:] Stavely & McCalla, Printers, 1853-4. [wrapper imprint]
57, [3] p. LCP (wrappers)

BULOS, -----. The Art of Wine Making.
See FISHER, S.I. Observations On The Character And Culture Of The European Vine .... To Which Is Added .... The Art Of Wine Making, By .... 1834.

841. BUMP, JAMES H.
J.H. Bump's Atmospheric Attempering Churn. The subscriber calls the attention of Dairymen and all other persons interested in the art of Butter making, to this improved Atmospheric Churn, patented by James H. Bump, of Morris, Otsego County, New-York, the 26th of October 1858 ....
Morris, N.Y.: A.S. Avery, Job Printer, [1859].
Brs. LCP

842. BURCHARD, NATHAN.
Address On Agriculture: Delivered At The Request Of The American Institute Of The City Of New York At Mechanics Hall, No. 472 Broadway, On Monday Evening, 6th September, 1847 ... Accompanied By Remarks On The Establishment Of An Agricultural College And Experimental Farm, Made At The New York Farmers Club, On Tuesday, 7th September, 1847, By Charles L. Fleischman ....
New York: W.B. Dean, Printer, 1847.
28 p. LCP (wrapper)

843. [BURDON, WILLIAM].
The Gentleman's Pocket-Farrier, Showing How To Use A Horse On A Journey. And What Remedies are proper for common Accidents that may befal [sic] him on the road. A New Edition, Improved and carefully revised by a Veterinary Surgeon. With Additions By Mr. Blyth ....
Philadelphia: Published by Thomas Desilver, 1823.
50 p.; front. LCP

844. [__________].
The Gentleman's Pocket-Farrier, Showing How To Use A Horse On A Journey .... A New Edition, Improved and carefully revised by a Veterinary Surgeon. With Additions By Mr. Blyth ....
Philadelphia: Published by Thomas Desilver, 1829.
50 p.; front. HSP in LCP; LCP

[__________]. Gentleman's Pocket-Farrier .... See also
Every Man His Own Farrier. Containing Ten Minutes' Advice .... [1853]. New edition of the work below.
Ten Minutes Advice .... [with] The Gentleman's Pocket Farrier .... 1787.

845. BURGER, JOHANN, 1773-1842.
The Economy Of Farming, Translated From The German ... With Many Additional Notes From The German Of Thaer, Veit, Schwertz, Sprengel, Petri, &c; And A Copious Index: By E. Goodrich Smith.
New York: Leavitt & Trow; Boston: Crocker & Brewster; Philadelphia: Geo. S. Appleton, 1843.
viii, 132, [4] p. LCP

846. BURGES, TRISTAM, 1770-1853.
Address, Delivered Before The American Institute Of The City Of New-York, At The Third Annual Fair, Held at Masonic Hall, October, 1830. Printed By Request Of The Institute.
New York: Published By John M. Danforth, 1830.
34 p. LCP; APS; HSP (wrapper)

847. __________.
Address To The Rhode-Island Society For The Encouragement Of Domestick Industry, Delivered At Pawtuxet, October 17, 1821.
Providence: Miller And Hutchens, Printers, August, 1822.
29, [2] p. LCP (wrappers); APS

848. BURKE, B.W.
A Compendium Of The Anatomy, Physiology And Pathology, Of The Horse: Being a clear and familiar Description of the various Organs and Parts, together with their Functions, of that useful and beautiful Animal. And also, comprising a View of the Diseases and Injuries ... Together With A Concise Examination of the Economy and Structure of The Foot, And Observations On Shoeing.
Philadelphia: Printed And Sold By James Humphreys, 1806.
292, [4 (pub. adv.] p.; 2 pls. HSP in LCP

849. Burlington Silk Record.
Burlington, N.J.
APS: Vol. 1, nos. 1-12 (Oct. 1840-Dec. 1841); vol. 2, nos. 1-6 (Jan.-Nov. 1842); vol. 3, nos. 1-2, 4 (Jan.-Feb., April 1843). Except for 3.3, all pub.

850. BURR, RICHARD.
Authentic Documents Relative To The Black River & St. Lawrence Countries; Being Extracts From Addresses, Reports, &c. Made To The Agricultural Society of Jefferson County, (N.Y.).
Philadelphia: 1820.
55 p. APS

851. BUSBY, JAMES.
Grapes And Wine. A Visit To The Principal Vineyards Of Spain and France; Giving A Minute Account Of The Different Methods Pursued In The Cultivation Of The Vine And The Manufacture Of Wine; With A Catalogue Of The Different Varieties Of Grape; An Attempt To Calculate The Profits Of Cultivating The Vine; An Estimate Of The Profits Of Malaga Fruit, &c. &c.
New York: C.S. Francis & Co.; Boston: J.H. Francis, 1848.
166, [2 (pub. adv.)] p. LCP

852. __________.
Journal Of A Recent Visit To The Principal Vineyards Of Spain And France. With Some Remarks On The Very Limited Quantity Of The Finest Wines Produced Throughout The World, And Their Consequent Intrinsic Value; An Attempt To Calculate The Profits Of Cultivating The Vine; A Catalogue Of The Different Varieties Of Grape; And An Estimate Of The Profits Of Malaga Fruits; Together With Observations Relative To The Introduction Of The Vine Into New South Wales.
Philadelphia: Jacob Snider, Jr., Wine Merchant [C. Sherman & Co. Printers (t.p. verso)], 1838.
xi, [1], 177 p. LCP

853. BUTLER, FREDERICK, 1766?-1843.
The Farmer's Manual; Being A Plain Practical Treatise On The Art Of Husbandry, Designed To Promote An Acquaintance With The Modern Improvements In Agriculture, Together With Remarks On Gardening, And A Treatise On The Management Of Bees.
Hartford: Published By Samuel G. Goodrich; Middletown: Clark & Lyman, Print, 1819.
224 p. LCP

854. __________.
The Farmer's Manual; Being A Plain Practical Treatise On The Art Of Husbandry ....
Weathersfield [Conn.]: Published By The Author; Middletown: Clark & Lyman, Printers, 1821.
Pp. [1-4], [3]-224. LCP

855. BYRN, MARCUS LAFAYETTE.
The Complete Practical Brewer, Or, Plain, Accurate And Thorough Instructions In The Art Of Brewing Ale, Beer, And Porter; Including The Process Of Making Bavarian Beer; Also, All The Small Beers, Such As Root-Beer, Ginger-Pop, Sarsaparilla-Beer, Mead, Spruce-Beer, Etc. Etc. Adapted To The Use Of Public Brewers, And Private Families ....
Philadelphia: Henry Carey Baird, Successor To E.L. Carey [Printed By T.K. & P.G. Collins (t.p. verso); Stereotyped By L. Johnson And Co. (colop.)], 1856.
199, [1], 24 (pub. adv.) p.; illus. LCP

856. __________.
The Complete Practical Distiller: Comprising The Most Perfect And Exact Theoretical And Practical Description Of The Art Of Distillation And Rectification; Including All Of The Most Recent Improvements In Distilling Apparatus; Instructions For Preparing Spirits From The Numerous Vegetables, Fruits, &c. .... With Numerous Engravings.
Philadelphia: Henry Carey Baird, (Successor To E.L. Carey) [Printed by T.K. & P.G. Collins (t.p. verso); Stereotyped By L. Johnson And Co. (colop.)], 1853.
198, 24 (pub. adv.) p.; illus. LCP

C**, M.
See CHAMBON, -----.

857. CABELL, NATHANIEL FRANCIS, 1807-1891.
Early History Of Agriculture In Virginia.
Washington, D.C.: Printed by Lemuel Towers, [1858].
41 p. 1st published in 5 parts in De Bow's Review, April-August 1858, "at the instance of the Virginia Agricultural Society" (footnote Part I). HSP: presentation copy from the author; in MS. after the title he has added: A Fragment. Wrapper.

858. CAIRD, JAMES, 1816-1892.
Prairie Farming In America. With Notes By The Way On Canada And The United States.
New York: D. Appleton And Company, 1859.
130, [2 (pub. adv.)] p. LCP

CALIFORNIA. GEOLOGICAL SURVEYS.
See
TRASK, JOHN BOARDMAN, 1824-1879. Report On The Geology Of Northern And Southern California, Embracing The ... Agricultural Resources .... [1856].
__________. Report On The Geology Of The Coast Mountains .... Embracing ... Resources In Agriculture .... [1854].

859. CALIFORNIA. STATE AGRICULTURAL SOCIETY.
Transactions.
Sacramento.
LCP: 1858; 1864-65; 1880.
Later known as: Reports.

CAMBRIA CO. [PA]. AGRICULTURAL SOCIETY.
See TAYLOR, GEORGE. Annual Address Before The Agricultural Societies Of Blair and Cambria Counties .... 1859.

860. CAMPBELL, PATRICK.
Travels In The Interior Inhabited Parts Of North America. In The Years 1791 and 1792. In which is given an account of the manners and customs of the Indians ... the mode of life and system of farming among the new settlers of both Canadas, New York, New England, New Brunswick, and Nova Scotia; interspersed with anecdotes of people, observations on the soil .... Illustrated With Copper-Plates.
Edinburgh: Printed For The Author, And Sold By John Guthrie, 1793.
Pp. [1-4], [iii]-x, [1]-387, [1]; front.; 2 pls., 1 fold.; fold. tbl. HSP in LCP

861. CAREY, HENRY CHARLES, 1793-1879.
The Harmony Of Interests, Agricultural, Manufacturing, And Commercial.
Philadelphia: J.S. Skinner; Office Of The Plough, The Loom, And The Anvil; Stereotyped By L. Johnson And Co., 1851.
235, [1], [3 (pub. adv.)] p. APS (wrappers); HSP (wrappers)

862. __________.
The Prospect: Agricultural, Manufacturing, Commercial, And Financial. At The Opening Of The Year 1851.
Philadelphia: J.S. Skinner; Office Of "Plough, Loom, And Anvil" [Printed by T.K. & P.G. Collins (t.p. verso)], 1851.
84 p. APS (wrappers); HSP (wrappers)

863. CAREY, JOHN, 1792-1875.
The Claims Of Agriculture. Speech Of Hon. John Carey, Of Ohio, Delivered in the U.S. House of Representatives, April 27, 1860. [cap. title]
[Washington: 1860].
7, [1] p. LCP

864. CAREY, MATHEW, 1760-1839.
Address Delivered Before The Philadelphia Society For Promoting Agriculture, At Its Meeting, On The Twentieth Of July, 1824. Printed By Order Of The Society.
Philadelphia: Printed By Joseph R.A. Skerret, 1824.
Pp. [i]-xii, [9]-80. LCP (wrapper); APS; HSP (wrappers)

865. __________.
Address Delivered Before The Philadelphia Society For Promoting Agriculture, At Its Meeting, On The Twentieth Of July, 1824. Printed By Order Of The Society. Second Edition.
Philadelphia: Printed By Joseph R.A. Skerrett, 1824.
66 p. LCP

866. __________.
Address Delivered Before The Philadelphia Society For Promoting Agriculture, At Its Meeting On The Twentieth Of July, 1824. Printed By Order Of The Society. Fifth Edition, Revised And Corrected.
Philadelphia: Mifflin & Parry, Printers, November, 1827.
71 p. LCP; HSP

CARR, ROBERT, 1778-1866, Nurseryman, Kingsessing, Pa.
See BARTRAM'S BOTANIC GARDEN, Kingsessing, Pa. Carr became proprietor in 1812 on the death of his father-in-law, John Bartram the younger.

867. CARTER, NATHANIEL HAZELTINE, 1787-1830.
Address Read Before The New-York Horticultural Society, At The Anniversary Celebration, Of The 28th Of August, 1827.
New York: Published At The Request Of The Society; Sleight & George, Printers, 1827.
22 p. APS

868. CARVER, JAMES.
Facts, Interesting To The Gentleman, the Farmer, The Grazier And The Butcher; Being An Account Of An Epidemic Disease, Which has lately been prevalent in England, And now prevailing in this City ... Together With The Causes, Symptoms and best mode of Treatment, By Which Numbers Have Been Saved. Also, An Account Of The Same Disease Among Oxen, Cows And Sheep. [wrapper title]
Philadelphia: Printed For The Author, 1817. [wrapper imprint]
24 p. Cap. title: Remarks and Observations On The Epidemic Catarrh .... LCP (wrappers); APS

869. __________.
A Treatise On The Age Of The Horse, Being a true delineation, with instructions how to tell his age, from a foal to the period of sixteen years; Taken From Actual Dissections ... Also, An Essay, On Founder, Contraction And Running Thrush .... Also, The Causes, Best Mode Of Treatment, And Cure By The Use Of The Hot And Cold Bath ....
Philadelphia: Published For The Author By Littell & Henry; Merritt, Printer, 1818.
93, [3] p.; 2 pls. HSP in LCP; LCP (with an extra pl. not belonging to this work bound at back); APS

870. __________.
Veterinary Science, Important To The Physician, the Student, and the Gentleman of Philadelphia: Giving An Account Of The Royal Veterinary College Of London, And A Similar Establishment In This City; Of A New Veterinary Forge and Horse Infirmary, For Instructing City And Country Smiths In that Important Branch of Domestic Science called The Shoeing Art; A Comparison Of The Two Systems Between Man and Horse. Also, Comparative Shoeing Explained ....
[Philadelphia:] T.S. Manning, Printer, 1817.
49 p. LCP (wrappers); APS; HSP

__________, ed.
See The Farrier's Magazine .... [pdcl.].

871. CARVER, WILLIAM.
The Practical Horse Farrier; Or, The Traveller's Pocket Companion. Shewing The Best Method To Preserve The Horse In Health; And, Likewise, The Cure Of The Most Prominent Diseases To Which This Noble Animal Is Subject, In The United States Of America ....
Philadelphia: Printed And Published By M'Carty & Davis, 1818.
111, [1] p. HSP in LCP

872. __________.
The Practical Horse Farrier; Or, The Traveller's Pocket Companion ....
Norristown, Pa.: Published By David Sower, Jun., 1818.
111, [1] p. LCP

873. __________.
The Practical Horse Farrier; Or, The Traveller's Pocket Companion .... Second Edition, Much enlarged, and embellished with three engravings.
Philadelphia: Printed And Published By M'Carty And Davis, 1820.
251 p.; illus. LCP

874. CASEY, JOSEPH P.
The Farmers' And Gardeners' Hive, Shewing The Expense And Profit Attending The Cultivation Of Three Hundred Acres Of Land, And So In Proportion For Any Other Quantity; And The Work Necessary To Be Done On A Farm And In A Garden, For Every Month In The Year. Also, A Treatise On The Cultivation Of The Peach. To Which Is Added, A Number Of Recipes, To Protect All Sorts Of Fruit Trees, Vegetables, &c. From All Sorts Of Diseases, Insects, Electricity ....
Baltimore: Printed by Wm. F. Redding, 1819.
24 p.; fold. tbl. LCP

875. __________.
A Treatise On The Culture And Growth Of Different Sorts Of Flower Roots, And Of Green House Plants Kept In Rooms, &c. To Which Is Added, A Table Of The Linnaean Classes Of Botany, With Their Orders And Examples.
Baltimore: Published by the Author; Printed By John D. Toy, 1821.
99, [3] p. LCP

876. CASS, LEWIS, 1782-1866.
Address Of Gen. Lewis Cass, Delivered Before the Kalamazoo County [Mich.] Agricultural Society, on Friday, October 11, 1850. Published By Request. [cap. title]
Washington: Towers, printer, [1850].
16 p. LCP

CATTARAUGUS CO. [N.Y.]. AGRICULTURAL AND HORTICULTURAL SOCIETY.
See MANLEY, JOHN, comp. Cattaruagus County: Embracing Its Agricultural Society, Newspapers .... 1857 (2 issues).

Cattle pastured and soiled .... [1798].
See BORDLEY, JOHN BEALE, 1727-1804.

CAYUGA CO. [N.Y.]. AGRICULTURAL SOCIETY.
See JOHNSON, BENJAMIN PIERCE, 1793-1869. An Address Before .... 1848.

877. CENTRAL ILLINOIS FARM, COAL AND LUMBER CO., Philadelphia.
Central Illinois Farm, Coal And Lumber Co. Office Corner Of Fourth And Library Streets, Philadelphia .... [Cap. title:] A Brief Statement Of The Organization, Plan Of Operations, And Prospects, Of The ....
[Philadelphia: 1855].
22 p.; 2 maps, 1 fold. The lands lay in Macon, Moultrie, Shelby, and Clay Cos. LCP

878. __________.
Central Illinois Farm, Coal, and Lumber Co. Projected And Conducted Upon True Mutual Joint Stock Principles. [cap. title]
[Philadelphia:] [Steam-Power Printing Establishment (colop.)], [1855].
4 p. Prospectus. LCP

879. Ceres.
Lebanon, Pa.
LCP: Vol. 2, nos. 1, 7 (Aug. 1840, Feb. 1841). Wrappers.

CHADERTON, WILLIAM.
See
[FELLENBERG INSTITUTION, Bolton Farm, Bucks Co., Pa.]. Classical, English, And Agricultural Institute .... [1830]. Incl. letters signed by Chaderton.
__________. Fellenberg Institution .... [1831]. Co-signed by Chaderton.

880. [CHAMBON, -----].
Le Commerce De L'Amerique Par Marseille, Ou Explication des Lettres-Patentes du Roi, portant Réglement pour le Commerce qu'il fait de Marseille aux Isles Françoises de l'Amérique .... Par Un Citadin. Nouvelle Edition, Ornée de Cartes & de Figures en taille douce. Tome Premier [-Second].
Leyden and Aix: chez Pierre Chambon, Libraire, 1782.
2 vols. Vol. I: [8], 615, [1] p.; extra eng. t.p.; 7 pls.; 7 maps, 6 fold. II: [8], 618, [1]] p.; extra eng. t.p.; 5 pls.; 3 fold maps. Incl. remarks on cotton in the Louisiana Territory. APS

881. [-----].
Traité Général Du Commerce De L'Amérique. Contenant L'Histoire Des Découvertes des Européens dans cette Partie du Monde, son etendue, ses productions .... Ouvrage utile aux Négocians, Jurisconsultes, Gens d'affaire, & autres. Orné De Figures Et Cartes Géographiques. Par M. C**. Ancien Receveur des Fermes du Roi. Tome Premier.
Amsterdam: Chez Marc-Michel Rey, Libraire; Marseille: Et Se Trouve chez Jean Mossy, Imprimeur & Libraire, 1783.
APS: Vol. 1 only (of 2): [8], 615, [1] p.; 7 pls.; 7 maps, 6 fold. Reissue with different title of Le Commerce De L'Amerique .... (above)

882. CHANDLER, ADONIRAM.
Address, Delivered At The Close Of The Nineteenth Annual Fair Of The American Institute, Of The City Of New-York, At Castle Garden, October 23, 1846.
New York: Van Norden & Amerman, Printers, 1846.
18 p. HSP (wrapper)

883. CHAPMAN, THOMAS.
Cyder-Maker's Instructor, Sweet-Maker's Assistant, And Victualler's and Housekeeper's Director. In Three Parts. Part I. Directs the Grower to make his Cyder in the Manner foreign Wines are made .... Part II. Instructs the Trader or House-Keeper to make Raisin Wines .... Part III. Directs the Brewer to fine his Beer and Ale in a short Time ....
London: Printed; And, Philadelphia: Re-printed by And. Steuart, [1762].
23, [1 (pub. adv.)] p. HSP in LCP

884. CHAPTAL DE CHANTELOUPE, JEAN ANTOINE CLAUDE, comte, 1756-1832.
Chymistry Applied To Agriculture. First American, Translated From The Second French, Edition.
Boston: Hilliard, Gray And Co. [Cambridge: Charles Folsom, Printer To The University (t.p. verso)], 1835.
xl, 365 p. LCP

885. CHAZOTTE, PETER STEPHEN, ca. 1770-ca. 1849.
Facts And Observations On The Culture Of Vines, Olives, Capers, Almonds, &c. In The Southern States, And Of Coffee, Cocoa, And Cochineal, In East Florida, By Peter Stephen Chazotte, Who Has For Upwards Of Ten Years Been Engaged In The Culture Of Vines, &c. In Southern France ....
Philadelphia: Printed And Published By J. Maxwell, 1821.
24 p. APS; HSP

886. The Cheap And Famous Farrier.
Ephrata [Pa.]: Printed and sold, 1795.
64 p.; illus. HSP in LCP

887. CHEEVER, SAMUEL.
Address Delivered Before The New York State Agricultural Society At Its Annual Meeting at Albany, Feb. 15, 1856. Published By Order Of The Society.
Albany: Fisk & Little, 1856.
30 p. LCP (wrapper)

888. CHELTENHAM NURSERIES, Montgomery Co., Pa.
Catalogue of Fruit and Ornamental Trees, Shrubs, Vines, &c. For Sale At The Cheltenham Nurseries, On Oak Lane .... Address -- Haines & Hacker, Cheltenham P.O., Montgomery Co., Pa. [wrapper title.]
Philadelphia: From Bryson's Printing Rooms, 1859. [wrapper imprint]
12 p. Wrapper headed: Robert B. Haines. William Hacker. LCP (wrappers)

889. CHENEY, JOHN MILTON.
An Address Delivered Before The Society Of Middlesex [Mass.] Husbandmen And Manufacturers, At Concord, Mass. Oct. 5, 1831.
Concord: Published By Herman Atwill, 1831.
22 p. LCP

CHENEY, WARD, & BROTHERS, eds.
See The American Silk Grower And Farmer's Manual [pdcl.].

890. CHERRY HILL NURSERY, West Chester, Pa.
Catalogue Of Fruit And Ornamental Trees, Shrubbery, Vines, Roses And Green-House Plants, Cultivated And For Sale At The Cherry Hill Nursery, West Chester, Pa. Josiah Hoopes, Proprietor.
Philadelphia: Merrihew & Thompson, Printers, 1856.
24 p. HSP (wrappers)

891. ________.
Catalogue Of Fruit And Ornamental Trees, Shrubbery, Vines .... Hoopes & Brother, Proprietors. Josiah Hoopes. Abner Hoopes.
Philadelphia: Robb, Pile & M'Elroy, Printers, 1859.
35 p. HSP (wrappers)

892. Chesapeak. [Cap. title p. 5:] A Letter From A Reputable Farmer To His Friend.
[Philadelphia: 1768].
LCP: pp. 1-18 (of 19 [?] p.). On mutual benefits to farmers and merchants of a joint local market in Maryland (not on a canal, as per Evans 10854)

893. CHESTER CO. [PA.]. AGRICULTURAL SOCIETY.
Annual Report Of The Managers Of The Chester County Agricultural Society For 1858. West Chester, October 10, 1858.
Philadelphia: Merrihew & Thompson, Printers, 1858.
56 p. HSP (wrappers)

________.
See also
BIDDLE, JOHN BARCLAY, 1815-1879. An Address Delivered Before .... 1856.
DARLINGTON, WILLIAM, 1782-1863. A Discourse Upon Agriculture: At A Meeting ... For The Purpose Of Forming An Agricultural Society .... 1847.
EMERSON, GOUVERNEUR, 1796-1874. Address Delivered Before .... 1853.
FISHER, SIDNEY GEORGE, 1809-1871. Address Delivered Before .... 1858.

CHESTER COUNTY [PA.]. CABINET OF NATURAL SCIENCE.
See
BROWNE, PETER ARRELL, 1782-1860. An Essay On Indian Corn. Read Before .... 1837.
DARLINGTON, WILLIAM, 1782-1863. A Discourse On ... True Grasses. Delivered ... Before .... 1841.

894. CHESTER CO. [PA]. HORTICULTURAL SOCIETY.
Constitution And By-Laws Of The Chester County Horticultural Society. Instituted Dec. 1845.
West Chester: Strickland And Bosee, Printers, 1846.
12 p. HSP

________.
See also
DARLINGTON, WILLIAM, 1782-1863. Address Before 1846
DILLINGHAM, WILLIAM HENRY, 1790-1854. An Address Delivered Before .... With The Transactions Of .... 1847.
LEE, ZACCHEUS COLLINS, 1805-1859. Address Before .... 1858.

895. CHILD, DAVID LEE, 1794-1874.
The Culture Of The Beet, And Manufacture Of Beet Sugar.
Boston: Published By Weeks, Jordan & Co.; Northampton: J.H. Butler [Press Of I.R. Butts (t.p. verso)], 1840.
156 p. LCP

896. CHILD, SAMUEL.
Every Man His Own Brewer, A Small Treatise, Explaining The Art And Mystery Of Brewing Porter, Ale And Table-Beer; Recommending And Proving The Ease And Possibility Of Every Man's Brewing His Own Porter, Ale And Beer, In Any Quantity. From one Peck to an hundred Bushels of Malt .... Improved And Calculated According To The Measures And Monies Of The United States.
Philadelphia: Printed for T. Condie; and H. Kammerer, 1796.
Pp. [1-2], 5-34; last 4 p. pub. adv. Preface dated August, 1796. Not in Evans or Bristol. LCP

897. ________.
Every Man His Own Brewer, A Small Treatise, Explaining The Art And Mystery Of Brewing Porter, Ale, And Table-Beer ... Second American Edition ....
Philadelphia: Printed for T. Condie, 1796.
Pp. [1-2], 5-34; last 4 p. pub. adv. Preface dated September, 1796. Evans 30189. HSP in LCP

898. CHORLTON, WILLIAM.
The American Grape Grower's Guide. Intended Especially For The American Climate. Being A Practical Treatise On The Cultivation Of The Grape-Vine In Each Department Of Hot House, Cold Grapery, Retarding House, And Out Door Culture. With Plans For The Construction Of The Requisite Buildings .... Every Department being fully Illustrated.
New York: C.M. Saxton, Barker & Co. [Edward O. Jenkins, Stereotyper And Printer (t.p. verso)], 1860.
Pp. [i]-xii, 11-204, 1-8 (pub. adv.); front.; illus. LCP

CHRETIEN DE ROVILLE, A.J., jt. auth.
See BENTZ, LOUIS, and ________.

CHRIST, JOHANN LUDWIG, 1739-1813.
See SOUDER, DAVID. The Rural Economist's Assistant, In The Management Of Bees: Principally taken from the German writings of .... 1807.

899. The Christian's, Scholar's, and Farmer's Magazine.
Elizabeth, N.J.
HSP in LCP: Vol. 1, nos. 1-2 (Apr./May-June/July 1789); no. 4 (Oct./Nov. 1789)
LCP: Vol. 1, nos. 1-6 (Apr./May 1789-Feb./Mar. 1790)

900. [CHRISTY, DAVID, b. 1802].
Cotton Is King: Or The Culture Of Cotton, And Its Relation To Agriculture, Manufactures and Commerce; To The Free Colored People; And To Those Who Hold That Slavery Is In Itself Sinful. By An American.
Cincinnati: Moore, Wilstach, Keys & Co. [William Overend & Co., Printers (t.p. verso)], 1855.
210, [1] p. LCP; HSP

901. [CHURCH, EDWARD, b. 1779].
Notice On The Beet Sugar. Containing--First, A Description Of The Culture And Preservation Of The Plant. Second, An Explanation Of The Process Of Extracting Its Sugar. Preceded By A Few Remarks On The Origin And Present State Of The Indigenous Sugar Manufactories Of France. Translated from the works of Dubrunfaut, De Domballe, and others. Second Edition.
Northampton: Published By J.H. Butler; Boston: Hilliard, Gray, & Co.; New York: D. Appleton & Co.; Philadelphia: William Marshall & Co.; Buffalo: T. & M. Butler [Northampton: John Metcalf, Printer (t.p. verso)], 1837.
55, [1], 4 (pub. adv.) p. LCP; APS

902. CINCINNATI. HORTICULTURAL SOCIETY.
A Brief History Of The Cincinnati Horticultural Society. Its Charter, Constitution And By-Laws; Officers From 1841 To 1859; Life, Annual And Honorary Members; And Catalogue Of Books In Library. Together With Its Transactions For The Past Year, And Premium Lists For The Year 1859. Prepared By Order Of Society.
Cincinnati: Printed At Office Of The "Cincinnatus," By Ongley & Shain, 1859.
Pp. [1-4], [3]-127, [1], [1-12 (adv.)]. LCP

903. __________.
The Charter, Constitution And By-Laws, Of The ... With A Report of its Transactions for 1843, '44, and '45, List Of Members, &c.
Cincinnati: L'Hommedieu & Co., Printers, 1846.
68, xix p. Wrapper headed: Part I. APS (wrappers)

__________.
See also ALLEN, ISAAC JACKSON. Address Of ... Delivered ... During The Exhibition Of 1855

904. The Cincinnatus: Devoted To Scientific Agriculture ....
Cincinnati.
LCP: Vol. 1, no. 3 (Mar. 1, 1856)

905. CLARKE, JOHN.
Sericon: Or, A Brief Treatise On The Art Of Rearing Silk-Worms, And Conducting Filatures.
Philadelphia: Published For The Author, 1841.
81, [1 (adv.)] p.; illus. APS (wrappers)

906. __________.
A Treatise On The Mulberry Tree And Silkworm. And On The Production And Manufacture Of Silk. Embellished With Appropriate Engravings.
Philadelphia: Thomas, Cowperthwait & Co. [Printed By T.K. And P.G. Collins; Stereotyped By L. Johnson (t.p. verso)], 1839.
364, 4 (pub. adv.) p.; illus. LCP; APS

907. __________.
A Treatise On The Mulberry Tree And Silkworm .... Second Edition.
Philadelphia: Thomas, Cowperthwait & Co. [Printed By T.K. And P.G. Collins (t.p. verso)], 1839.
364, 4 (pub. adv.) p; illus. HSP

908. [CLARKE, THOMAS COTTRELL, 1801-1874].
Pocket Manual for Producing Silk. [Cap. title continues:] Containing ample Directions for growing the Mulberry tree, managing the Eggs, feeding the Worms, securing the Cocoons, and Reeling the Silk, comprising the whole process of Silk making. Compiled from authentic sources.
Philadelphia: Published by Thomas C. Clarke, [1840].
23, [1] p.; illus. Author's name p. 3. HSP

909. CLATER, FRANCIS, 1756-1823.
Every Man His Own Cattle Doctor; Or, A Practical Treatise On The Diseases Of Horned Cattle: Being A Concise And Familiar Description Of All The Diseases Incident To Oxen, Cows, And Sheep; With the most simple and effectual Method of curing each Disorder ....
Philadelphia: Printed And Published By Abm. Small, 1815.
256, [7], [1 (pub. adv.)] p. LCP

910. __________.
Every Man His Own Cattle Doctor; Or, A Practical Treatise On The Diseases Of Horned Cattle ....
Albany: 1817.
191 p. LCP

911. __________.
Every Man His Own Cattle Doctor. Containing The Causes, Symptoms, And Treatment Of All The Diseases Incident To Oxen, Sheep, And Swine; And A Sketch Of The Anatomy And Physiology Of Neat Cattle. Edited, Revised, And Almost Rewritten By William Youatt .... With Numerous Additions, Embracing An Essay On The Use Of Oxen, And The Improvement In The Breed Of Sheep, &c. By J.S. Skinner .... With Numerous Cuts And Illustrations.
Philadelphia: Lea And Blanchard [J. Fagan, Stereotyper; J. & W. Kite, Printer (t.p. verso)], 1844.
Pp. [5]-251; illus. LCP

CLAVIÈRE, ÉTIENNE, 1735-1793, jt. auth.
See BRISSOT DE WARVILLE, JEAN PIERRE, 1754-1793, and __________.

912. [CLUNY, ALEXANDER].
The American Traveller: Containing Observations On The Present State, Culture and Commerce Of The British Colonies in America, And the further Improvements of which they are capable; With An Account of the Exports, Imports and Returns of each Colony respectively .... Together With The Amount of the Revenue arising to Great-Britain therefrom. In A Series Of Letters, Written originally to the Right Honourable the Earl of ******** By an Old and Experienced Trader.
[Philadelphia:] Printed [by Crukshank and Collins] in the Year 1770.
89, [1 (adv.)] p. Imprint from Evans 11603. HSP in LCP; LCP (with 2nd leaf, i.e. ded. and Tbl. of Contents, bound preceding t.p.)

913. ________.
Le Voyageur Américair, Ou Observations Sur l'Etat actuel, la Culture, le Commerce des Colonies Britanniques en Amérique, les Exportations & Importations respectives entre elles & la Grande-Bretagne .... Adressées par un Négociant experiencé, en forme de lettres, au très-honorable Comte de .... Traduit de l'Anglois, Augmenté d'un Précis sur l'Amérique Septentrionale & la République des Treize-Etats-Unis. Par M. J^{n}. M.........
Amsterdam: Chez J. Schuring, 1783.
viii, 264 p.; 1 fold. map; 3 fold. tbls. LCP

914. COBB, JONATHAN HOLMES, 1799-1882.
A Manual Containing Information Respecting The Growth Of The Mulberry Tree, With Suitable Directions For The Culture Of Silk, In Three Parts.
Boston: Carter, Hendee And Babcock [Boston Classics Press: I.R. Butts (t.p. verso)], 1831.
68 p.; 3 pls., 2 col. LCP

915. ________.
A Manual Containing Information Respecting The Growth Of The Mulberry Tree ....
Boston: Carter, Hendee And Babcock; London: R.J. Kennett [Boston Classics Press: I.R. Butts (t.p. verso)], 1831.
68 p.; 3 pls., 2 col. APS (as above but with slip bearing the London publisher's name pasted below printed imprint)

916. ________.
A Manual Containing Information Respecting The Growth Of The Mulberry Tree .... In Three Parts. New Edition.
Boston: Carter, Hendee And Co. [Boston: I.R. Butts (t.p. verso)], 1833.
Pp. [1-2], [5]-98; col. front.; 2 pls., 1 col. LCP; APS

917. ________.
A Manual Containing Information Respecting The Growth Of The Mulberry Tree .... Fourth Edition Enlarged.
Boston: Weeks, Jordan And Company [Tuttle, Dennett & Chisholm, Printers (t.p. verso)], 1839.
[4], 162, 2 (pub. adv.) p.; 3 pls. on 1 fold. front. sheet; illus. APS

918. COBBETT, WILLIAM, 1763-1835.
The American Gardener; Or, A Treatise On the Situation, Soil, Fencing and Laying-Out of Gardens; on the Making and Management of Hot-Beds and Green-Houses; and on the Propagation and Cultivation of the several sorts of Vegetables, Herbs, Fruits and Flowers.
London: Published By C. Clement [Printed by J. Edgerley (t.p. verso)], 1821.
[254] p.; 4 pls. HSP in LCP; APS

919. ________.
The American Gardener; Or, A Treatise on the Situation, Soil, Fencing and Laying-Out of Gardens .... Stereotype Edition.
London: Published By C. Clement [Printed by C. Clement (t.p. verso)], 1821.
[254] p.; 4 pls. LCP; APS

920. ________.
The American Gardener .... Stereotype Edition.
London: Published By C. Clement, 1821.
[254] p.; 4 pls. APS

921. ________.
The American Gardener: A Treatise On The Situation, Soil, Fencing, And Laying-Out Of Gardens .... First American Stereotype Edition.
Concord, N.H.: Published By L. Hamilton; Boston: Saxton & Pierce; New York: Saxton & Miles [Stereotyped And Printed By Morrill, Silsby, & Co. (t.p. verso)], 1842.
271 p.; illus. LCP

922. ________.
Cottage Economy; Containing Information Relative To The Brewing Of Beer, Making Of Bread, Keeping Of Cows, Pigs, Bees, Ewes, Goats, Poultry, And Rabbits, And Relative To Other Matters Deemed Useful In The Conducting Of The Affairs Of A Labourer's Family .... To Which Is Added The Poor Man's Friend; Or, A Defence Of The Rights Of Those Who Do The Work, And Fight The Battles.
New York: Published By John Doyle; Stereotyped By Connor & Cooke, 1833.
158, 68 p. Sep. ½ t.p. and pagination for Poor Man's Friend. LCP

923. ________.
The Emigrant's Guide; In Ten Letters, Addressed To The Tax-Payers Of England; Containing Information Of Every Kind, Necessary To Persons Who Are About To Emigrate; Including Several authentic and most interesting Letters from English Emigrants, now in America, to their Relations in England.
London: Printed By Mills, Jowett, And Mills; Published By The Author, 1829.
153 p. LCP (with 12 p. pub. adv. at front); APS

924. ________.
The Emigrant's Guide; In Ten Letters, Addressed To The Tax-Payers Of England .... Including Several authentic and most interesting Letters ... and an Account of the Prices of Houses and Land, recently obtained from America .... A New Edition.
London: Printed By Mills, Jowett, And Mills; Published By The Author, 1830.
12 (pub. adv.), 162 p. APS; HSP

925. ________.
Il Giardiniere Americano Ossia Trattato Della situazione del Suolo, Assiempamento ed Impiantagione de Giardini; della Costruzione e Maneggio dei Letti e delle Stufe; e della Propagazione e Coltivazione delle varie sorta di Vegetabili, Erbe, Frutta e Fiori. Recato in Italiano da Luca Regini di Cefalonia.
Corfu: 1826.
Pp. [1-6], [3]-245; 4 pls. LCP

926. ________.
The Woodlands: Or, A Treatise On the preparing of ground for planting; on the planting; on the cultivating ... Describing The usual growth and size and the uses of each sort of tree, the seed of each ... The Trees Being arranged in Alphabetical Order, and the List of them, including those of America as well as those of England ....
London: Printed And Published By William Cobbett, 1825.
[344] p.; illus. LCP; APS

927. ________.
A Year's Residence In The United States Of America. Treating of the Face of the Country, the Climate, the Soil, the Products, the Mode of Cultivating the Land, the Prices of Land, of Labour, of Food, of Raiment .... In Three Parts. Part I. Containing, -I. A Description of the Face of the Country .... -II. An Account of the Author's agricultural experiments in the Cultivation of the Ruta Baga ....
London: Printed For Sherwood, Neely And Jones [J. M'Creery, Printer (t.p. verso)], 1818 (Pt. I); 1819 (II).
APS: 1st 2 vols. only (of 3); paginated continuously: pp. [1-4], [1]-186; [187]-434. Vol. II has subtitle: Part II. Containing, -III. Experiments as to Cabbages. -IV. Earthburning. -V. Transplanting Indian Corn .... For Part III see below.

928. ________.
A Year's Residence In The United States Of America; Treating Of ....
Belfast: Printed At The Ulster Register Office, From An American Copy, 1818.
144 p. This is Part I (all pub. in this edn.) LCP

929. ________.
A Year's Residence In The United States Of America; Treating Of .... In Three Parts. Part I .... [- II].
New York: Printed For The Author By Clayton And Kingsland, 1818 (Part I); Sold At No. 53 Vesey-Street ....; Clayton And Kingsland, Printers, 1819 (II-III).
3 vols. paginated continuously: pp. [1]-134; [135]-304; [1-4], [307]-432. T.pp. for Parts II-III do not incl. subtitles. LCP; APS (Part III only of this edn.; see above for I-II); HSP

930. ________.
A Year's Residence, In The United States Of America. Treating of .... In Three Parts. Second Edition. Part I .... [- III ....].
London: Printed For Sherwood, Neely, And Jones [John M'Creery, Printer (t.p. verso)], 1819.
3 pts. in one: pp. [1-4], 1-186; [187]-434; [435]-610; fold. map. Each part has sep. t.p. APS

931. ________.
A Year's Residence In The United States Of America; Treating Of .... In Three Parts. Part I ....
New York: Printed For The Author By Clayton And Kingsland, 1819.
1st part of 3; 134 p. Reissue with new t.p. of 1818 edn. of Part I (above). See Gaines 77b. HSP

932. ________.
A Year's Residence, In The United States Of America. Treating of .... In Three Parts. Third Edition. Part I .... [-III ....].
London: Printed for J.M. Cobbett [B. Bensley (t.p. verso)], 1822.
3 parts in one: pp. [1-4], [i]-vi, [1]-100; [101]-246; [247]-360; fold. map. Each part has sep. ½ t.p. APS

933. ________.
A Year's Residence In The United States Of America. Treating Of .... In Three Parts. Third Edition. Part I .... [- III ....].
Andover: Printed By B. Bensley; London: And Published By The Author, 1828.
3 parts in one; with sep. ½ t.p. for each part. 12 (pub. adv.), 370 p.; fold. front. map. LCP

________.
See also
BIRKBECK, MORRIS, 1764-1825. Extracts .... [with] Reply To William Cobbett, Esq. July 31st, 1819. Replies to Cobbett's criticisms in A Year's Residence .... Part III.
FORSYTH, WILLIAM, 1737-1804. A Treatise On The Culture And Management Of Fruit Trees .... 1802 etc. With notes and introduction by Cobbett.

934. COCK, MICAJAH R.
The American Poultry Book; Being A Practical Treatise On The Management Of Domestic Poultry.
New York: Harper & Brothers, 1843.
179 p.; illus. LCP

935. COLE, SAMUEL W., 1796-1851.
The American Fruit Book; Containing Directions For Raising, Propagating, And Managing Fruit Trees, Shrubs, And Plants; With A Description Of The Best Varieties Of Fruit, Including New And Valuable Kinds ....
Boston: Published By John P. Jewett; New York: C.M. Saxton [Stereotyped by George A. Curtis, New England Type And Stereotype Foundery (t.p. verso)], 1849.
288 p. incl. double-sided front.; illus. T.p. headed: A Book For Every Body. LCP

936. ________.
The American Veterinarian, Or Diseases Of Domestic Animals, Showing The Cause, Symptoms, And Remedies, And Rules For Restoring And Preserving Health, By Good Management; With Directions For Training And Breeding.
Boston: John P. Jewett & Co. [Stereotyped by George A. Curtis; New England Type And Stereotype Foundery (t.p. verso)], 1847.
288 p. incl. front.; illus. T.p. headed: A Book For Every Farmer. LCP

937. COLMAN, HENRY, 1785-1849.
An Address Before The Hampshire, Franklin, And Hampden [Mass.] Agricultural Society; Delivered In Greenfield, Oct. 23, 1833. By Henry Colman. Published At The Request Of The Society.
Greenfield, Mass.: Printed By Phelps And Ingersoll, 1833.
28 p. LCP

938. __________.
An Address Delivered Before The Massachusetts Agricultural Society [i.e. Society for Promoting Agriculture], At the Brighton Cattle Show, October 17th, 1821.
Boston: Wells And Lilly, 1822.
[2], 17 p. LCP

939. __________.
An Address To The Essex [Mass.] Agricultural Society, Delivered At Andover, Massachusetts, 29th Sept. 1831. At Their Annual Cattle Show. Published At The Request Of The Society.
[Andover:] Gazette Office, 1832.
Pp. [1-2], [5]-22. LCP

940. __________.
Agricultural Addresses Delivered At New Haven, Norwich, And Hartford, Connecticut, At The County Cattle Shows, In The Year 1840. Published By The Request Of The Three Societies.
Boston: Dutton And Wentworth, Printers, 1840.
72 p. Incl. (pp. 3-24): Address Before The Agricultural And Horticultural Societies Of New Haven County ....; (pp. 25-42): An Address Delivered ... Before The Agricultural And Mechanic Institute Of New London And Windham Counties, Ct.; and (pp. 43-63); An Address Delivered ... Before The Hartford County Agricultural Society .... LCP

941. __________.
Agriculture And Rural Economy. From Personal Observation. Sixth Edition, With Additions.
Boston: Phillips, Sampson, & Company [Stereotyped At The Boston Stereotype Foundry (t.p. verso)], 1857.
2 vols. Vol. I: xxvi, 492 p.; front.; 4 pls.; illus. II: xxiv, 588 p.; front.; 4 pls.; illus. LCP (2 vols. in 1)

942. __________.
Letter To The Farmers Of Massachusetts On The Subject Of An Agricultural Survey Of The State By The Authority Of The Legislature.
Boston: Weeks, Jordan And Company, 1837.
18 p. LCP

943. __________.
First Report On The Agriculture Of Massachusetts. County Of Essex. 1837.
Boston: Dutton And Wentworth, State Printers, 1838.
Pp. [1-4], [3]-139; 2 fronts. LCP; APS

944. __________.
Second Report On The Agriculture Of Massachusetts. County Of Berkshire, 1838.
Boston: Dutton And Wentworth, State Printers, 1839.
194 p. LCP

945. __________.
Third Report Of The Agriculture Of Massachusetts, On Wheat And Silk.
Boston: Dutton And Wentworth, Printers To The State, 1840.
252 p. T.p. headed: Senate. No. 36. LCP (wrapper)

946. __________.
Fourth Report Of The Agriculture Of Massachusetts. Counties Of Franklin And Middlesex.
Boston: Dutton And Wentworth, State Printers, 1841.
xxix, [3], 528 p. LCP

947. __________.
Report On The Cultivation Of Spring Wheat, Prepared In Obedience To An Order Of The Senate Of Massachusetts, 16 March, 1838.
Boston: Dutton And Wentworth, State Printers, 1838.
27 p. T.p. headed: Senate. No. 77. LCP

948. __________.
Treatise On Bone Manure.
Boston: Printed By Tuttle, Dennett & Chisholm, 1838.
7, [1] p. LCP

__________.
See also
MASSACHUSETTS. AGRICULTURAL SURVEY OF THE STATE. Report .... With Notes By Henry Colman, Commissioner .... 1840.
MONROE CO. [N.Y.]. AGRICULTURAL SOCIETY. Annual Report Of The Transactions ... Containing The Address ... By .... 1842.

949. COLUMBIAN HORTICULTURAL SOCIETY, Washington, D.C.
Constitution And By-Laws Of The Columbian Horticultural Society, In The District Of Columbia. August, 1833.
Washington: Printed By Gales And Seaton, 1833.
8 p. LCP

__________.
See also WATTERSTON, GEORGE, 1783-1854. An Address Delivered Before .... 1834.

COLUMBIAN INSTITUTE FOR THE PROMOTION OF ARTS AND SCIENCES, Washington, D.C.
See CUTBUSH, EDWARD. An Address, Delivered Before .... 1817.

COLUMELLA, *pseud.*
See MOORE, CLEMENT CLARK, 1779-1863.

950. COLVIN, RICHARD.
A Small Treatise Containing Some Important Facts With A Few Practical Directions For Managing The Bee And Hive, Compiled By Permission From Langstroth On The Honey Bee.
Baltimore: Printed By Forrester & Bringman, 1859.
48 p.; illus. APS (with double set of wrappers, yellow outer and blue inner)

Le Commerce De L'Amerique Par Marseille .... 1782.
See CHAMBON, -----.

951. COMMISSIONERS FOR THE CULTIVATION OF THE VINE, Philadelphia.
Culture of the Vine. An address of the Commissioners for the cultivation of the Vine to the public.
[Philadelphia: ca. 1810?].
[2] p. Signed at end: Timothy Pickering, and 12 others. APS

952. The Complete Family Brewer; Or The Best Method Of Brewing Or Making Any Quantity Of Good Strong Ale And Small Beer, In The Greatest Perfection, For The Use Of Private Families; From A Peck Of Malt To 60 Bushels. Together With Directions for chusing good Malt, Hops, Water, Brewing-Vessels, &c. .... To Which Is Added, The Method of making Porter, as well as a Cheap and Wholesome Beer, from Treacle or Molasses ....
Philadelphia: Printed And Published By B. Graves, 1805.
24, 12 p. Sep. pagination for section cap.-titled: An Account, Of The Operations In Brewing, As Well As The Vessels Made Use Of Therein. HSP in LCP

953. The Complete Farrier, Or, Gentleman's Travelling Companion. Comprising A General Description Of The Perfections And Imperfections Of That Noble Animal The Horse, With A Concise Account Of His Diseases ... And Advice With Respect To Purchase .... Compiled from the best authorities ... the result of the observations of a Gentlemen of known experience in the United States.
Philadelphia: Published By Bradford And Inskeep, 1809.
vi, [10], 248 p.; fold. front.; 2 fold. pls. HSP in LCP; LCP

954. The Complete Farrier, Or, Gentleman's Travelling Companion .... Second Edition.
Philadelphia: Published By Bradford And Inskeep, 1810.
vi, [10], 248 p.; fold. front.; 2 fold. pls. LCP

955. The Complete Florist: A Manual Of Gardening, Containing Practical Instruction For The Management Of Greenhouse Plants, And For The Cultivation Of The Shrubbery, The Flower Garden, And The Lawn. With Descriptions Of Those Plants And Trees Most Worthy Of Culture, In Each Department. With Additions, And Amendments, Adapted To The Climate Of The United States.
Philadelphia: Lea And Blanchard [J. Fagan, Stereotyper; J. & W. Kite, Printers (t.p. verso)], 1844.
108, [12 (pub. adv.)] p. LCP

956. The Complete Florist: A Manual Of Gardening ....
Philadelphia: Lea And Blanchard [Printed by T.K. & P.G. Collins (t.p. verso)], 1847.
108 p. LCP

957. COMSTOCK, F.G.
A Practical Treatise On The Culture Of Silk, Adapted To The Soil And Climate Of The United States.
Hartford: Wm. G. Comstock, 1836.
108 p.; illus. Last 2 p. adv. LCP

958. ________.
A Practical Treatise On The Culture Of Silk .... Second Edition, Revised And Improved.
Hartford: P.B. Gleason And Co.; For Sale By The Booksellers Generally, 1839.
96 p. LCP

COMSTOCK, FERRE & CO., Nurserymen, Wethersfield, Conn.
See ALMANACS. WETHERSFIELD, CONN. 1850. The Gardener's Almanac; contains: Descriptive Catalogue Of Garden Seeds Cultivated And Sold At ....

CONNECTICUT agricultural societies.
See the following individual societies:
CONNECTICUT. SOCIETY FOR PROMOTING AGRICULTURE IN THE STATE [founded 1794].
CONNECTICUT. STATE AGRICULTURAL SOCIETY [founded 1852].

CONNECTICUT RIVER VALLEY. AGRICULTURAL SOCIETY.
See SANBORN, EDWIN DAVID, 1808-1885. An Address Delivered Before .... 1850.

959. CONNECTICUT. SOCIETY FOR PROMOTING AGRICULTURE IN THE STATE [founded 1794].
Transactions.
New Haven.
APS: 1802. All pub.

________.
See also ALMANACS. NEW HAVEN. 1815. An Agricultural and Economical Almanack .... Published under the patronage of the ....

960. CONNECTICUT. STATE AGRICULTURAL SOCIETY [founded 1852].
Transactions. With: Reports of the County Societies (1854-55); Reports of the Annual Meeting (1858-59).
Hartford.
LCP; APS: 1854-55, 1858-59.

________.
See also
JOHNSON, SAMUEL WILLIAM, 1830-1909. First Annual Report Of ..., Chemist To The .... 1858.
________. Essays On Peat ... By ..., Chemist To The .... 1859.

Considerations On ... Establishing An Agricultural College .... 1819.
See DE WITT, SIMEON.

Conversations On Vegetable Physiology .... 1830 etc.
See MARCET, JANE HALDIMAND, 1769-1858.

961. COOK, GEORGE HAMMELL, 1818-1889.
Report On The Geology And Agricultural Resources Of The Southern Division Of The State [of N.J.]. Printed by order of the House of Assembly.
Trenton: Printed At the "True American" Office, 1857.
30 p. LCP (wrapper); HSP (wrapper)

962. COOK, ZEBEDEE, 1786-1858.
An Address, Pronounced Before The Massachusetts Horticultural Society. In Commemoration Of Its Second Annual Festival, The 10th Of September, 1830.
Boston: Printed By Isaac R. Butts, 1830.
Pp. [1-4], [3]-54. LCP

963. COOKE, GEORGE.
The Complete English Farmer: Or, Husbandry made perfectly easy, In All Its Useful Branches. Containing what every Farmer ought to know and Practice. Among the various Articles treated of in this Work, are the following, viz. The Management and Qualities of the different kinds of Grass, and making Hay. Description of the different Soils ....
London: Printed; Boston: Re-printed by Daniel Kneeland, for Henry Knox, [1772?].
142 p. Dated by Bristol B3444. HSP

964. COOPER, JOHN W., b. 1803?
A Treatise On Cocking, Giving A History Of The Various Breeds Of Imported And American Game Fowls; The Manner Of Feeding and Training Them for the Pit, A Scientific Mode Of Gafting For Battle, Together With Rules of the Pit and for Breeding, And The Proper Treatment Of The Various Diseases Incident To Game Fowls.
Media, Delaware Co., Pa.: Cooper & Vernon, Prs., 1859.
96 p.; col. front.; fold. pl. HSP

COPE, CALEB, 1797-1888.
See THOMAS, M., & SONS, Auctioneers, Philadelphia. Catalogue Of Splendid And Rare Green And Hot-House Plants ... To Be Sold By Public Auction, At ... The Country Seat Of .... 1857.

COPE, F.J. Physiology In Its Application To Grape Culture ....
See [SPANGLER, ANDREW M., pub.]. Both Sides Of The Grape Question .... 1860. Incl. Cope's essay.

965. COPELAND, ROBERT MORRIS.
Country Life: A Handbook Of Agriculture, Horticulture, And Landscape Gardening.
Boston: John P. Jewett And Company; Cleveland, Ohio: Henry P.B. Jewett [Illustrations Drawn By Warren, Engraved By Andrew; Electrotyped By Cowles & Company; Cambridge: Printed By H.O. Houghton And Company (t.p. verso), 1859.
x, 813 p.; col. fold. front; 8 pls; illus. LCP

966. COPPINGER, JOSEPH.
The American Practical Brewer And Tanner: In Which Is Exhibited The Whole Process Of Brewing without boiling. Brewing strong Beer ....
New York: Printed By Van Winkle And Wiley, 1815.
246, [2] p.; illus. Errata slip tipped in before t.p. LCP

967. The Cottage Bee Keeper; Or Suggestions For The Practical Management Of Amateur, Cottage and Farm Apiaries, On Scientific Principles. With An Appendix Of Notes, Chiefly Illustrative. By A Country Curate.
New York: C.M. Saxton, Agricultural Book Publisher [E.O. Jenkins, printer (t.p. verso)], 1851.
xii, 119, [1], 11+[13] (pub. adv.) p.; illus. T.p. headed: Saxton's Cottage And Farm Library. LCP

968. Cotton, From The Pod To The Factory: A Popular View Of The Natural and Domestic History Of The Plant, The Adaptation And Improvement Of The Raw Material, With The Rise And Progress Of The Cotton Factory To Its Present State Of Perfection; And A Brief Account Of Bleaching And Dyeing.
New York: Homans & Ellis [R. Craighead, Printer (t.p. verso)], 1845.
iv, 77, [1], [2 (pub. adv.)] p.; front. Wrapper headed: The Shilling Library. No. 2. LCP (wrappers)

Cotton Is King .... 1855.
See CHRISTY, DAVID, b. 1802.

969. [COULTON, MISS -----].
Our Farm Of Four Acres And The Money We Made By It. From the Twelfth London Edition. With An Introduction By Peter B. Mead, Editor Of The Horticulturist.
New York: C.M. Saxton, Barker & Company, 1860.
126, 6 (pub. adv.) p. LCP

970. COURTAULD, GEORGE.
Address To Those Who May Be Disposed To Remove To The United States of America, On the advantage of Equitable Associations Of Capital and Labour, In The Formation Of Agricultural Establishments In The Interior Country. Including Remarks On Mr. Birkbeck's Opinions upon this Subject.
Sudbury: Printed And Sold By J. Burkitt; And Sold By D. Eaton; And R. Hunter, 1820.
40 p. APS

971. ________.
[Circular letter dated June 22, 1820; begins:] Sir, My object in publishing a late address to those who may be disposed to remove to the United States, was to shew that in the formation of new settlements in the interior country ....
[London: 1820].
3 p. Announces a meeting to form an "equitable association." APS

972. COXE, TENCH, 1755-1824.
An Addition, Of December 1818, To The Memoir, Of February And August 1817, On The Subject Of The Cotton Culture, The Cotton Commerce, And The Cotton Manufacture Of The United States, Most Respectfully submitting a suggestion, For Consideration Only, Of A Specific Measure For Securing To The Planters Of Cotton A Market For Their Crops. [cap. title]
[Philadelphia: 1818].
16 p. Note p. 1: See former Memoir, 16 pages; or extract from it in Rees's Cyclopedia, Philadelphia edition [1805-22], title "United States," sect. Agriculture U.S.: half vol. 77 [i.e. vol. 39; unpaged]. LCP; APS

973. __________.
An Addition, Of December 1818, To The Memoir .... [cap. title]
[Philadelphia: 1818].
16 p. Copy without footnotes pp. 1 and 16. HSP

974. __________.
An Address To An Assembly Of The Friends Of American Manufactures, Convened For The Purpose Of Establishing A Society For The Encouragement Of Manufactures And The Useful Arts, Read In The University Of Pennsylvania, On Thursday The 9th Of August, By Tench Coxe And Published At Their Request.
Philadelphia: Printed By R. Aitken & Son, 1787.
30 p. APS

975. [__________].
A Memoir, Of February, 1817, Upon the subject of the Cotton Wool Cultivation, the Cotton Trade, and the Cotton Manufactories of the United States of America. [cap. title]
[Philadelphia: 1817].
16 p. Author's name p. 16. HSP in LCP

976. [__________].
A Plan For encouraging Agriculture, and increasing the Value of Farms in the Midland and more Western Counties of Pennsylvania, applicable to several other Parts of that State, and to many Parts of the United States. [cap. title]
[Philadelphia: 1793].
11 p. HSP

977. [__________].
Proposals For Establishing A Number Of Farms, Like Those Of New England, New York, New Jersey, Pennsylvania, and Delaware. On The South Side Of The Western Districts Of North Carolina, For The Mutual Benefit Of The Settlers, And Of The Trade Of Charleston, Wilmington, (N.C.) Fayetteville, Georgetown, (S.C.) Augusta, And Savannah; And Of The Planters on the Sea-coast of the two Carolinas, Who raise Sugar, Rice, Indigo, Cotton, and Tobacco ... and may not raise Grain or Cattle.
[Philadelphia:] 1816.
16 p. LCP; APS; HSP

[__________.] A Series Of Tables Of The Several Branches Of American Manufactures .... Together with Returns of certain doubtful Goods, Productions of the Soil and agricultural Stock .... [1813].
See __________. A Statement Of The Arts And Manufactures .... 1814. Incl. the above.

978. __________.
A Statement Of The Arts And Manufactures Of The United States of America, For The Year 1810 ....
Philadelphia: Printed By A. Cornman, Junr., 1814.
Pp. [1-12], v-xiv (i.e. lxiv), [1-2], 1-46, 1-169 (p. lxiv misnumbered xiv). Consists of the following parts written at intervals between 1812 and 1814 (see J.E. Cooke, Tench Coxe, 1978, pp. 498-500): A Collection Of Facts [essays, pp. v-lxiv]; Tabular Statements Of The Several Branches Of American Manufactures, Exhibiting Them I. By States, Territories and Districts [pp. 1-46], II. By Counties, Cities and Towns [pp. 1-169]. Each series of tables has sep. t.p. T.p. for Tabular Statements has imprint: Philadelphia: Printed By A. Cornman, Junr., 1813. HSP in LCP; LCP (also holds tabular portions bound sep.); APS (tabular portions only); HSP

[__________.] Tabular Statements Of The Several Branches Of American Manufactures .... 1813.
See __________. A Statement Of The Arts And Manufactures .... 1814. Incl. the above.

979. __________.
A View Of The United States Of America, In A Series Of Papers, Written At Various Times, Between The Years 1787 And 1794. Interspersed With Authentic Documents: The Whole Tending To Exhibit The Progress And Present State Of Civil And Religious Liberty, Population, Agriculture, Exports, Imports, Fisheries, Navigation, Ship-Building, Manufactures, And General Improvement.
Philadelphia: Printed For William Hall; and Wrigley & Berriman, 1794.
Pp. [i]-v, [1-3], [7]-513; 6 fold. tbls. LCP

980. COXE, WILLIAM, 1762-1831.
A View Of The Cultivation Of Fruit Trees, And The Management of Orchards and Cider; With Accurate Descriptions Of The Most Estimable Varieties Of Native And Foreign Apples, Pears, Peaches, Plums, and Cherries, Cultivated In The Middle States Of America ....
Philadelphia: Published By M. Carey And Son; D. Allinson, Printer, Nov. 1, 1817.
253, [15] p.; 77 pls. HSP in LCP; LCP; APS

981. [CRÈVECOEUR, MICHEL GUILLAUME ST. JEAN DE, 1735-1813].
Letters From An American Farmer; Describing Certain Provincial Situations, Manners, And Customs, Not Generally Known; And Conveying Some Idea Of The Late And Present Interior Circumstances Of The British Colonies In North America. Written For The Information Of A Friend In England, By J. Hector St. John [pseud.], A Farmer In Pennsylvania.
Dublin: Printed By John Exshaw, 1782.
[12], 256 p.; 2 fold. maps. LCP

982. [__________].
Letters From An American Farmer; Describing Certain Provincial Situations ....
London: Printed For Thomas Davies; And Lockyer Davis, 1782.
[16], 318, [2 (pub. adv.)] p.; 2 fold. maps. HSP in LCP; LCP; APS (lacking ½ t.p.)

983. [__________].
Letters From An American Farmer: Describing Certain Provincial Situations .... A New Edition, With An Accurate Index.
London: Printed For Thomas Davies; And Lockyer Davis, 1783.
[16], 326, [2 (pub. adv.)] p.; 2 fold. maps. APS

984. [__________].
Letters From An American Farmer, Describing Certain Provincial Situations ....
Philadelphia: From The Press Of Mathew Carey, March 4, 1793.
240 p. HSP in LCP; LCP; APS; HSP

__________. Letters From An American Farmer ....
See also [AYSCOUGH, SAMUEL, 1745-1804]. Remarks On Letters From .... 1783.

985. [__________].
Lettres D'Un Cultivateur Américain, Écrites A W.S. Ecuyer, Depuis l'Arnée 1770, jusqu'à 1781. Traduites de l'Anglois par ***. Tome Premier [- Second].
Paris: Chez Cuchet, Libraire, 1784.
2 vols. Vol. I: xxiv, iv, 422, [2] p. II: [4], iv, 400, [2] p. LCP

986. __________.
Lettres D'Un Cultivateur Américain addressées à W[m.] S...on Esq[r.] depuis l'Année 1770 jusqu'en 1786. Traduites de l'Anglois. Tome I [- III].
Paris: Chez Cuchet Libraire, 1787.
3 vols. Vol. I: xxxii, 478, [4] p.; eng. t.p.; front.; 2 pls.; fold. map. II: 438, [6] p.; eng. t.p.; 3 fold. maps. III: 592, [1] p.; eng. t.p.; fold. pl.; fold. map. HSP in LCP (lacking all 5 maps); LCP; APS

__________. Lettres D'Un Cultivateur Américain ....
See also [CUTLER, MANASSEH, 1742-1823]. Description Du Sol .... 1789. Incl. Extraits des Lettres ....

987. CROLL, M. MARGARETHA, comp.
[Eine Sammlung Auserlesenen Recepte Heilsamer Mittel bey Krankheiten der Menschen und des Viehe zu gebrauchen.]
[(Ephrata:) 1826].
LCP: pp. [3]-23; lacking t.p.

988. CROPPER, BENSON & CO., merchants, Liverpool.
Circular On The Cultivation Of Cotton. [cap. title]
[Philadelphia?:] [Printed At The Office Of The Aurora (colop.)], [1822].
16 p. Incl. statistics on American cotton. Dated at head: Liverpool, 9th Mo. 27, 1822. LCP; APS; HSP

989. CULLEY, GEORGE, 1735-1813.
Observations On Live Stock: Containing Hints For Choosing And Improving The Best Breeds Of The Most Useful Kinds Of Domestic Animals. To Which Is Added, Some Observations Respecting The Cattle Imported Into America, In 1792, By Robert Heaton, Of Throg's Neck, Westchester County, State Of New-York.
New York: Printed And Published By D. Longworth, 1804.
Pp. [1-8], [5]-103. LCP

990. The Cultivator ... Published By The New York State Agricultural Society.
Albany.
LCP: Vols. 1, 3-5 (Mar. 1834-Feb. 1835; Mar. 1836-Feb. 1839); ns. vols. 1-5 (Jan. 1844-Dec. 1848). Also: 2nd edn. vols. 1-3 (Mar. 1834-Feb. 1837) pub. 1838.

991. CURTIS, JOHN.
America And The Corn Laws; Or, Facts And Evidence, Showing The Extensive Supply Of Food Which May Be Brought From America, And The Effects Of The Restrictive System On The British And American Trade. Printed For The National Anti-Corn-Law League.
Manchester: J. Gadsby, 1841.
35 p. LCP

992. CURTIS, SAMUEL, 1747-1822.
A Valuable Collection Of Recipes, Medical And Miscellaneous. Useful In Families, And Valuable To Every Description Of Persons.
Amherst, N.H.: Printed By Elijah Mansur, 1819.
70, [2] p. Incl. veterinary and agricultural remedies. LCP (wrappers)

993. CURTIS & LINCOLN, Nurserymen, Boston.
Catalogue Of A Fine Collection Of Floricultural, Vegetable And Agricultural Seeds, Comprising Assortments of the most approved varieties in cultivation. To Be Had Of Curtis & Lincoln ... 348 Washington Street, Boston. Daniel T. Curtis. Calvin A. Lincoln ....
Boston: Cross & Freeman, Printers, 1854.
33 p. LCP (with pp. 5-6 following pp. 7-8; wrappers)

994. CUSHING, CALEB, 1800-1879.
Address. Delivered September 26, 1850, At Salem, Before The Essex [Mass.] Agricultural Society. Published By Order Of The Society. December, 1850.
Salem: Printed at Office of Salem Gazette and Essex Co. Mercury, 1850.
18 p. LCP

995. CUTBUSH, EDWARD.
An Address, Delivered Before The Columbian Institute, For The Promotion Of Arts And Sciences, At The City Of Washington, On The 11th January, 1817. Published By The Request Of The Columbia Institute.
Washington: Printed By Gales and Seaton, 1817.
29 p. On the natural products and agriculture of the United States. LCP; APS

996. [CUTLER, MANASSEH, 1742-1823].
Description Du Sol, Des Productions, &c. &c De Cette Portion Des États-Unis, Située entre la Pensylvanie, les rivières de l'Ohio & du Scioto, & le lac Erié. Traduite d'une Brochure imprimée à Salem, en Amérique, en 1787.
Paris: 1789.
30 p. Incl. (pp. 24-29): Extraits des Lettres d'un Cultivateur Américain, par M. S' John de Crèvecoeur .... HSP in LCP

997. [__________].
An Explanation Of The Map Which Delineates That Part Of The Federal Lands, Comprehended between Pennsylvania West Line, the Rivers Ohio and Sioto, and Lake Erie; confirmed to the United States by sundry Tribes of Indians, in the Treaties of 1784 and 1786, and now ready for Settlement.
Salem: Printed By Dabney And Cushing, 1787.
24 p. Incl. (pp. 23-24): Extracts From the Letters of M. St. John De Crevecoeur .... HSP in LCP; LCP

998. [DABNEY, JOHN, 1752-1819, comp.].
An Address To Farmers On The following Interesting Subjects: 1. The character of a complete Farmer. 2. The Importance of Manure. 3. Labor. 4. Exchanging Work. 5. The profits of a Nursery. 6. The advantages of an Orchard 7. The management of Cyder .... To Which Is Added An Appendix, Containing the most approved methods for the management and improvement of Tillage .... (Extracted principally from a variety of Authors ....)
Newburyport: Printed, By Blunt And March, For John Dabney, Salem, 1796.
[1], 61 p. LCP

999. [__________].
An Address To Farmers ....
Newburyport: Printed, By Blunt And March, For John Dabney, Salem, 1796.
[1], 64 p. As above but with Postscript (pp. 63-64). HSP

1000. DADD, GEORGE H., b. 1813.
The Advocate Of Veterinary Reform And Outlines Of Anatomy And Physiology Of The Horse; Also, A General History Of The Rise And Progress Of The Veterinary Science In England, With Practical Observations On Feeding, Watering, Grooming, Shoeing, &c., &c.: Written With A View Of Reforming The Veterinary Science, And Teaching Every Man How To Prevent Disease In Animals. Containing, Also, A Veterinary Dictionary, Selected From The Works Of R. White, V.S. ....
Boston: Published By The Author [Stereotyped At The Boston Stereotype Foundry (t.p. verso)], 1850.
307 p.; 2 fold. pls.; illus. LCP

1001. __________.
The American Reformed Cattle Doctor; Containing The Necessary Information For Preserving The Health And Curing The Diseases Of Oxen, Cows, Sheep, And Swine, With A Variety Of Original Recipes, And Valuable Information In Reference To Farm And Dairy Management ....
Boston: J.G. Tillotson And Company [Stereotyped At The Boston Stereotype Foundry (t.p. verso)], 1851.
359, [1]+8 (pub. adv.) p. incl. front. LCP

1002. __________.
The Anatomy And Physiology Of The Horse: With Anatomical And Questional Illustrations. Containing, Also, A Series Of Examinations Of Equine Anatomy And Physiology .... To Which Is Added, Glossary of Veterinary Technicalities, Toxological Chart, and Dictionary of Veterinary Science.
Boston: Published By John P. Jewett And Company; Cleveland, Ohio: H.P.B. Jewett; New York: Sheldon, Blakeman And Company [Boston: Lithotyped By The American Stereotype Company (t.p. verso)], 1857.
[2], 291 p.; front; 19 pls., each with unnumbered leaf of explanatory text. LCP

1003. __________.
Dadd On The Nature And Treatment Of The Diseases Of Cattle, With Descriptions And Illustrations Of Various Organs And Functions Of The Animal Economy. Containing Also, Useful And Practical Information On Breeding, Ventilation, And Diet.
Boston: John P. Jewett And Company; New York: C.M. Saxton; Cleveland, Ohio: Henry P.B. Jewett [Boston: Lithotyped By Cowles And Company; Printed by Geo. C. Rand and Avery (t.p. verso)], 1859.
395 p.; front.; 3 pls.; illus. LCP

1004. __________.
The Modern Horse Doctor: Containing Practical Observations On The Causes, Nature, And Treatment Of Disease And Lameness In Horses. Embracing The Most Recent And Approved Methods, According To An Enlightened System Of Veterinary Therapeutics, For The Preservation And Restoration Of Health. With Illustrations. Eleventh Thousand.
Boston: Published By John P. Jewett And Company; Cleveland, Ohio: Jewett, Proctor & Worthington; New York: Sheldon, Lamport & Blakeman [Stereotyped At The Boston Stereotype Foundry (t.p. verso)], 1856.
432 p. incl. front.; illus. LCP

1005. DANA, SAMUEL LUTHER, 1795-1868.
An Essay On Manures, Submitted To The Trustees Of The Massachusetts Society For Promoting Agriculture, For Their Premium.
New York: Published By C.M. Saxton; Also, Stringer & Townsend, H. Long & Brother, W.F. Burgess, Dewitt & Davenport, Wilson & Co., Dexter & Brother; Boston: Redding & Co.; Philadelphia: W.B. Zieber, Lindsay & Blakiston, 1852.
60 p. LCP

1006. [__________].
Manures: A Prize Essay. Published By The Massachusetts Society for the Promotion of Agriculture.
Boston: Printed By Samuel N. Dickinson, 1843.
40 p. Author's name t.p. verso. LCP

1007. __________.
Manures: A Prize Essay.
Lowell [Mass.]: Published By Bixby & Whiting, 1844.
47 p. LCP

1008. __________.
A Muck Manual, For Farmers.
Lowell [Mass.]: Daniel Bixby [W. Schouler, Printer (t.p. verso)], 1842.
242, [2] p. LCP

1009. __________.
A Muck Manual For Farmers: A Treatise On The Physical And Chemical Properties Of Soils; The Chemistry Of Manures; Including Also The Subjects Of Composts, Artificial Manures And Irrigation. Fourth Edition, With a new Chapter on Bones and Superphosphates.
New York: A.O. Moore, Agricultural Book Publisher (Late C.M. Saxton & Co.), 1859.
312 p. LCP

1010. DARLINGTON, WILLIAM, 1782-1863.
Address Before The Chester County Horticultural Society, At Their First Annual Exhibition, In The Borough Of West Chester, Sept. 11, 1846.
West Chester, Penn'a: 1846.
15 p. LCP; HSP

1011. __________.
Address Delivered Before The Philadelphia Society For Promoting Agriculture, At the Annual Exhibition, held Oct. 17, 1844. [cap. title]
[Germantown, Pa.:] [Printed At The Office Of The Germantown Telegraph (colop.)], [1844].
8 p. HSP

1012. __________.
Address Delivered Before The Philadelphia Society For Promoting Agriculture, At the Annual Exhibition, held October 17, 1844.
[Philadelphia:] 1844.
16 p. LCP; APS

1013. __________.
Agricultural Botany: An Enumeration And Description Of Useful Plants And Weeds, Which Merit The Notice, Or Require The Attention, Of American Agriculturists.
Philadelphia: J.W. Moore; New York: Mark H. Newman & Co. [Lancaster, Pennsylvania: E.C. Darlington, Printer (t.p. verso)], 1847.
lviii, 270 p. LCP; APS

1014. __________.
American Weeds And Useful Plants: Being A Second and Illustrated Edition Of Agricultural Botany: An Enumeration and Description .... Revised, With Additions, By George Thurber ....
New York: A.O. Moore & Company, Agricultural Book Publishers [Stereotyped And Printed By Edward O. Jenkins (t.p. verso)], 1859.
xvi, 460, 12 (pub. adv.) p.; col. front.; illus. LCP

1015. __________.
A Discourse On The Character, Properties, and Importance to Man, Of The Natural Family Of Plants Called Gramineae, Or True Grasses. Delivered As A Lecture Before The Class Of The Chester County Cabinet Of Natural Sciences, Feb. 19, 1841.
Westchester, Penn.: 1841.
22 p. LCP; APS

1016. __________.
A Discourse Upon Agriculture: At A Meeting Of The Citizens Of Oxford, And Vicinity, Chester County, Penn., Assembled For The Purpose Of Forming An Agricultural Society, September 4, 1847.
[Oxford, Pa.:] 1847.
16 p. LCP; HSP

1017. DAUBENTON, LOUIS JEAN MARIE, 1716-1799.
Advice To Shepherds And Owners Of Flocks, On The Care And Management Of Sheep. Translated From The Original French Of M. Daubenton, By A Gentleman Of Boston. To which are added, Explanations Of The Plates, And A Table Of Contents.
Boston: Printed By Joshua Belcher, 1811.
[2], 154 p.; front.; 2 pls. LCP

1018. DAVIDGE, FRANCIS HATHORN.
Address Delivered Before The Horticultural Society Of Maryland, At Its Third Anniversary Celebration, June 4th, 1835.
Baltimore: Printed By Jos. Robinson, 1835.
32 p. LCP (wrapper); APS

1019. DAVIE, WILLIAM RICHARDSON, 1756-1820.
An Address Delivered Before The South Carolina Agricultural Society, At Their Anniversary Meeting, Held In Columbia, On The 8th Of December, 1818. Together With The Report Of The Curators For The Preceding Year, And A List Of The Officers And Members.
Columbia, S.C.: Printed At The Telescope Press, 1819.
29, [2] p. APS

1020. DAVY, HUMPHRY, 1778-1829.
Elements Of Agricultural Chemistry, In A Course Of Lectures For The Board Of Agriculture [of Great Britain].
New York: Published By Eastburn, Kirk & Co.; Boston: Ward & Lilly, 1815.
viii, 332, lxxiv, [6] p.; 8 pls., 2 fold. Incl. sep.-paginated Appendix: An Account Of The Results Of Experiments On The Produce And Nutritive Qualities Of Different Grasses And Other Plants, Used As The Food Of Animals. Instituted By John, Duke Of Bedford; by George Sinclair, Gardener to his Grace. HSP in LCP; LCP

1021. _________.
Elements Of Agricultural Chemistry, In A Course Of Lectures For The Board Of Agriculture. With An Appendix, Containing A Series Of Experiments To Test The Value Of The Grasses Cultivated In Great Britain. Second American Edition. To Which Is Added, Practical Minutes On Some Of The Manures Mentioned In The Lectures.
Hartford: Hudson And Co., Printers, 1819.
viii, 304 p. LCP

1022. DAVY, JOHN TANNER.
Davy's Devon Herd Book; Containing The Names Of The Breeders, The Ages, And Pedigrees Of The Devon Cattle, With The Prizes They Have Gained. Arranged By John Tanner Davy .... American Edition, Edited By Sanford Howard .... Volumes I. And II.
New York: Printed By James D. Torrey, 1855.
2 vols. paginated continuously: 196 p.; front.; 2 pls. Vol. II has ½ t.p. at p. 75. LCP

_________. Devon Herd Book ....
See also HOWARD, SANFORD. To The Breeders Of Devon Cattle .... Now collecting materials for the Third Volume of .... 1856.

1023. DEAN, AMOS, 1803-1868.
An Attempt To Present The Claims Of Long Lake To The Consideration Of All Those Who Are In Search After Good Land At A Low Price.
Albany: Printed By Joel Munsell, 1846.
35 p. The lands lay in Hamilton Co., N.Y. HSP

1024. DEANE, SAMUEL, 1733-1814.
The New-England Farmer; Or, Georgical Dictionary: Containing A Compendious Account Of The Ways And Methods In Which the most Important Art Of Husbandry, In All Its Various Branches, Is, Or May Be, Practised To The Greatest Advantage In This Country.
Worcester, Massachusetts: Printed By Isaiah Thomas; Sold at his bookstore, and by him and Company in Boston, 1790.
viii, 335 p. HSP in LCP; LCP

1025. _________.
The Newengland Farmer; Or Georgical Dictionary. Containing A Compendious Account .... The Second Edition, Corrected, Improved, And Enlarged, By The Author.
Worcester, Massachusetts: Printed At The Press Of Isaiah Thomas, By Leonard Worcester, for Isaiah Thomas, 1797.
viii, 396 [i.e. 397] p. P. 397 misnumbered 396. LCP; APS

1026. _________.
The New-England Farmer; Or Georgical Dictionary. Containing A Compendious Account .... Third Edition, Corrected, Improved, Greatly Enlarged And Adapted To The Present State Of The Science Of Agriculture.
Boston: Wells And Lilly, 1822.
xi, [1], 532 p. LCP

1027. DEARBORN, HENRY ALEXANDER SCAMMEL, 1783-1851.
An Address Delivered Before The Massachusetts Horticultural Society, On The Celebration Of Their First Anniversary, September 19, 1829.
Boston: From The Press Of Isaac R. Butts, 1829.
28 p. LCP; APS (wrappers)

1028. _________.
An Address Delivered Before The Massachusetts Society For Promoting Agriculture, At Brighton, Oct. 14, 1835.
Boston: N.E. Farmer Office - Geo. C. Barrett [Tuttle And Weeks, Printers (t.p. verso)], 1835.
27 p. LCP

1029. [DEBAR, JOSEPH HUBERT DISS].
For The Benefit Of The Catholic Hospital and Orphan Asylum Of Wheeling, Va. Distribution of Farm Lots in St. Clara Colony, Doddridge County, Virginia. A Chance Of A Home For One Dollar! [cap. title]
[N.p.: ca. 1855].
4 p. Incl. description of the lands. Author's name at end. LCP

1030. DE BOW, JAMES DUNWOODY BROWNSON, 1820-1867.
The Industrial Resources, Etc., Of The Southern And Western States: Embracing A View Of Their Commerce, Agriculture, Manufactures, Internal Improvements, Slave And Free Labor, Slavery Institutions, Products, Etc., Of The South, Together with Historical And Statistical Sketches Of The Different States And Cities Of The Union ... And Returns Of The Census Of 1850, On Population, Agriculture And General Industry, Etc., With An Appendix. In Three Volumes. Vol. I [- III].
New Orleans, New York, and Charleston: Published At The Office Of De Bow's Review [Pudney & Russell, Printers (t.p. verso vol. I)], 1852 (vols. I-II); 1853 (III).
3 vols. Vol. I: vii, [1], 464, [5 (pub. adv.)] p. II: iv, 560 p. III: iv, 584, x, [2 (pub. adv.)] p. LCP

1031. De Bow's Review And Industrial Resources, Statistics, etc. Devoted To Commerce, Agriculture, Manufactures, Internal Improvements ....
New Orleans.
LCP: Vols. 12-29; vol. 30, nos. 1-4 (Jan. 1852-April 1861); + scattered issues to 1870.
HSP: Vols. 1-4 (Jan 1846-Dec. 1847); 8-15 (Jan. 1850-Dec. 1853); 28-30, ns. 1-6 (Jan. 1860-Dec. 1868)
Also known as: Commercial Review Of The South And West (1846-1850) etc.

1032. DE CORDOVA & FRAZIER, General land agents, Austin, Texas.
De Cordova & Frazier's Texas Land Register And Emigrants' Guide. [cap. title]
[Austin, Texas: 1855].
23 p. With dateline following title: City of Austin, Texas, December 1, 1855. LCP

1033. DEGELOS, PIERRE A.
Statement Of Sugar Made In Lousiana In 1828 And 1829.
New Orleans: Printed By Benjamin Levy, Bookseller And Stationer, [1830].
Brs. LCP

1034. DEIGENDESCH, JOHANNES.
Nachrichters: Oder Nutzliches und aufrichtiges Ross-Artzney-Büchlein. In Welchem die meisten innerliche Kranckheiten und äusserliche Zustände der Ross aufs deutlichste beschrieben und erkläret werden: Samt Beyfügung der darzu gehörigen und nöthigsten Artzney-Mittel, und approbirte Recepte .... Auch wird gelehret einige Composita selbsten zu machen .... Welchem annoch beygefüget ein Anhang von Rind-Viehs Artzneyen ...
Germantown [Pa.]: Gedruckt bey Christoph Saur, 1770.
208, [8] p. HSP in LCP

1035. __________.
Nachrichters: Oder Nutzliches und aufrichtiges Ross-Artzney-Büchlein ....
Germantown [Pa.]: Gedruckt bey Christoph Saur, 1770.
208, [8], 12 p. Copy with final 12-p. sep.-paginated section, headed: Dritter Anhang; incl. further comments on the horse and notes on geese and chickens. LCP

1036. __________.
Nachrichters: oder Nutzliches und aufrichtiges Ross-Artzney-Büchlein ....
Germantown [Pa.]: Zum zweyten mal gedruckt bey Christoph Saur, 1771.
208, [8], 12 p. With Dritter Anhang. LCP

1037. __________.
Nachrichters: Oder Nutzliches und aufrichtiges Ross-Artzney-Büchlein ....
Germantown: Zum dritten mal gedruckt und zu finden bey Peter Leibert, 1791.
221, [7] p. LCP

DELAWARE. GEOLOGICAL SURVEY.
See BOOTH, JAMES CURTIS, 1810-1888. Memoir Of The Geological Survey .... Including The Application ... To Agriculture. 1841.

DELAWARE. HORTICULTURAL SOCIETY.
See EMERSON, GOUVERNEUR, 1796-1874. Address Delivered Before .... 1851.

1038. DELAWARE. STATE AGRICULTURAL SOCIETY.
Proceedings Of The Agricultural Convention Of The State Of Delaware, Held At Dover, On Wednesday, January 17th, 1849.
Wilmington, Del.: Evans & Vernon, Printers, 1849.
16 p. Establishing the Society. HSP

1039. DELAWARE CO. [PA.]. INSTITUTE OF SCIENCE.
Report Of The Delaware County Institute Of Science. [Cap. title; begins:] The committee appointed to conduct the second Annual Exhibition ....
[Chester, Pa.: 1847].
16 p. Items exhibited were chiefly agricultural. APS

1040. __________.
Report Of The .... [Cap. title; begins:] The Committee of Arrangement appointed to conduct the Third Annual Exhibition ....
[Chester: 1848].
16 p.; fold front. APS (lacking front.); HSP

1041. __________.
Report Of The Committee Appointed To Conduct The 4th Annual Exhibition Of The ... Made September 22, 1849. Ordered That Five Hundred Copies Be Printed For Distribution.
Chester: Printed By Y.S. Walter, 1849.
24 p. APS

1042. __________.
Report Of The Committee Appointed To Conduct The Fifth Annual Exhibition Of The ... Read September 21st, 1850. Also, The Address Of James Broomall, Esq. Delivered At The Close Of The Exhibition. Ordered That Five Hundred Copies Be Printed For Distribution.
Chester: Printed By Y.S. Walter, 1850.
27 p. HSP

1043. __________.
Report Of The Committee Appointed To Conduct The Sixth Annual Exhibition Of The ... Made October 4, 1851. Also, The Address Of Charles D. Manley, Esq., Delivered At The Close Of The Exhibition. Ordered That Six Hundred Copies Be Printed For Distribution.
Chester: Printed By Y.S. Walter, 1851.
32 p. APS; HSP

1044. __________.
Report Of The Committee Appointed To Conduct The Seventh Annual Exhibition Of The ... Made October 2, 1852. Ordered That Six Hundred Copies Be Printed For Distribution.
Chester: Printed By Y.S. Walter, 1852.
26 p. APS; HSP

1045. __________.
Report Of The Committee Appointed To Conduct The Eighth Annual Exhibition Of The .... Made October 1, 1853, Ordered That Six Hundred Copies Be Printed For Distribution.
Chester: Printed By Y.S. Walter, 1853.
19 p. APS

__________.
See also BOWEN, JOHN S. Address, Delivered At The .... 1848.

1046. Delaware Register And Farmer's Magazine. By William Huffington.
Dover.
LCP; HSP: Vols. 1-2 (Feb. 1838-Jan. 1839). All pub.

1047. DENNIS, JONATHAN, JR.
Dennis' Silk Manual: Containing Complete Directions For Cultivating The Different Kinds Of Mulberry Trees, Feeding Silk Worms, And Manufacturing Silk To Profit, Adapted To The Wants Of The American Cultivator .... With A Supplement Of Extracts, From Various Authors, In Relation To The Profit Of Raising Silk. In Three Parts ....
New York: Press Of Mahlon Day & Co.; James Egbert, Printer, 1839.
Pp. [1-4], [7]-107; illus. APS (wrappers; with 1st and last text leaves pasted to wrappers)

1048. DENNY, NATHANIEL P.
Address Delivered Before The Worcester [Mass.] Agricultural Society, September 25, 1822, Being Their Anniversary Cattle Show And Exhibition Of Manufactures. Printed by order of the Trustees.
Worcester: Printed By William Manning, [1822].
24, [2] p. LCP

1049. DENNY, SAMUEL.
The Farmer: A Practical Treatise On Agriculture.
Brookfield [Mass.]: E. & G. Merriam, Printers, 1824.
45 p. LCP

1050. [DERBY, ELIAS HASKET].
[M]erino Sheep .... to the quality and breed of MERINO SHEEP, imported in the Ship ... [Com]mander, belonging to Gen. ELIAS HASKET DERBY, of Salem, Massachu[setts] ... purchase with considerable difficulty, and obtained permission of egress for ... six hundred and fifty arrived in safety in New-York.
[New York?: 1810].
Brs. APS (with upper left corner damaged)

Descriptions Du Sol, Des Productions .... 1789.
See CUTLER, MANASSEH, 1742-1823.

A Description Of East-Florida .... 1769.
See STORK, WILLIAM.

A Description Of South Carolina .... 1761.
See GLEN, JAMES, 1701-1777.

1051. DES MOINES NAVIGATION COMPANY, Des Moines, Iowa.
To The Western Emigrant! Iowa Lands In The Valley Of The Des Moines River. One Million Acres Offered For Sale On Long Credit! By The Des Moines Navigation Company.
[Des Moines:] 1856.
[2] p.; illus. Signed at end: Geo. R. Perkins, Land Commissioner; and dated: Fort Des Moines, Iowa, June 16th, 1856. LCP

1052. [DeWITT, SIMEON].
Considerations On The Necessity Of Establishing An Agricultural College, And Having More Of The Children of Wealthy Citizens, Educated For The Profession Of Farming.
Albany: Printed By Websters And Skinners, 1819.
42 p. LCP; APS

D'HOMERGUE, JEAN.
See HOMERGUE, JEAN D'.

DICKEHUT, H.F., ed.
See Magazine Of Gardening And Botany [pdcl.].

DICKINSON, DANIEL STEVENS, 1800-1866.
See QUEENS CO. [N.Y.]. AGRICULTURAL SOCIETY. Annual Report ... Containing The Address .... 1844.

1053. DICKINSON, SAMUEL FOWLER, 1775-1838.
An Address Delivered At Northampton [Mass.], Before The Hampshire, Hampden And Franklin Agricultural Society, October 27, 1831.
Amherst: J.S. & C. Adams, Printers, 1831.
LCP: 38 p. only (of 40; lacking last leaf)

1054. DILLINGHAM, WILLIAM HENRY, 1790-1854.
An Address Delivered Before The Chester County Horticultural Society, At West Chester, Pa. September 10, 1847. With The Transactions Of The Society For The Years 1846, 1847.
Philadelphia: John C. Clark, Printer, 1847.
48 p. Incl. (pp. 37-48): Abstracts of Transactions and Committee Reports. LCP; APS; HSP (wrappers)

1055. DIX, JOHN ADAMS, 1798-1879.
Address Delivered Before The N.Y. State Agricultural Society, At Albany, October 7, 1859.
Albany: Printed By C. Van Benthuysen, 1859.
40 p. HSP (wrappers)

1056. DODGE, ALLEN WASHINGTON, 1804-1878.
Address Before The Barnstable [Mass.] County Agricultural Society. October 7th, 1846.
Barnstable: Press Of Barnstable Patriot, 1847.
18 p. LCP (wrappers)

1057. The Domestic Animal's Friend, Or The Complete Virginia And Maryland Farrier, Being A Copious Selection From The Best Treatises On Farriery Now Extant In The United States, In Five Parts .... To Which Is Added A Number Of Receipts ....
Winchester, Va.: Printed And Published By J. Foster, 1818.
Pp. [1-4], [3]-436; 2 fold. pls. LCP

1058. DONNANT, DENIS FRANÇOIS, b. 1769.
Statistical Account Of The United States Of America. Translated From The French, By William Playfair: With an Addition on the Trade to America, For The Use Of Commercial Men, By The Same. Illustrated By A Divided Circle, Representing The Proportional Extent of the different States ....
London: Printed For Messrs. Greenland And Norris; By J. Whiting [Glendinning, Printer (colop.)], 1805.
72 p.; col. fold. front. LCP

DORCHESTER [MD.]. AGRICULTURAL SOCIETY.
See MUSE, JOSEPH E. An Address Delivered .... 1837.

1059. DOUGLAS, STEPHEN ARNOLD, 1813-1861.
Address Of The Hon. Stephen A. Douglas, At The Annual Fair Of The New-York State Agricultural Society, Held At Rochester, September, 1851.
Washington: Printed by Jno. T. Towers, 1852.
24 p. LCP; HSP

1060. DOWNING, ANDREW JACKSON, 1815-1852.
The Architecture Of Country Houses; Including Designs For Cottages, Farm Houses, And Villas. With Remarks On Interiors, Furniture, And The Best Modes Of Warming And Ventilating. With Three Hundred and Twenty Illustrations.
New York: D. Appleton & Co.; Philadelphia: Geo. S. Appleton [R. Craighead, Printer and Stereotyper (t.p. verso)] 1850.
x, [2], 484 p.; 32 pls.; illus. Hitchcock 325 calls for 36 pls., but this may be an error; 4 unnumbered pls. follow Pl. XXXII, but the numeration of the preceding pls. is erratic. Incl. suggestions and designs for various farm buildings. LCP

1061. __________.
The Architecture Of Country Houses ....
New York: D. Appleton & Company, 1853.
x, [2], 484 p.; 32 pls.; illus. Numeration of pls. slightly altered (e.g. XXXII is now XXXIII) but still incorrect. LCP

1062. __________.
The Architecture Of Country Houses .... Fourteenth Thousand.
New York: D. Appleton & Co., 1856.
x, [2], 484 p.; 32 pls.; illus. Pl. numeration as in above edn. LCP

1063. __________.
Cottage Residences; Or A Series Of Designs For Rural Cottages And Cottage-Villas, And Their Gardens And Grounds. Adapted To North America. Illustrated By Numerous Engravings.
New York And London: Wiley And Putnam [New York: William Osborn, Printer (t.p. verso)], 1842.
[4 (pub. adv.)], [4], 187, [1], [4 (pub. adv.)] p.; 12 pls., 1 col.; illus. LCP

1064. __________.
Cottage Residences; Or A Series Of Designs .... Part I. Illustrated By Numerous Engravings. Second Edition.
New York and London: Wiley And Putnam [New York: William Osborn, Printer (t.p. verso)], 1844.
[2], 187, [1], [2 (pub. adv.)] p.; front.; 11 pls., 1 col.; illus. LCP

1065. __________.
The Fruits And Fruit Trees Of America: Or The Culture, Propagation, And Management, In Garden And Orchard, Of Fruit Trees Generally; With Descriptions Of All The Finest Varieties Of Fruit, Native And Foreign. Cultivated In This Country. Illustrated With Many Engravings.
New York & London: Wiley And Putnam [R. Craighead's Power Press; Stereotyped By T.B. Smith (t.p. verso)], 1845.
xiv, 594 p.; illus. LCP

1066. __________.
The Fruits And Fruit Trees Of America; Or, The Culture, Propagation, And Management ....
New York & London: Wiley And Putnam, 1847.
xiv, 594 p.; col. front.; 69 col. pls.; illus. LCP

1067. __________.
The Fruits And Fruit Trees Of America; Or, The Culture, Propagation, And Management .... Revised And Corrected By Charles Downing.
New York: Wiley & Halsted [R. Craighead, Printer And Stereotyper (t.p. verso)], 1857.
xix, [1], 760 p.; illus. LCP

1068. __________.
The Fruits And Fruit Trees Of America; Or, The Culture, Propagation, And Management .... Revised And Corrected By Charles Downing. Third Thousand, With Corrections.
New York: Wiley & Halsted [R. Craighead, Printer And Stereotyper (t.p. verso)], 1858.
xix, [1], 760 p.; illus. LCP

1069. __________.
Rural Essays. Edited, With A Memoir Of The Author, By George William Curtis; And A Letter To His Friends, By Frederika Bremer.
New York: George P. Putnam And Company [John F. Trow, Printer And Stereotyper (t.p. verso)], 1853.
lxxi, [1], 557 p.; extra eng. t.p.; front.; 8 pls.; illus. This "... comprises, with one or two exceptions, all his editorial papers in the 'Horticulturist'" (p. iii). LCP; HSP

1070. __________.
A Treatise On The Theory And Practice Of Landscape Gardening, Adapted To North America; With A View To The Improvement Of Country Residences. Comprising Historical Notices And General Principles Of The Art, Directions For Laying Out Grounds And Arranging Plantations .... With Remarks On Rural Architecture. Illustrated By Engravings.
New York & London: Wiley And Putnam; Boston: C.C. Little & Co. [New York: Hopkins & Jennings, Printers (t.p. verso)], 1841.
[4], 451, [1] p.; front.; illus. LCP; HSP

1071. __________.
A Treatise On The Theory And Practice Of Landscape Gardening, Adapted To North America .... Second Edition, Enlarged, Revised, And Newly Illustrated.
New York & London: Wiley And Putnam [New York: William Osborn, Printer (t.p. verso)], 1844.
Pp. [i]-xiv, [9]-497; front.; 15 pls.; illus. LCP

1072. __________.
A Treatise On The Theory And Practice Of Lansdscape Gardening, Adapted To North America .... With Remarks On Rural Architecture. Fourth Edition, Enlarged, Revised, And Newly Illustrated.
New York: George P. Putnam; London: Longman, Green & Longmans [New York: R. Craighead, Printer (t.p. verso)], 1849.
532, 4 (pub. adv.) p; front.; 18 pls.; illus. LCP

1073. __________.
Treatise On The Theory And Practice Of Landscape Gardening .... Sixth Edition, Enlarged, Revised, And Newly Illustrated. With A Supplement, Containing Some Remarks About Country Places .... By Henry Winthrop Sargent.
New York: A.O. Moore & Co. Agricultural Book Publishers [Wynkoop, Hollenbeck & Thomas, Printers (t.p. verso)], 1859.
576 p.; front.; 51 pls.; illus. LCP

__________, ed.
See The Horticulturist .... [pdcl.].

1074. DOWNING, JOSEPH.
A Treatise On The Disorders Incident To Horned Cattle, Comprising A Description Of Their Symptoms, And The Most Rational Methods of Cure, Founded On Long Experience. To Which Are Added, Receipts For Curing The Gripes, Staggers And Worms, In Horses; And An Appendix, Containing Instructions For The Extracting Of Calves.
Baltimore: Published By E.J. Coale; Philadelphia: And Anthony Finlay; Merritt, Printer, 1813.
95, [1] p. LCP

DOWNINGTOWN [PA.]. SOCIETY FOR THE ACQUISITION AND PROMOTION OF NATURAL KNOWLEDGE.
See KERSEY, JESSE, 1768-1846. Lectures On Agriculture, Delivered Before .... 1828.

DOYLE, MARTIN, pseud.
See HICKEY, WILLIAM, 1787?-1875.

1075. DRAKE, DANIEL, 1785-1852.
A Discourse On Intemperance; Delivered At Cincinnati, March 1, 1828, Before The Agricultural Society Of Hamilton County, And Subsequently Pronounced, By Request, To A Popular Audience.
Cincinnati: Looker & Reynolds, Printers, 1828.
96 p. HSP in LCP; LCP

1076. DREER, HENRY A., Nurseryman, Philadelphia.
H.A. Dreer's Abridged Catalogue Of Green and Hot-House Plants, Dahlias, Roses, And Other Plants Suitable For Bedding Out, Fruit Trees, Ornamental Plants And Trees. For Sale At His Nursery and Green Houses, 35th and Elm St. .... Orders Received At His Seed And Horticultural Warehouse, No. 117 Chestnut St., Near 4th, Philada. Where also may be found, constantly on hand, an extensive variety of Garden, Field And Flower Seeds, Of The Best Quality. Also, Garden Implements, Books ....
[Philadelphia:] A.C. Bryson & Brother, [1857].
24 p. Dated from addresses given in Philadelphia directories. LCP

1077. __________.
H.A. Dreer's Descriptive Catalogue Of Garden Seeds: Also a List of Flower Seeds, Implements, Books, &c. Warehouse: 117 Chestnut St. ....
[Philadelphia:] King & Baird Prs., [1857].
23, [1] p. Dated from Philadelphia directories. LCP

1078. DROWNE, WILLIAM, 1793-1874.
Compendium Of Agriculture, Or The Farmer's Guide, In The Most Essential Parts Of Husbandry And Gardening; Compiled From The Best American And European Publications, And The Unwritten Opinions Of Experienced Cultivators. By William Drown [sic], With The Aid And Inspection Of Solomon Drown, M.D.
Providence: Printed By Field & Maxcy, 1824.
288 p. LCP

DU BOSQUE, -----. The History Of America .... 1770.
See [POSTLETHWAYT, MALACHY, 1707-1767.] The History Of America .... 1770.

DUCOUEDIC, P.
See DU COUEDIC DE VILLENEUVE, PIERRE LOUIS, 1743-1822.

1079. DU COUEDIC DE VILLENEUVE, PIERRE LOUIS DE, 1743-1822.
The Pyramidal Bee-Hive. A Plain and Natural Method Of Preserving And Perpetuating The Population Of Bees, And Of Receiving Annually, From Each Family, A Box Full Of Wax And Pure Honey .... Also, The Art Of Converting Honey Into White, Inodorous Sugar, And Of Making Hydromel, Sirups, &c. .... Abridged, And Translated From The French By Silas Dinsmoor.
Philadelphia: Carey, Lea & Carey, 1829.
103 p.; front. APS

1080. DUDGEON, THOMAS.
A Nine Years Residence, And A Nine Months Tour On Foot, In The States Of New York And Pennsylvania, For The Use Of Labourers, Farmers, And Emigrants.
Edinburgh: Printed For The Author, And Sold By The Booksellers, 1841.
52 p. HSP

1081. DUFOUR, JOHN JAMES.
The American Vine-Dresser's Guide, Being A Treatise On The Cultivation Of The Vine, And The Process Of Wine Making, Adapted To The Soil And Climate Of The United States.
Cincinnati: Printed By S.J. Browne, At The Emporium Office, 1826.
Pp. [1-4], [3]-314, [1-3]. LCP

1082. DUMAS, JEAN BAPTISTE ANDRÉ, 1800-1884.
Manures, Amelioraters And Stimulants: Modes Of Manufacture And Application. Translated By Campbell Morfit .... The Nutrition Of Plants. Translated By Campbell Morfit And Barnet Phillips. The Action Of Salt Upon Vegetation; And Its Use In Agriculture. By M. Becquerel And J.B. Lawes.
New York: Greely & McElrath, 1848.
40 p. LCP

1083. DU PONCEAU, PETER STEPHEN, 1760-1844.
The History Of The Silk Bill, In A Letter From Peter S. Du Ponceau, To David B. Warden, Esq. ....
Philadelphia: Printed By A. Waldie, 1837.
12 p. LCP; APS

1084. __________.
Letter--Peter S. Du Ponceau--Silk. Letter from Peter S. Du Ponceau, upon the subject of the bill (H.R. No. 294) to promote the growth and manufacture of Silk. May 10, 1832. Read, and committed to the Committee .... [cap. title]
[Washington: 1832].
5 p. Title headed: 22d Congress, 1st Session. (Doc. No. 232.) Ho. of Reps. LCP

__________, jt. auth.
See HOMERGUE, JEAN D', and __________.

1085. EASTMAN, MOSES.
A Discourse Delivered Before The Hillsborough Agricultural Society, At Amherst, N.H. September 25, 1822.
Amherst: Printed By Richard Boylston, For The Society, 1822.
24 p. LCP

1086. EASTWOOD, B.
A Complete Manual For the Cultivation Of The Cranberry, With A Description Of The Best Varieties, By B. Eastwood, "Septimus," Of The New York Tribune.
New York: C.M. Saxton & Co., Agricultural Book Publishers [Edward O. Jenkins, Printer (t.p. verso)], 1856.
120 p.; extra eng. t.p.; 9 pls. LCP

1087. __________.
A Complete Manual For the Cultivation Of The Cranberry ....
New York: C.M. Saxton, Barker & Co. [Edward O. Jenkins, Printer (t.p. verso)], 1860.
120 p.; extra eng. t.p.; 9 pls. LCP

1088. [EATON, AMOS, 1776-1842].
A Geological And Agricultural Survey Of Rensselaer County, In The State Of New-York. To Which Is Annexed, A Geological Profile, Extending From Onondaga Salt Springs, Across Said County, To Williams College In Massachusetts. Taken Under The Direction Of The Honorable Stephen Van Rensselaer.
Albany: Printed By E. And E. Hosford, 1822.
70 p.; 1 section; 1 fold. tbl. Author's name p. vii. LCP (wrapper); APS; HSP

1089. [__________].
A Geological And Agricultural Survey Of The District Adjoining The Erie Canal, In The State Of New-York; Taken Under The Direction Of The Hon. Stephen Van Rensselaer. Part I. Containing A Description Of The Rock Formations ....
Albany: Printed By Packard & Van Benthuysen, 1824.
163 p.; 2 fold. sections, 1 col. Author's name p. 10. LCP (with adv. slip tipped in at front); HSP (lacking 1st 26 p.)

1090. [__________, and THEODRIC ROMEYN BECK, 1791-1855].
A Geological Survey Of The County Of Albany, Taken under the direction Of The Agricultural Society Of The County.
[Albany: 1820].
Pp. [1-2], [5]-55, [1]; 2 fold. charts. "The present ... is the first attempt yet made in this country, to collect and arrange geological facts, with a direct view to the improvement of agriculture," p. 5. Authors' names p. 5. LCP (lacking ½ t.p.); HSP

1091. The Eccaleobion, A Treatise On Artificial Incubation, And The General Management Of Poultry.
Philadelphia: Printed For The Author, And Sold At The Exhibition Room, In The Masonic Hall, Chestnut Street, 1844.
12 p. HSP: copy with the above imprint on a cancel slip pasted over the original imprint, which is similar except that the address of the Room is: Broadway, Corner Of Chambers Street [i.e. New York].

1092. [EDDOWES, RALPH, 1751-1833].
Account Of The Wheat Moth Or Virginia Fly, As It Appeared In France, In the Year 1755; And Which Damaged The Grain In The Township Of Lower Dublin, In The Harvest Of 1802, 1803, and 1804.
Philadelphia: Printed By Archibald Bartram, For Thomas Dobson, 1805.
8 p. LCP; APS

1093. EDWARD, DAVID BARNETT, 1790-1870.
The History Of Texas; Or, The Emigrant's, Farmer's, And Politician's Guide To The Character, Climate, Soil And Productions Of That Country: Geographically Arranged From Personal Observation And Experience.
Cincinnati: Stereotyped And Published By J. A. James & Co. [Printed by James & Gazlay (t.p. verso)], 1836.
336 p.; col. fold. front. map. LCP

1094. [Ein und vierzig] 41 Bewährte Recepte; für Handwerker und Hausväter. Erste Auflage.
Millgrove [Pa.]: gedruckt von Samuel und Salomon Siegfrieds, 1834.
32 p. LCP

1095. ELDRIDGE, JOHN S.
Address On National Industry, Delivered Before The Norfolk [Mass.] Agricultural Society, At Dedham, September 29th, 1858.
Boston: J.M. Hewes, Printer, 1858.
20 p. HSP (wrappers)

1096. ELEY, JAMES NORMAN.
The American Florist; Or A Guide To The Management & Cultivation Of Plants In Conservatories, Greenhouses, Rooms, And Gardens; To Which are added, Directions For The Culture Of Annual, Biennial, And Perennial Flowers, Trees, Shrubs, Bulbs, Etc., Etc.
Hartford: Printed By Elihu Geer, 1845.
Pp. [i]-viii, [13]-183. LCP

1097. ELIOT, JARED, 1685-1763.
An Essay Upon Field-Husbandry In New-England As it is or may be Ordered.
N. London: Printed and Sold by T. Green, 1748.
[4], iv, 25, [3] p. LCP

1098. __________.
A Continuation Of The Essay Upon Field-Husbandry .... [The Second Part].
N. London: Printed and Sold by T. Green, 1749.
[4], 34, [2] p. LCP

1099. __________.
A Continuation Of The Essay Upon Field-Husbandry .... [The Third Part].
New London: Printed, and Sold by T. Green, 1751.
[4], 32 p. LCP

1100. __________.
A Continuation Of The Essay Upon Field-Husbandry .... The Fourth Part.
New York: Printed and Sold by J. Parker and W. Weyman, 1753.
[2], 33 p. HSP in LCP; LCP (lacking ½ t.p.); APS (lacking ½ t.p. and last leaf)

1101. __________.
A Continuation Of The Essay Upon Field-Husbandry .... The Fifth Part.
New York: Printed and Sold by J. Parker and W. Weyman, 1754.
44 p. LCP

1102. __________.
The Sixth Essay On Field-Husbandry ....
New Haven: Printed by J. Parker, And Company, 1759.
34 p. LCP

1103. __________.
Essays Upon Field-Husbandry In New-England, As it is or may be Ordered.
Boston: Printed and Sold by Edes and Gill, 1760.
[6], 166 p. Six essays and appendix. LCP

1104. ELIOT, THOMAS DAWES, 1808-1870.
Anniversary Address Delivered Before The American Institute Of The City Of New-York, At The Broadway Tabernacle, October 17, 1845. During The Eighteenth Annual Fair.
New York: James Van Norden & Co., Printers, 1845.
19 p. LCP

1105. ELLIOTT, FRANKLIN REUBEN, 1817-1878.
Eliott's Fruit Book; Or, the American Fruit-Grower's Guide In Orchard And Garden. Being A Compend Of The History, Modes Of Propagation, Culture, &c, Of Fruit Trees And Shrubs, With Descriptions Of Nearly All The Varieties Of Fruits Cultivated In This Country: Notes Of Their Adaptation To Localities And Soils, And Also A Complete List Of Fruits Worthy Of Cultivation.
New York: C.M. Saxton, Agricultural Book Publisher [E.O. Jenkins, Printer and Stereotyper; Cleveland: Brainard & Burridge, Engravers (t.p. verso)], 1854.
503, [3], 10 (pub. adv.) p.; illus. LCP

1106. ELLIS, JOHN, 1710-1776.
Directions For Bringing Over Seeds And Plants, From The East-Indies And Other Distant Countries, In A State Of Vegetation: Together With A Catalogue of such Foreign Plants as are worthy of being encouraged in our American colonies .... To Which Is Added, The Figure and Botanical Description of .... Venus's Fly-Trap.
London: Printed; and sold by L. Davis, Printer to the Royal Society, 1770.
[2], 41 p.; front; 1 pl. Sep. ½ t.p. for section on Venus's Fly-Trap (p. 35); title continues: In a Letter To Sir Charles Linnaeus .... LCP (with pl. col.); PHS in APS (with pl. uncol.)

1107. __________.
Some Additional Observations On The Method Of Preserving Seeds from Foreign Parts, For the Benefit of Our American Colonies. With An Account Of The Garden at St. Vincent, Under The Care Of Dr. George Young.
London: Printed for W. Bowyer and J. Nichols, 1773.
15 p.; 1 pl. APS; PHS in APS

1108. ELLSWORTH, HENRY LEAVITT, 1791-1858.
Sketches Of An Address Delivered Before The Hartford County [Conn.] Agricultural Society, On The Twenty-Fourth Of March, 1818. Published By Request.
Hartford: George Goodwin & Sons, Printers, 1818.
23 p. LCP

__________.
See also Illinois In 1837; A Sketch Descriptive Of The Situation .... Together With A Letter On The Cultivation Of The Prairies, By .... 1837.

1109. ELLSWORTH, HENRY WILLIAM, 1814-1864.
The American Swine Breeder, A Practical Treatise On The Selection, Rearing And Fattening Of Swine.
Boston: Weeks, Jordan And Company, 1840.
304 p.; 2 fronts.; illus. Half t.p.: Farmers' Series. Vol. I .... LCP

1110. __________.
Valley Of The Upper Wabash, Indiana, With Hints On Its Agricultural Advantages: Plan Of A Dwelling, Estimates Of Cultivation, And Notices Of Labor-Saving Machines.
New York: Published By Pratt, Robinson, And Co. [Washington, D.C.: Printed By Blair And Rives (t.p. verso)], 1838.
xii, 175, p.; col. fold. front. map; 3 fold. pls. LCP (2 copies: 1. with wrappers; 2. in cloth)

1111. ELLWANGER & BARRY, Nurserymen, Rochester, N.Y.
Descriptive Catalogue Of Fruits, Ornamental Trees, Flowering Shrubs And Plants, Cultivated And For Sale By Ellwanger & Barry, At The Mount Hope Botanic Garden and Nurseries, Saint Paul-Street .... Rochester, N.Y.
Rochester: Printed By Canfield & Warren, 1845.
42, [1] p. T.p. headed: 1845 & 1846. LCP

1112. __________.
Ellwanger & Barry, Catalogue Of Dahlias, Verbenas, Fuchsias, Geraniums, Chrysanthemums, And Other Select Green-House and Bedding-Out Plants, Cultivated And Sold At The Mount Hope Nurseries, Rochester, N.Y.
Rochester: Press Of Curtis & Butts, Daily Union Office, 1854.
16 p. T.p. headed: No. 3.-1854. LCP

1113. __________.
Wholesale Catalogue For Autumn Of 1854 And Spring Of 1855. Mount Hope Nurseries .... [cap. title]
[Rochester: 1854].
Pp. [29]-35. Title headed: No. 4. LCP

1114. EMERSON, GOUVERNEUR, 1796-1874.
Address Delivered Before The Agricultural Society Of Chester County, Pennsylvania, September 17, 1853. Published By Order Of The Society.
Philadelphia: T.K. And P.G. Collins, Printers, 1853.
24 p. LCP (wrappers); HSP (wrappers)

1115. __________.
Address Delivered Before The Agricultural Society Of Kent County, Delaware, October 15, 1857. Published By Order Of The Society.
Philadelphia: National Merchant Print, 1857.
20 p. APS

1116. __________.
An Address Delivered Before The Agricultural Society of Newcastle
County, Delaware, At Its Annual Exhibition Held in Wilmington, September 11,
1855. Published By Order Of The Society.
Philadelphia: T.K. And P.G. Collins, Printers, 1855.
12 p. LCP (wrappers)

1117. __________.
Address Delivered Before The Delaware Horticultural Society, At Wilmington, On the 24th of September, 1851. Published By Order Of The Society.
Philadelphia: T.K. Ard P.G. Collins, Printers, 1851.
22 p. LCP (wrappers); HSP (wrappers)

1118. __________.
Address Delivered Before The Society For Promoting Agriculture Of The County Of Philadelphia, At Their Annual Exhibition, At The Rising Sun Tavern, October 6, 1848. Together With The Reports of Committees.
Philadelphia: John C. Clark, Printer, 1849.
20 p. LCP; HSP

1119. __________.
Results Of Extensive Experience In The Use Of Superphosphate of Lime, &c. Communicated To The Agricultural Society of Kent County, Delaware, February 2, 1859. Published By Order Of The Society.
Philadelphia: Butler & Myers, Printers, 1859.
11 p. LCP

1120. __________.
Results Of Extensive Experience In The Use Of Superphosphate of Lime, &c. ....
Philadelphia: H.L. Butler, Printer, 1859.
12 p.; with Additional Note p. 12. APS

EMERY BROTHERS, Proprietors of agricultural works.
EMERY, HORACE L.
See ALBANY AGRICULTURAL WORKS, Albany, N.Y.

1121. EMMONS, EBENEZER, 1799-1863.
Agriculture Of New-York: Comprising An Account Of The Classification, Composition And Distribution Of The Soils And Rocks, And The Natural Waters Of The Different Geological Formations .... Volume I [- V].
Albany: Printed By C. Van Benthuysen & Co., 1846 (Vol. I); 1849 (II); 1851 (III-[IV]); 1854 (V).
5 vols. Vol I: pp. [1-6], [v]-xi, [1], 371; front.; eng. general t.p.; 18 pls., 4 col. fold., 9 tinted; 2 maps, 1 col. fold.; 3 col. fold. sections; illus. II: viii, 343, [1], 50 p.; eng. general t.p.; 28 pls., 16 col.; 14 col. charts. III: viii, 340 p.; eng. general t.p.; illus. IV: vii p.; eng. general t.p.; 100 col. pls. (i.e. for Vol. III; numbered 1-81). V: viii, 272 p. + [47] explanatory leaves interleaved with pls.; eng. general t.p.; 50 pls., 47 col. This is Division V of the Natural History of New-York. LCP; APS (5 vols. in 2)

1122. __________.
Report Of The North-Carolina Geological Survey. Agriculture Of The Eastern Counties; Together With Descriptions Of The Fossils Of The Marl Beds. Illustrated by Engravings.
Raleigh: Henry D. Turner [Holden And Wilson, Printers to the State (t.p. verso)], 1858.
xvi, 314 p.; errata slip tipped in at back; illus. LCP; HSP

__________, ed.
See American Journal of Agriculture and Science [pdcl.].

1123. ENGELMAN, DANIEL, Nurseryman, Philadelphia.
A List of a great number of Bulbous Flowers, Shrubs, Plants, and Flower Seeds, sold by Daniel Engelman, Florist, &c. &c. from Haarlem in Holland, (since 1795 in Philadelphia,) now living at the place of Mr. George Honey, between Arch and Race-streets and between Second and Third-street from Schuylkill, Philadelphia.
[Philadelphia:] Robert Cochran, Printer, [1811].
Brs. LCP; HSP (with "now living ... Honey" crossed out and with MS. addition at bottom of: Now Stands at the S.E. Corner of the old Court house Market Street)

1124. ENGLISH, WILLIAM HAYDEN, 1822-1896.
Address Of Hon. William H. English, Delivered At Livonia, Indiana, Before The Washington And Orange District Agricultural Society, October 15, 1857. Published By Request Of The Society.
Washington: Printed At The Congressional Globe Office, 1858.
8 p. LCP

1125. ENGLISH, GOODRICH & DIXON, Mfrs. of agricultural machinery, New Philadelphia, Ohio.
Combined Iron Cylinder Grain & Seed Drill! Patented by Coggeshall & Warner, and Manufactured at the New Philadelphia Agricultural Machine Works, by English, Goodrich & Dixon.
[New Philadelphia, Ohio: 1858].
Brs. Illus. LCP

Essai Sur La Manière D'Améliorer L'Education Des Chevaux .... 1795.
See MOREAU DE SAINT MÉRY, MÉDÉRIC LOUIS ELIE, 1750-1819.

1126. An Essay On The Climate of the United States: Or, An Inquiry Into The Causes Of The Difference In Climate Between The Eastern Side Of The Continent Of North America And Europe. With Practical Remarks on the Influence of Climate on Agriculture, and particularly the cultivation of the Vine.
Philadelphia: Published By Hopkins And Earle; Fry and Kammerer, Printers, 1809.
42 p. HSP in LCP; LCP

1127. An Essay On The Culture Of Silk, And Raising White Mulberry Trees, The Leaves Of Which Are The Only Proper Food of The Silk-Worm. By A Citizen Of Philadelphia.
Philadelphia: Printed And Sold By Joseph Crukshank, 1790.
8 p. Possibly by Peletiah Webster (1725-1795); see Sabin 102410. HSP in LCP; LCP

An Essay On The Manner Of Improving The Breed Of Horses In America. 1795.
See MOREAU DE SAINT MÉRY, MÉDÉRIC LOUIS ELIE, 1750-1819.

1128. Essays On The Spirit of Legislation, In The Encouragement Of Agriculture, Population, Manufactures, And Commerce. Containing Observations On The Political Systems At Present Pursued In Various Countries Of Europe, For The Advancement Of Those Essential Interests. Interspersed With Various Remarks On The Practice Of Agriculture. Societies Of Agriculture. Rewards. Bounties. The Police. Luxury .... Translated From The Original French, Which gained the Premiums offered by the OEconomical Society of Berne, in Switzerland, for the best Composition on this subject.
Newark [N.J.]: Printed By Pennington & Dodge, For William Reid, New York, 1798.
479, [1], vii, [5] p. LCP

1129. Essays On The Spirit of Legislation, In The Encouragement Of Agriculture, Population, Manufactures, And Commerce. Translated From The Original French ....
Newark [N.J.]: Printed For William Reid, By Pennington & Gould, 1800.
479, [1], vii p. HSP in LCP; LCP (lacking 2nd leaf, the Translator's Preface)

1130. ESSEX [MASS.]. AGRICULTURAL SOCIETY.
Account Of Premiums Awarded In 1825, And A List Of Premiums Offered In 1826; Accompanied With Remarks & Explanations For The Information Of The Farmers Of The County. Published By Order Of The Trustees. February, 1826.
Salem: W. & S.B. Ives, Printers Observer Office, 1826.
33 p. LCP

1131. __________.
Agricultural Essays, By The Essex Agricultural Society.
Danvers: Printed At The Courier Office, 1847.
43 p. LCP (wrappers)

1132. __________.
The Constitution And Catalogue Of The Officers And Members Of The Essex Agricultural Society. 1849.
Salem: Printed At The Gazette Office, 1849.
iv, 43 p. LCP

1133. __________.
Essay On The Cultivation Of Mulberry Trees, And The Culture Of Silk. Extracted From The Transactions Of The Essex Agricultural Society.
Salem: Press Of Foote And Brown, 1832.
Pp. [1-2], [75]-106; col. fold. pl.; fold. tbl. Errata slip pasted t.p. verso. LCP (wrappers); HSP

1134. __________.
Order Of Arrangements, List Of Premiums And Committees, For The Cattle Show And Exhibition Of The ... At Lawrence, On Wednesday And Thursday, 29th And 30th Of September, 1852.
[Salem:] Salem Gazette Press, [1852].
24 p. LCP

1135. __________.
Order Of Arrangements, List Of Premiums And Committees, For The Cattle Show And Exhibition Of The ... At Haverhill, On Wednesday And Thursday, 26th And 27th Of September, 1855.
Haverhill: Printed at the Essex Banner Office, [1855].
25 p. LCP

1136. __________.
Order Of Arrangements, Premiums And Committees, For The Annual Exhibition Of The ... At South Danvers, Wednesday And Thursday, September 25 And 26, 1860.
Salem: Printed At The Office Of The Gazette And Mercury, 1860.
22 p. LCP (with t.p. corrected in MS. to read ... Tuesday And Wednesday ....; wrappers)

1137. __________.
Transactions.
Salem.
LCP: Vol. 2, nos. 5, 7 (1835, 1837)

1138. __________.
The Trustees' Account Of The Cattle Show, And Other Exhibitions, At Topsfield, Oct. 5, 1820.
Salem: Printed By John D. Cushing, 1821.
56 p. LCP

_________.
See also
ABBOT, ABIEL, 1770-1828. An Address Delivered Before .... 1822.
CUSHING, CALEB, 1800-1879. Address. Delivered .... 1850.
GAGE, NATHANIEL, 1800-1861. An Address Before .... 1838.
HOWES, FREDERICK. An Address To .... 1824.
KELLEY, E.G. An Address ... Delivered Before .... 1857.
KING, DANIEL PUTNAM, 1801-1850. An Address To The .... 1836.
MOSELEY, EBENEZER, 1781-1854. An Address To .... 1835.
NICHOLS, ANDREW, 1785-1853. An Address, To .... 1821.
PICKERING, TIMOTHY, 1745-1829. Address To The .... 1818.
_________. A Discourse, Read Before .... 1820.
SALTONSTALL, LEVERETT, 1783-1845. An Address Before The .... 1843.
STONE, EDWIN MARTIN, 1805-1883. An Address, Delivered .... 1846.

1139. ESSEX CO. [N.J.]. AGRICULTURAL SOCIETY.
Constitution And By-Laws Of The Agricultural Society, Of The County Of Essex, State Of New Jersey.
Newark: Printed By Kinney And Cameron, (Publishers of the Eagle), 1823.
10 p. HSP

1140. EVANS, EVAN.
Patent Cyder, or Tobacco Press [title to illus.; text headed:] Explanation.
[Philadelphia?: 1807].
Brs. Illus. Incl. 1-paragraph propectus by Oliver Evans of his foundry and machine shop. LCP (with MS. note by Oliver Evans)

1141. EVANS, GURDON.
The Dairyman's Manual: Being A Complete Guide For The American Dairyman. With Numerous Illustrations.
Utica [N.Y.]: John W. Fuller & Company; Hartford: O.D. Case & Co. [R. Northway & Co. Printers (t.p. verso)], 1851.
Pp. [1]-4, [i]-vi, [1-2], [5]-235; front.; 6 pls.; illus. LCP

1142. EVANS, OLIVER, 1755-1811.
License. Know all Men, that I, Oliver Evans, of the City of Philadelphia, have received of [Thomas Hixon of Maysville Kentucky state] the sum of [Seven Hundred & twenty dollars] in full, for License hereby granted to the said [Hixon his] Heirs, Executors ... to construct, make, and use, my Patented Machine and Patented Improvements in the art of manufacturing flour or meal .... In witness whereof I have hereunto set my hand and seal this [Eighteenth] day of [July 181[8].
[Philadelphia: 1818].
Brs. With pl. 8 from Evan's Millwright's Guide on verso. LCP (with blanks in printed text completed in MS., as indicated in brackets)

1143. _________.
Oliver Evans To His Counsel, Who Are Engaged In Defence Of His Patent Rights, For The Improvements He Has Invented: Containing A Short Account Of Two Out Of Eighty Of His Inventions ... And Two Of His Patents, With Explanations.
[Philadelphia: 1817].
55, [1] p. LCP; APS; HSP

1144. _________.
To The Public.
[Philadelphia: 1795].
Brs. Proposal for publishing by subscription The Young Mill-wright and Miller's Guide. Incl. Design Of The Work; Principal Articles Of The Contents; and testimonial statement by Davd. Rittenhouse, R. Patterson, J. B. Bordley, et al. HSP (text mutilated at beginning)

1145. _________.
The Young Mill-Wright & Miller's Guide. In Five Parts--Embellished With Twenty Five Plates. Containing, Part I.--Mechanics and Hydraulics .... Part II.--Rules for applying the theories to practice .... Part III.--Directions for constructing and using all the authors patented improvements in mills. Part IV.--The art of manufacturing meal and flour in all its parts .... Part V.--The Practical Mill-Wright .... Appendix. Containing rules for discovering new improvements--exemplified in improving the art of thrashing and cleaning grain, hulling rice ....
Philadelphia: Printed For, And Sold By The Author, 1795.
[8], 160, 178, 90, 10, [12] p.; 26 pls., 2 fold. Sep. pagination for Pts. I, II-IV, V. HSP in LCP (with last section, subscribers' names, misbound); LCP; APS

1146. _________.
The Young Mill-Wright & Miller's Guide ....
Philadelphia: Printed For, And Sold By The Author, 1795.
[8], 160, [4], 178, 90, 10, [12] p. Copy with 4-p. unpaginated Tbl. of Contents section preceding Pt. II. LCP

1147. _________.
The Young Mill-Wright's & Miller's Guide. In Five Parts ....
Octoraro [Pa.]: Printed By Francis Bailey, And Sold By The Principal Booksellers In The United States, 1807.
LCP: viii, 364 p.; 25 pls. only, 2 fold. (of 26; lacking pl. [2]XIII; pl. [2]XI supplied in photostat)

1148. _________.
The Young Mill-Wright's & Miller's Guide. In Five Parts .... Third Edition.
Philadelphia: Printed By M. Carey And Son, 1818.
364 p.; 26 pls., 2 fold. LCP

1149. _________.
The Young Mill-Wright And Miller's Guide. In Five Parts .... Fourth Edition.
Philadelphia: M. Carey & Son, 1821.
374 p.; 26 pls., 2 fold. LCP

1150. ________.
The Young Mill-Wright And Miller's Guide .... The Eighth Edition, With Additions And Corrections; By Thomas P. Jones ... And A Description Of An Improved Flour-Mill, With Engravings By C. & O. Evans, Engineers.
Philadelphia: Carey, Lea & Blanchard [Griggs & Co., Printers (t.p. verso)], 1834.
392 p.; errata slip at back; 28 pls., 2 fold. LCP; APS (with 4 p. pub. adv. at back); HSP

1151. ________.
The Young Mill-Wright And Miller's Guide .... The Ninth Edition, With Additions And Corrections; By Thomas P. Jones ....
Philadelphia: Carey, Lea & Blanchard, 1836.
392 p.; 28 pls., 2 fold. LCP

1152. ________.
The Young Mill-Wright And Miller's Guide; Illustrated By Twenty-Eight Descriptive Plates .... Fourteenth Edition.
Philadelphia: Blanchard And Lea [Wm. S. Young, Printer (t.p. verso)], 1853.
Pp. [i]-viii, [17]-400, 1-6 (pub. adv.); 28 pls., 2 fold. HSP

________.
See also
EVANS, EVAN. Patent Cyder, or Tobacco Press. [1807].
UNITED STATES. CITIZENS. Memorial ... Praying Relief From The Oppressive Operation Of .... 1813.

1153. EVERETT, DAVID, 1770-1813.
Common Sense In Dishabille: Or, The Farmer's Monitor, Containing A Variety Of Familiar Essays, On Subjects Moral & Economical. To Which Is Added, A Perpetual Calendar, Or Economical Almanack.
Worcester, Massachusetts: Printed By Isaiah Thomas, Jun. For Isaiah Thomas; Sold at the Worcester Bookstore, March, 1799.
120 p. LCP

1154. EVERETT, EDWARD, 1794-1865.
Address Delivered Before The American Institute Of The City Of New York, At Their Fourth Annual Fair, October 14, 1831.
New York: Van Norden And Mason, 1831.
50 p. LCP; HSP (wrapper)

1155. Every Man His Own Farrier. Containing Ten Minutes' Advice How To Buy A Horse. To Which Is Added Directions How To Use Your Horse At Home Or On A Journey; And What Remedies Are Proper For All The Diseases To Which He Is Liable.
Philadelphia: Published By Leary & Getz, [1853]. [cover date]
98, 46 (pub. adv.) p. (prelims. erratically paginated); front.; illus. This is a new edition of: Ten Minutes Advice To Every Gentleman Going to Purchase a Horse .... (q.v.) [with] The Gentleman's Pocket Farrier .... 1787. The 2nd part is attributed to William Burdon (q.v.). LCP

1156. Every Man His Own Gardener; An Account Of Every Vegetable Production Cultivated For The Table, By The Plough And The Spade.
New York: Published By Wm. H. Graham [Stereotyped by J. Van Norden & Co. (t.p. verso)], 1846.
92 p. LCP

An Explanation Of The Map Which Delineates That Part Of The Federal Lands .... 1787.
See CUTLER, MANASSEH, 1742-1823.

1157. FALKNER, FREDERIC.
The Farmer's Manual; A Practical Treatise On The Nature And Value Of Manures, Founded From Experiments On Various Crops. With A Brief Account Of All The Most Recent Discoveries In Agricultural Chemistry.
New York: D. Appleton & Co.; Philadelphia: Geo. S. Appleton [John F. Trow, Printer (t.p. verso)], 1843.
153, [1], [14]+24 (pub. adv.) p. LCP

The Farm: A Pocket Manual .... 1858.
See WELLS, SAMUEL ROBERTS, 1820-1875.

Farm Journal And Progressive Farmer.
See Pennsylvania Farm Journal ....

1158. Farmer and Gardener. Devoted to Agriculture, Horticulture and Rural Affairs Generally.
Philadelphia.
HSP: Ns. vol. 2, nos. 1-6, 8-9 (July-Dec. 1860, Feb.-March 1861); vol. 3, no. 5 (Oct. 1861); vol. 6, no. 1 (Jan. 1864)
Later (1864) known as: Pennsylvania Farmer and Gardener.

1159. The Farmer's And Horsemen's True Guide, Pointing Out The Perfections And Imperfections Of The Horse, With Cures For His Diseases, And For Neat Cattle And Sheep. Together With Useful Recipes In Gardening, Horticulture And Domestic Economy. From Practical Experience. Well Tried And Approved By Farmers, Farriers, &c.
Oswego: Published And Sold By A.P. Searing; Printed At The Gazette Office, 1834.
132 p. LCP

1160. The Farmers' Cabinet, and American Herd Book.
Philadelphia.
LCP: Vols. 1-11; vol. 12, nos. 5, 8, 12 (July 1, 1836-July 1847; Dec. 15, 1847; Mar. 15, July 15, 1848). Also: another copy of 3:3 with additional matter in publisher's statement at end.
APS: Vols. 1-4 (July 1, 1836-July 1840); vol. 5, nos. 2, 10-11 (Sept. 15, 1840; May 15-June 15, 1841); vol. 6, nos. 1-3, 5 (Aug. 15-Oct. 15, Dec. 15, 1841)
HSP: Vols. 1-4; vol. 6, nos. 6, 8, 9 (Jan., Mar., Apr. 1842); vol. 7, nos. 1, 5 (Aug., Dec. [frag.] 1842); vol. 11, no. 7 (Feb. 1847); vol. 12, no. 4 (Nov. 1847)
Vols. 1-4 known as: The Farmers' Cabinet; Devoted To Agriculture, Horticulture, And Rural Economy.

1161. Farmer's Museum.
Walpole, Mass.
LCP: scattered weekly numbers from vols. 3, 5, 7-9 (Oct. 20, 1795-April 13, 1802)

1162. Farmer's Register.
Conway, Mass.
LCP: Vol. 1, nos. 23, 25 (Aug. 18, Sept. 1, 1798)

1163. The Farmer's Register; A Monthly Publication. Edmund Ruffin, Editor And Proprietor.
Shellbanks, Va.; Petersburg, Va.
LCP; APS: Vol. 7 (1839)

1164. The Farmer's Year Book Of Useful And Entertaining Knowledge. 1834.
Boston: Charles Ellms, Agent; And Willard Felt & Co., [1833].
47, [1] p.; illus. T.p. headed: Vol. I. No. I. LCP

1165. FARNHAM, ELIZA WOODSON, 1815-1864.
California, In-Doors And Out; Or, How we Farm, Mine, and Live generally in the Golden State.
New York: Dix, Edwards & Co. [Miller & Holman, Printers & Stereotypers (t.p. verso)], 1856.
xiv, [2], 508, [6 (pub. adv.)] p. LCP

1166. The Farrier's Magazine, Or The Archives Of Veterinary Science .... By James Carver.
Philadelphia.
LCP: [Vol. 1, no. 1] ([June] 1818). All pub.?

1167. FAUX, WILLIAM.
Memorable Days In America: Being A Journal Of A Tour To The United States, Principally Undertaken To Ascertain, By Positive Evidence, The Condition And Probable Prospects Of British Emigrants; Including Accounts Of Mr. Birkbeck's Settlement In The Illinois: And intended to shew Men and Things as they are in America.
London: Printed For W. Simpkin And R. Marshall [J. M'Creery (t.p. verso)], 1823.
xvi, 488 p.; front. LCP; HSP

1168. FEARON, HENRY BRADSHAW, b. ca. 1770.
Sketches of America. A Narrative Of A Journey Of Five Thousand Miles Through The Eastern And Western States Of America; Contained In Eight Reports Addressed To The Thirty-Nine English Families By Whom The Author Was Deputed, In June 1817, To Ascertain Whether Any, And What Part Of The United States Would Be Suitable For Their Residence. With Remarks On Mr. Birkbeck's "Notes" And "Letters."
London: Printed For Longman, Hurst, Rees, Orme, And Brown [Printed by A. Strahan (t.p. verso)], 1818.
vii, [1], 462 p. LCP

1169. __________.
Sketches of America. A Narrative Of A Journey Of Five Thousand Miles .... Second Edition.
London: Printed For Longman, Hurst, Rees, Orme, And Brown [Printed by A. Strahan (½ t.p. verso)], 1818.
xi, [1], 454 p. HSP

1170. __________.
Sketches of America. A Narrative Of A Journey Of Five Thousand Miles .... Third Edition.
London: Printed by Strahan and Spottiswoode; For Longman, Hurst, Rees, Orme, And Brown, 1819.
xv, [1], 454 p. LCP; APS

1171. __________.
Skizzen von America, entworfen auf einer Reise durch die Vereinigten Staaten in den Jahren 1817 und 1818. Nach dem Englischen. Dritter Wegweiser fur Alle, die nach den Vereinigten Staaten auswandern und sich dort ansiedeln wollen. (Aus dem Ethnographischen Archiv besonders abgedruckt.)
Jena: in der Bran'schen Buchhandlung, 1819.
iv, [2], 276 p. LCP

FEATHERSTONHAUGH, GEORGE W.
See SCHENECTADY CO. [N.Y.]. AGRICULTURAL SOCIETY. Proceedings .... With The President's Address .... 1822.

1172. [FELLENBERG INSTITUTION, Bolton Farm, Bucks Co., Pa.].
Classical, English, And Agricultural Institute, At Bolton Farm, Bucks County, Pa. [cap. title]
[Bristol?: 1830].
12 p. Title headed: (Circular.) Opening letter signed: William Chaderton. LCP; APS

1173. __________.
Fellenberg Institution, At Bolton Farm, Near Bristol, Pennsylvania. [Cap. title; text begins:] The undersigned have associated themselves, and have opened at Bolton Farm, an Institution to be conducted on the principles of the justly celebrated School of Emanuel Fellenberg, at Hoffwyl, in Switzerland ....
[Bristol?: 1831].
2 p. Signed at end: William Chaderton; F.A. Ismar (q.v.). APS

1174. FENNELL, JAMES, 1766-1816.
Description Of The Principles And Plan Of Proposed Establishments Of Salt Works; For The Purpose Of Supplying The United States With Home Made Salt.
Philadelphia: Printed By John Bioren, 1798.
Pp. [1]-12, [17]-60; fold. front. HSP in LCP; LCP (lacking front.); APS (lacking front.)

1175. FENWICK, JOHN, 1618-1683.
Friends, These are to Satisfie you, or any other who are Sober, and are any wise minded to go along with me, and Plant within my Colony; That we shall no doubt find, but that New Cesarea or New Jersey ... is a Healthy Pleasant, and Plentiful Country .... [cap. title]
[London:] 1675.
2 p. Dated: This 8th. of the 1st Month, 1675. HSP in LCP

1176. [FERNAGUS DE GELONE, -----].
Manuel-Guide Des Voyageurs Aux États-Unis De L'Amérique Du Nord, Ou l'on trouve tous les renseignements nécessaires et fondés sur des faits pour y commercer avec sûreté et profit .... Addressé Particulièrement Aux Commerçans, Aux Agriculteurs, et aux personnes de toutes les professions .... Par M.F.D.G.*** ....
Paris: Chez Pillet, Imprimeur-Libraire [De L'Imprimerie De Pillet (½ t.p. verso)], 1818.
[6], 196 p. APS

1177. FERRIS, JACOB.
The States And Territories Of The Great West; Including Ohio, Indiana, Illinois, Missouri, Michigan, Wisconsin, Iowa, Minesota [sic], Kansas, And Nebraska; Their Geography, History, Advantages .... Also, Pointing Out The Best Districts For Agriculture, Commerce, Lumbering, And Mining Operations. With A Map And Numerous Illustrations.
New York and Auburn: Miller, Orton, And Mulligan; Buffalo: E.F. Beadle [Buffalo: C.E. Fulton, Stereotyper (t.p. verso)], 1856.
LCP: 352, [4 (pub. adv.)] p.; 4 pls.; illus. (lacking fold. map; see Graff 1309)

1178. FERRIS, W.L., Nurseryman, Throg's Neck, N.Y.
Catalogue Of Fruit And Ornamental Trees, Shrubs, Vines, Roses, Green-House And Stove Plants, Cultivated And For Sale At The Nursery Of W.L. Ferris, Throg's Neck, Westchester County, New York. 1856.
New York: R. Craighead, Printer, 1856.
29 p. LCP (wrappers)

1179. FESSENDEN, THOMAS GREEN, 1771-1837.
The American Kitchen Gardener; Containing Practical Directions For The Culture Of Vegetables. Also, Garden Fruits, Strawberry, Raspberry, Gooseberry, Currants, Melons, &c., &c. Revised From The 35th Edition, And Adapted To The Use Of Families, By A Practical Gardener.
New York: C.M. Saxton, Agricultural Book Publisher [S.W. Benedict, Stereotyper And Printer (t.p. verso)], 1852.
120 p. LCP

1180. __________.
The New American Gardener; Containing Practical Directions On The Culture Of Fruits And Vegetables; Including Landscape And Ornamental Gardening, Grape-Vines, Silk, Strawberries, &c. &c.
Boston: Published By J.B. Russell; Cambridge, Mass.: Sold Also By Hilliard & Brown; New York: And G. Thorburn & Son [Stereotyped at the Boston Type and Stereotype Foundry (t.p. verso)], 1828.
306, [1], [5 (adv.)] p. LCP

1181. __________.
The New American Gardener; Containing Practical Directions .... Fourth Edition.
Boston: Published By Carter And Hendee; Also By John B. Russell; New York: Sold By G. Thorburn And Son [Stereotyped at the Boston Type and Stereotype Foundry (t.p. verso)], 1830.
306, [1], [4 (adv.)] p. LCP

1182. __________.
The Register Of Arts, Or A Compendious View Of Some Of The Most Useful Modern Discoveries And Inventions.
Philadelphia: Published By C. And A. Conrad, And Co.; Baltimore: Conrad Lucas & Co.; Petersburg: Somervill And Conrad; Norfolk: Bonsal Conrad & Co., 1808.
xi, [1], 404 p.; 2 pls., 1 fold. Incl. many agricultural items. LCP (with 9 p. pub. adv. at back); APS

1183. FIELD, THOMAS WARREN, 1820-1881.
Pear Culture. A Manual For The Propagation, Planting, Cultivation, And Management Of The Pear Tree. With Descriptions And Illustrations Of The Most Productive Of The Finer Varieties ....
New York: A.O. Moore, Agricultural Book Publisher, 1858.
Pp. [i]-viii, [13]-286, 1-12 (pub. adv.); front.; 2 pls.; illus. LCP

1184. FISHER, S.I.
Observations On The Character And Culture Of The European Vine, During A Residence Of Five Years In The Vine Growing Districts Of France, Italy And Switzerland. To Which Is Added, The Manual Of The Swiss Vigneron ... By Mons. Brun Chappuis. And The Art Of Wine Making, By Mons. Bulos. Translated From The French.
Philadelphia: Key And Biddle, 1834.
Pp. [i]-vii, [1], [13]-244. LCP

1185. FISHER, SIDNEY GEORGE, 1809-1871.
An Address Delivered Before The Agricultural Society Of New Castle County, Delaware, At Their Annual Exhibition, Held At The Society's Farm, Near Wilmington, October 17, 1860. Published By Order Of The Society.
Philadelphia: C. Sherman & Son, Printers, 1860.
26 p. LCP (wrappers); APS; HSP (wrappers)

1186. __________.
Address Delivered Before The Chester County Agricultural Society, At Their Annual Exhibition Held At West-Chester, September 26, 1857. Published By Order Of The Society.
Philadelphia: Merrihew & Thompson, Printers, 1857.
27 p. LCP (wrappers); HSP (wrappers)

1187. __________.
Address Delivered Before The Montgomery County [Pa.] Agricultural Society At Their Annual Exhibition Held At Springtown, October 7th, 1859. Published By Order Of The Society.
Philadelphia: James B. Chandler, Printer, 1859.
26 p. LCP (wrapper); HSP (wrapper)

1188. __________.
Address Delivered Before The Philadelphia Society For Promoting Agriculture, At Their Annual Exhibition Held At The Rising Sun Village, October 17, 1850.
Germantown, Pa.: Printed At The Office Of The Telegraph, 1850.
22 p. This address also incl. in: PHILADELPHIA. SOCIETY FOR PROMOTING AGRICULTURE. Proceedings Of The Annual Exhibition .... 1850 (q.v.). LCP (wrapper); HSP (wrappers)

1189. [_________].
A Report On The Cultivation Of Native Grapes For Fruit And For Wine. Read To The Philadelphia Society for Promoting Agriculture, April 2, 1856. Printed By Order Of The Society.
Philadelphia: Inquirer Printing Office, 1856.
15 p. Author's name at end. APS (wrappers); HSP (wrappers)

1190. FISKE, OLIVER, 1762-1836.
An Address Delivered Before The Worcester [Mass.] Agricultural Society, October 20, 1831: Being Their Thirteenth Anniversary Cattle Show and Exhibition of Manufactures.
Worcester: From M.W. Grout's Power Press, 1831.
16 p. LCP

1191. FITCH, ASA, 1809-1879.
An Essay Upon The Wheat-Fly, And Some Species Allied To It.
Albany: Printed By Carroll And Cook, 1845.
[4], 32 p.; 1 pl. T.p. headed: From the American Quarterly Journal of Agriculture and Science, Vol. II. No. 2. LCP

1192. _________.
First Report On The Noxious, Beneficial And Other Insects, Of the State Of New York, Made To The State Agricultural Society, Pursuant To An Appropriation For This Purpose From The Legislature Of The State.
Albany: C. Van Benthuysen, Printer To The Legislature, 1855.
[2], 180 p.; illus. LCP

1193. _________.
The Hessian Fly, Its History, Character, Transformations, And Habits. Originally Published In The American Journal Of Agriculture And Science, Vols. IV., V.
Albany: Printed By Joel Munsell, 1846.
63 p.; 1 pl.; leaf of pl. description tipped in after t.p. LCP (wrappers); APS (wrappers); HSP (without inserted leaf)

1194. _________.
The Hessian Fly .... As Published In The Transactions Of The New-York State Agricultural Society, Vol. VI.
Albany: Printed By C. Van Benthuysen And Co., 1847.
60, [1] p.; 1 pl. LCP (wrappers)

1195. _________.
The Most Pernicious Species Of United States Insects, And The Curculio, Two Addresses Delivered At The Annual Meeting Of The New York State Agricultural Society, A.D. 1859 And 1860.
Albany: Printed By C. Van Benthuysen, 1860.
28 p.; illus. Wrapper title: Two Addresses On Insects And Curculio. LCP (wrappers)

FLEISCHMAN, CARL LUDWIG, b. ca. 1800. Remarks On The Establishment Of An Agricultural College .... 1847.
See BURCHARD, NATHAN. Address ... Accompanied By Remarks ... By Charles L. Fleischman .... 1847.

1196. FLINT, CHARLES LOUIS, 1824-1889.
Grasses And Forage Plants. A Practical Treatise, Comprising Their Natural History; Comparative Nutritive Value; Methods Of Cultivation, Cutting, And Curing; And The Management Of Grass Lands In The United States And British Provinces. With One Hundred And Seventy Illustrations. Fifth Edition. Revised And Enlarged.
Boston: Crosby, Nichols, Lee, & Co. [Cambridge: University Press; Printed by Welch, Bigelow, and Company (t.p. verso)], 1860.
398, [2 (pub. adv)] p. incl. front.; illus. LCP

1197. _________.
Milch Cows And Dairy Farming; Comprising The Breeds, Breeding, and Management .... The Selection Of Milch Cows ... And The Production of Milk, Butter, And Cheese ... With A Treatise Upon The Dairy Husbandry Of Holland; To Which Is Added Horsfall's System Of Dairy Management.
New York: Published By A. O. Moore; Boston: A. Williams & Co.; Philadelphia: J.B. Lippincott & Company; Cincinnati: Rickey, Mallory & Co. [Printed By George C. Rand & Avery; Stereotyped by Hobart & Robbins, New England Type and Stereotype Foundery (t.p. verso)], 1858.
416 p. incl. front.; illus. LCP

1198. _________.
A Practical Treatise On Grasses And Forage Plants; Comprising Their Natural History, Comparative Nutritive Value, Methods Of Cultivating, Cutting, And Curing, And The Management Of Grass Lands.
New York: G.P. Putnam & Co.; London: N. Trubner & Co. [Boston: William White, Printer (t.p. verso)], 1857.
iv, 236 p; front.; 1 pl.; illus. LCP

_________.
See also
MASSACHUSETTS. STATE BOARD OF AGRICULTURE. Abstract Of Returns Of The Agricultural Societies .... 1859.
_________. The Agriculture Of Massachusetts .... Prepared By .... 1854.
_________. Annual Reports .... Prepared By ..., Secretary .... 1854 etc.

1199. FLINT, JAMES.
Letters From America, Containing Observations On The Climate And Agriculture Of The Western States, The Manners Of The People, The Prospects Of Emigrants, &c. &c.
Edinburgh: Printed For W. & C. Tait; London: Longman, Hurst, Rees, Orme, And Brown [Edinburgh: Printed by Balfour & Clarke (t.p. verso)], 1822.
viii, 330 p.; illus. HSP in LCP

1200. The Floral Magazine and Botanical Repository. Published by D. & C. Landreth.
Philadelphia.
HSP in LCP; LCP; APS: Vol. 1 (1832). Col. pls. All pub.

FLORIDA. INSTITUTE OF AGRICULTURE, ANTIQUITIES AND SCIENCE, Tallahassee.
See GADSDEN, JAMES. Oration, Delivered ... To The .... 1827.

1201. The Florist And Horticultural Journal, A Monthly Magazine Of Horticulture, Agriculture .... H.C. Hanson, Editor.
Philadelphia.
LCP: Vols. 2-3; vol. 4, nos. 1-9 (Jan. 1853-Sept. 1855)
APS: Vol. 3; vol. 4, nos. 1-9 (Jan. 1854-Sept. 1855)
HSP: Vol. 1, nos. 2-12; vols. 2-3; vol. 4, nos. 1-3 (June 1852-March 1855)
Vol. 1 known as: The Philadelphia Florist And Horticultural Journal .... R. Robinson Scott, Editor. Col. pls.

1202. FOLLEN, CHARLES CHRISTOPHER.
Suggestions On Landscape Gardening; By Charles Follen, Architect And Landscape Gardener. Thorough Drainage; By J. Herbert Shedd, Civil Engineer.
Boston: Phillips, Sampson and Co. [Press Of The Franklin Printing House (t.p. verso)], 1859.
20, [3 (adv.)] p. LCP (wrappers)

1203. FOOTE, HENRY G.
Address Of The St. Lawrence County [N.Y.] Agricultural Society Delivered At The Eighth Annual Fair, At Canton, September 30th, 1859. [wrapper title]
Ogdensburgh: Abbott & O'Brian, Democrat Caloric Print, 1859.
32 p. LCP (wrappers)

1204. For Sale. [Heading; text begins:] The quantity of 389,565 acres of Farm Lands, in Sixty adjoining and adjacent tracts, fit for wheat, rye and other grain, cattle, tobacco, and cotton ....
[North Carolina: 1801].
2 p. Describes farm lands in Rutherford Co., North Carolina. LCP

1205. FORBES, ALEXANDER, 1778-1862.
California: A History Of Upper And Lower California From Their First Discovery To The Present Time Comprising An Account Of The Climate, Soil, Natural Productions, Agriculture, Commerce, &c. .... With An Appendix Relating To Steam Navigation In The Pacific ....
London: Smith, Elder And Co. [Chichester: Printed by William Hayley Mason (colop.)], 1839.
352 p.; front.; 9 pls.; fold. map. LCP

1206. FORSTALL, EDMUND J.
Agricultural Productions Of Louisiana, Embracing Valuable Information Relative To The Cotton, Sugar And Molasses Interests, And The Effects Upon The Same Of The Tariff Of 1842, By Edward [sic] J. Forstall.
[New Orleans?:] Tropic Print., 1845.
43 p. LCP

FORSTALL, EDWARD J.
See FORSTALL, EDMUND J.

1207. FORSYTH, WILLIAM, 1737-1804.
An Epitome Of Mr. Forsyth's Treatise On The Culture And Management Of Fruit-Trees. Also Notes On American Gardening And Fruits; With Designs For Promoting The Ripening Of Fruits, And Securing Them As Family Comforts: And Further Of Economical Principles In Building Farmers' Habitations. By An American Farmer [i.e. John Beale Bordley].
Philadelphia: Published By Wm. Poyntell & Co. Proprietors Of The Classic Press, 1804.
[4], 186, [6] p.; 15 pls., 13 fold. Gaines 63a calls Wm. Cobbett the "American Farmer"; but see the introductory essay to this catalogue. HSP in LCP (lacking pls.); LCP; HSP (lacking pls.)

1208. ________.
A Treatise On The Culture And Management Of Fruit Trees; In Which A New Method Of Pruning And Training Is Fully Described. Together With Observations On The Diseases, Defects, and Injuries ... A Particular Method Of Cure .... To Which Are Added, An Introduction And Notes, Adapting The Rules Of The Treatise To The Climates and Seasons of the United States of America. By William Cobbett.
Philadelphia: Printed For J. Morgan, 1802.
xii, 259 p.; 13 pls., 11 fold. LCP; APS

1209. ________.
A Treatise On The Culture And Management Of Fruit Trees .... To Which Are Added, An Introduction And Notes .... By William Cobbett.
Albany: Printed For And Sold By D. & S. Whiting, At The Albany Bookstore; Boston: Sold Also By Thomas & Andrews; N. York: A. & A. Stansbury; Troy: O. Penniman & Co.; Utica: S.P. Goodrich; Oxford: J. Glover & Co.; Canandaigua: H. Chapin, 1803.
280 p.; 13 pls., 11 fold. LCP

41 Bewährte Recepte .... 1834.
See [Ein und vierzig] 41 Bewährte Recepte ....

1210. FOSTER, EDEN BURROUGHS, 1813-1882.
An Address On Agriculture: Delivered Before The Hillsboro [N.H.] Agricultural And Mechanical Society, At Amherst, March 5, 1851.
Manchester: Union Press, - Wm. H. Gilmore & Co., Printers, [1851].
24 p. LCP

1211. FOWLER, JOHN.
Journal Of A Tour In The State Of New York, In The Year 1830; With Remarks On Agriculture In Those Parts Most Eligible For Settlers: And Return To England By The Western Islands, In Consequence Of Shipwreck In The Robert Fulton.
London: Whittaker, Treacher, And Arnot [Liverpool: Printed By E. Smith And Co. (t.p. verso)], 1831.
Pp. [1-4], [3]-333. LCP; HSP

1212. FRANCIS, JOHN WAKEFIELD, 1789-1861.
An Address Delivered Before The New-York Horticultural Society, At Their Anniversary, On The Eighth Of September, 1829. Published At The Request Of The Society.
New York: Printed By E. Conrad, 1830.
31 p. LCP

Frank; Or, Dialogues Between A Father And Son .... 1840.
See PEDDER, JAMES, 1775-1859.

1213. FRANKLIN, AUGUSTUS.
The American Farrier: Adapted For The Convenience Of The Farmer .... [sic] Gentleman .... [sic] And Smith; Being A Sure Guide To Prevent And Cure All Maladies And Distempers That Are Incident To Horses .... And Also, For The Diseases Incident To Cattle.
Strasbourg [Pa.]: Printed & Sold By Brown & Bowman, 1803.
290 p. LCP

1214. FRANKLIN CO. [MASS.]. AGRICULTURAL SOCIETY.
Franklin County Agricultural Society. 1860. [Cap. title; text begins:] The Eleventh Exhibition of the Franklin County Agricultural Society, will be holden at Greenfield, Thursday, September 27, 1860 ....
[Greenfield: 1860].
[2] p. LCP

1215. FREEMAN, SAMUEL.
The Emigrant's Hand Book, And Guide To Wisconsin, Comprising Information Respecting Agricultural And Manufacturing Employment, Wages, Climate, Population &c.; Sketches Of Milwaukee ... Table Of Routes From New-York ....
Milwaukee: Sentinel and Gazette Power Press Print., 1851.
148 p.; pp. 103-48 adv. LCP (wrappers)

1216. FREEOFF, P.R., Nurseryman, Auburn, N.Y.
Select List Of Florists' Flowers: Consisting Of Roses, Dahlias, Verbenas, Fuchsias, Geraniums, Salvias, Chrysanthemums, &c., Cultivated And Sold By P.R. Freeoff, At The Commercial Garden And Nursery, Auburn, N.Y.
Auburn: Knapp & Peck, Printers, Daily Advertiser Office, 1854.
20 p. LCP (wrappers)

FREITAG, EBERHARD.
See FREITAG, JOHN EBERHARD, 1764-1846.

1217. FREITAG, JOHN EBERHARD, 1764-1846.
Der Deutsche Pferd-Arzt. Herausgegeben von Doctor Eberhard Freitag, in Bethlehem.
Easton, Pennsylvania: Gedruckt bey Christ. Jac. Hütter, 1809.
v, [1], 90, vi [i.e. iv] p. P. ^{2}iv misnumbered vi. HSP in LCP; LCP

1218. [FRIENDS OF DOMESTIC INDUSTRY].
Report On The Production And Manufacture Of Cotton.
Boston: Printed By J.T. & E. Buckingham, 1832.
20 p. T.p. headed: New-York Convention [i.e. of the Friends of Domestic Industry, in October, 1831]. LCP

1219. FROST, JOHN, 1800-1859.
The Omnibus; Or, General Repository of Instruction and Entertainment. Embracing Articles On Agriculture, Commerce, Manufactures, Political Economy, Science .... With Many Hundred Valuable Receipts And Practical Directions, Of Use To Everybody.
Philadelphia: E.S. Morris, 1857.
136, [1 (pub. adv.)] p. (with unnumbered adv. p. throughout text); illus. LCP

1220. FRY, WILLIAM HENRY, 1815-1864.
A Complete Treatise On Artificial Fish-Breeding: Including The Reports On The Subject Made To The French Academy And The French Government; And Particulars Of The Discovery As Pursued In England. Translated And Edited By W.H. Fry. Illustrated With Engravings.
New York: D. Appleton And Company, 1854.
Pp. [1-2 (pub. adv.)], [i]-x, [1-3], [7]-188, [1-4 (pub. adv.)]; illus. LCP; APS

1221. FURMAN, GABRIEL, 1800-1854.
Address Delivered Before The Queens County Agricultural Society, At Its Third Anniversary, At Jamaica, Thursday, October 10th, 1844.
Jamaica: Printed At The Office Of "The Long Island Farmer," By C.S. Watrous, 1845.
24 p. LCP

G***, F.D.
See FERNAGUS DE GELONE, -----.

1222. GADSDEN, JAMES, 1788-1858.
Oration, Delivered By Col. James Gadsden, To The Florida Institute Of Agriculture, Antiquities And Science, At Its First Public Anniversary, On Thursday, the 4th January, 1827.
Tallahassee: Printed At The Office Of The Florida Intelligencer, 1827.
[16] p. APS

1223. GAGE, NATHANIEL, 1800-1861.
An Address Before The Essex [Mass.] Agricultural Society, At Topsfield, September 27, 1837. At Their Annual Cattle Show. Published By Order Of The Society.
Salem: Printed At The Gazette Office, 1838.
Pp. [1-2], [5]-27. LCP

1224. The Gardener's Monthly And Horticultural Advertiser. Edited By Thomas Meehan.
Philadelphia.
HSP: Vol. 1, no. 4 (April 1859); vol. 9, nos. 4, 6, 11 (April, June, Nov. 1867)

1225. GARDINER, JOHN, and DAVID HEPBURN.
The American Gardener, Containing Ample Directions For Working A Kitchen Garden, Every Month In The Year; And copious instructions for the cultivation of Flower Gardens, Vineyards, Nurseries, Hop-Yards, Green Houses, And Hot Houses.
Washington: Printed By Samuel Smith, For The Authors, 1804.
204, [4] p. LCP

1226. __________.
The American Gardener, Containing Ample Directions .... A New Edition Much Enlarged. To Which Is Added A Treatise On Gardening, By A Citizen Of Virginia [i.e. John Randolph]. Also, A Few Hints On The Cultivation Of Native Vines, And Directions For Making Domestic Wines.
Georgetown, D.C.: Published By Joseph Milligan [Wm. A. Rind & Co. Prints. (t.p. verso)], 1818.
347, [1] p. LCP; HSP

1227. GARDINER LYCEUM, Gardiner, Maine.
An Address To The Public From The Trustees Of The Gardiner Lyceum.
Hallowell: Printed By Goodale, Glazier & Co., 1822.
8 p. The object of the Lyceum was "to give to mechanics and farmers such a scientific education, as would enable them to become skilful in their professions" (p. 2). LCP; HSP

1228. __________.
Catalogue Of The ... And An Address Of The Trustees.
[Hallowell: Printed By Goodale, Glazier, & Co. (t.p. verso)], [1823].
16 p. APS

1229. __________.
Catalogue Of The Officers And Students Of ... With An Address To The Public. October 1824.
[Gardiner: P. Sheldon, Printer (t.p. verso)], [1824].
16 p. APS

1230. __________.
Laws Of The ... And Regulations For Its Government.
Gardiner, Me.: Printed By P. Sheldon, 1825.
24 p. APS

1231. GARDNER, AUGUSTUS KINSLEY, 1821-1876.
Report Of A Committee Appointed By The Academy Of Medicine, Upon The Comparative Value Of Milk, Formed From The Slop Of Distilleries And Other Food, With Chemical And Microscopical Analyses. Read March 1, 1848. (Extracted from the Transactions of the N.Y. Academy of Medicine.) [wrapper title]
New York: R. Craighead, Printer, 1851. [wrapper imprint]
19 p. Deals chiefly with diseases of cattle and the effect on milk. LCP (wrapper)

1232. GARDNER, DANIEL PEREIRA, d. 1853, ed.
The Farmer's Dictionary: A Vocabulary Of The Technical Terms Recently Introduced Into Agriculture And Horticulture From Various Sciences, And Also A Compendium Of Practical Farming: The Latter Chiefly From The Works Of The Rev. W.L. Rham, Loudon, Low, And Youatt, And The Most Eminent American Authors. With Numerous Illustrations.
New York: Harper & Brothers, Publishers, 1846.
876, 8+2+2 (pub. adv.) p.; illus. LCP

1233. GASKINS, JAMES.
The Farmer's Guide, Or A New Theory Of Agriculture; Founded On Philosophical And Practical Principles, And Adapted To All Climates.
Baltimore: Printed By Samuel Sands, 1838.
215 p. LCP

1234. GENET, EDMOND CHARLES, 1765-1834.
Address On The Means Of Opening New Sources Of Wealth For The Northern States. Delivered On The 19th October, 1821, Before The Agricultural Society Of The County Of Rensselaer, And The Public Attending Their Proceedings.
Troy: Printed By Francis Adancourt, 1821.
56 p.; 1 pl. LCP

1235. GENTLE, ANDREW.
Every Man His Own Gardener; Or, A Plain Treatise On The Cultivation Of Every Requisite Vegetable In The Kitchen Garden, Alphabetically Arranged. With Directions For The Green & Hothouse, Vineyard, Nursery .... Intended principally for the inexperienced Horticulturist.
New York: Sold ... by Alexander Smith, Nursery and Seedsman; Brooklyn: and by the Author, 1841.
105, [2] p. LCP

The Gentleman's Pocket-Farrier .... 1823 etc.
See BURDON, WILLIAM.

GEORGIA. CENTRAL AGRICULTURAL SOCIETY.
See SCOTT, THOMAS FIELDING, 1807-1867. Address On Education, Delivered Before .... 1852.

1236. GERHARD, FREDERICK.
Illinois As It Is; Its History, Geography, Statistics, Constitution ... Climate, Soil, Plants, Animals, State Of Health, Prairies, Agriculture, Cattle-Breeding, Orcharding, Cultivation Of The Grape, Timber-Growing, Market-Prices, Lands And Land-Prices ....
Chicago, Ill.: Keen And Lee; Philadelphia: Charles Desilver [Stereotyped By J. Fagan, Philadelphia (t.p. verso)], 1857.
451 p. + pp. [3]-6 (pub. adv.); front.; 2 pls.; 3 col. fold. maps. LCP

1237. GHOLSON, THOMAS SAUNDERS, 1809-1868.
Valedictory Address Of Thos. S. Gholson, Esq., Delivered Before The Union Agricultural Society Of Virginia and North Carolina, October 27, 1854.
[Petersburg, Va.:] Printed At The Office Of The Southern Farmer, 1854.
14 p. LCP; HSP (wrapper)

1238. GIBBONS, WILLIAM, 1781-1845.
Essay On The Culture Of Silk; Read Before The Agricultural Society Of Newcastle County, State Of Delaware, And Published At the Request Of The Society; With A Journal Annexed.
Philadelphia: Rackliff & Jones, Printers, 1836.
32 p. HSP (wrappers)

1239. [GILROY, CLINTON G.]
The History of Silk, Cotton, Linen, Wool, And Other Fibrous Substances; Including Observations On Spinning, Dyeing And Weaving. Also An Account Of The Pastoral Life Of The Ancients, Their Social State And Attainments In The Domestic Arts. With Appendices On Pliny's Natural History; On The Origin And Manufacture Of Linen And Cotton Paper; On Felting, Netting, &c. Deduced From Copious And Authentic Sources. Illustrated By Steel Engravings.
New York: Harper & Brothers, 1845.
xxii, [2], 464 p.; front.; 9 pls. LCP

1240. Gleanings From The Most Celebrated Books On Husbandry, Gardening, And Rural Affairs. From The London Second Edition Of 1803. Interspersed with Remarks and Observations by a Gentleman of Philadelphia [i.e. John Beale Bordley].
Philadelphia: Printed And Sold By James Humphreys, 1803.
viii, 365, [1] [2 (pub. adv.)] p.; fold. pl. For attribution see the introductory essay to this catalogue. HSP in LCP (lacking pl.); LCP; APS

1241. [GLEN, JAMES, 1701-1777].
A Description Of South Carolina; Containing, Many curious and interesting Particulars relating to the Civil, Natural and Commercial History of that Colony, viz. The Succession of European Settlers there ... The Nature of the Climate ... The Culture and Produce of Rice, Indian Corn, and Indigo; the Process of extracting Tar and Turpentine .... To which is added, A very particular Account of their Rice-Trade for Twenty Years, with their Exports of Raw Silk and Imports of British Silk Manufactures for Twenty-five Years.
London: Printed for R. and J. Dodsley, 1761.
viii, 110 p. HSP in LCP (with 1 leaf pub. adv. at back); LCP (lacking ½ t.p.)

1242. GLENN, JAMES.
A Treatise On The Breeding Of Animals, And Other Matters Interesting To Farmers.
Utica: O. Hutchinson, Publisher; R. Northway, Printer, 1842.
17 p. LCP

1243. GOLD, THOMAS, 1759-1827.
Address Of Thomas Gold, Esq. President of the Berkshire Agricultural Society, And Member Of The Massachusetts Society, Delivered Before The Berkshire Association, For The Promotion Of Agriculture and Manufactures, At Pittsfield, Oct. 3d, 1816. Together With The Report of the Committee on Agriculture, And The Address Of Elkanah Watson, Esq. Previous To Declaring The Premiums Reported.
Pittsfield [Mass]: Printed By Phinehas Allen, [1816].
22, [2] p. LCP

1244. GOLD, THOMAS RUGGLES, 1764-1827.
Address Of Thomas R. Gold, Esquire, Delivered Before The Agricultural Society Of The County Of Oneida, At The Request Of The Officers Of The Society, (And Published In Pursuance Of A Resolution Of The Society) At Whitestown, October 18, 1820.
Utica: Printed For The Society, By William Williams, 1820.
21 p. HSP

1245. GOODWIN, ISAAC, 1786-1832.
Address Delivered Before The Worcester Agricultural Society, October 13, 1824; Being Their Anniversary Cattle Show and Exhibition of Manufactures. Published by order of the Trustees.
Worcester: Printed By William Manning, 1824.
32 p. LCP

1246. GOWEN, JAMES.
Address Delivered Before The Lancaster County [Pa.] Agricultural Society, At Its Annual Meeting, On The Thirteenth Of January, 1852. Published By Order Of The Society.
Germantown, Pa.: Printed At The Office Of The Telegraph, 1852.
28 p.; errata slip at front. HSP (wrappers)

1247. __________.
Address Delivered Before The Lancaster County Agricultural Society ....
Philadelphia: Brown's Steam Power Book And Job Printing Office, 1852.
28 p. LCP; APS

1248. __________.
Address Delivered Before The Mercer County [Pa.] Agricultural Society, At Its Annual Meeting, On The Twentieth Of September, 1853. Published By Order Of The Society.
Philadelphia: Brown's Steam Power Book And Job Printing Office, 1853.
27 p.; errata slip at back. HSP (wrappers)

GRAFTON CO. [N.H.]. AGRICULTURAL SOCIETY.
See PATTERSON, JAMES WILLIS, 1823-1893. An Address Delivered Before .... 1857.

1249. GRANNISS, GEORGE B., Land agent, New York.
For Sale: 40,000 Acres Of Valuable Land Very Desirable For Large Or Small Farms, Tanneries, Mill Sites And Manufactories; Having An Abundance Of Magnificent Timber For Lumber, Hemlock Bark For Tanneries, And Excellent Water Power .... Situated Contiguous To, And Immediately Surrounding Laporte, The County Seat Of Sullivan County, Pennsylvania. Offices Of Sale: New York, George B. Granniss ... Philadelphia, Dr. Charles Neil ... Baltimore, Luther J. Cox ....
New York: Samuel T. Callahan, Printer, 1857.
17 p.; fold. front. map. LCP (wrapper); HSP (wrappers)

1250. GRATZ, SIMON, JOSEPH GRATZ, and JACOB GRATZ.
4,830 Acres Land, In Valuable And Improved Farms for Sale, In The towns of Edmeston and Pittsfield, Otsego County, New-York.
[Philadelphia: 1820].
Brs. LCP

1251. GRAY, ALONZO, 1808-1860.
Elements Of Scientific And Practical Agriculture, Or The Application Of Biology, Geology And Chemistry To Agriculture And Horticulture. Intended As A Text-Book For Farmers And Students In Agriculture.
Andover [Mass.]: Allen, Morrill And Wardwell; New York: Dayton And Newman, 1842.
Pp. [i]-xix, [1], [17]-368; illus. LCP

1252. GREELEY, HORACE, 1811-1872.
What The Sister Arts Teach as to Farming. An Address Before The Indiana State Agricultural Society At Its Annual Fair, Lafayette, Indiana October 13th, 1853. [wrapper title]
New York; Philadelphia; Boston; London: Fowlers And Wells, Publishers, 1853. [wrapper imprint]
33 p. incl. wrapper. LCP (wrappers)

1253. GREEN, ROLAND.
A Treatise On The Cultivation Of Ornamental Flowers; Comprising Remarks On The Requisite Soil, Sowing, Transplanting, And General Management: With Directions For The General Treatment Of Bulbous Flower Roots, Green House Plants, &c.
Boston: Published By John B. Russell; New York: And G. Thorburn & Son; Press of I.R. Butts & Co., 1828.
59, [1] p. LCP

1254. GRISCOM, DAVID J., Nurseryman, Woodbury, N.J.
Evergreen Nursery, Woodbury, N.J. David J. Griscom, Proprietor. Descriptive Catalogue of Foreign And Native Evergreen And Deciduous Trees, Fruits, &c. [wrapper title]
[Philadelphia:] Crissy & Markley Printers, [1855].
22 p.; front.; 1 pl. Title headed: 1855-6. LCP (wrappers)

1255. Guano. A Treatise On The History, Economy As A Manure, And Modes Of Applying Peruvian Guano, In The Culture Of The Various Crops Of The Farm And The Garden.
Boston: Published By C.L. Bartlett, Agent of the Peruvian Government [Steam Press of C.C.P. Moody (t.p. verso)], 1860.
48 p. LCP

Guano, Its Origin .... 1845.
See BARTLETT, EDWIN.

1256. GUÉNON, FRANÇOIS, 1796-1855.
A Treatise On Milch Cows, Whereby The Quality And Quantity Of Milk Which Any Cow Will Give Will Be Accurately Determined By Observing Natural Marks .... Translated For The Farmers' Library, From The French, By N.P. Trist .... With Introductory Remarks And Observations, On The Cow And The Dairy. By John S. Skinner ....
New York: Published By Greeley & McElrath, 1846.
Pp. [1]-32, [1-2], [785]-797, 904-919, 1014-1029, [1]; illus. LCP

1257. ________.
A Treatise On Milch Cows, Whereby The Quality And Quantity Of Milk .... Translated For The Farmers' Library .... With Introductory Remarks And Observations, On The Cow And The Dairy. By John S. Skinner ....
New York: Published By Greeley & McElrath, 1847.
Pp. [1]-32, [785]-797, 904-919, 1014-1029, [1]; illus. LCP

1258. ________.
A Treatise On Milch Cows .... Translated For The Farmers' Library .... With Introductory Remarks And Observations, On The Cow And The Dairy, By John S. Skinner. Forty-Fifth Thousand - With Additions.
New York: Thomas L. McElrath And Company, 1854.
88, 12 (pub. adv.) p.; illus. LCP (wrappers)

1259. GUERNSEY, ORRIN, and JOSIAH F. WILLARD, eds. and comps.
History Of Rock County [Wis.], And Transactions Of The Rock County Agriculture Society And Mechanics' Institute. Published By The Rock County Agricultural Society ....
Janesville, Wis.: Wm. M. Doty And Brother, Printers, 1856.
xii, 350 p.; 6 pls. on col. paper. APS

H., C.W.
See HOSKYNS, CHANDOS WREN, 1812-1876.

1260. HAINES, B. & C.S., Livestock breeders, Elizabeth, N.J.
Catalogue Of Pure Bred Short Horns, Owned By B. & C.S. Haines, Elizabethtown, New Jersey. 1855. [wrapper title]
New York: Edward O. Jenkins, Printer, [1855]. [wrapper imprint]
25 p. incl. 5 pls. LCP (wrappers)

1261. [HAINES, CHARLES GLIDDEN, 1792-1825.]
An Examination Into The Expediency Of Establishing A Board Of Agriculture In The State Of New-York. Published By The New-York Corresponding Association For The Promotion Of Internal Improvements.
Brooklyn: E. Worthington, Printer, 1819.
64 p. Haines is the addressee of the appended letters, pp. 45-64. LCP; APS; HSP

1262. [HALL, FAYRER].
The Importance Of The British Plantations In America To This Kingdom; With The State of their Trade, and Methods for Improving it; As Also A Description of the several Colonies there.
London: Printed for J. Peele, 1731.
[6], 114 p. HSP in LCP

1263. HALL, HARRISON, 1785-1866.
Hall's Distiller, Containing 1. Full and particular directions for mashing and distilling all kinds of Grain .... 3. A Treatise on Fermentation .... 4. Directions for making Yeast .... 6. Instructions for making all kinds of Cordials, Compound Waters, &c; also for making Cider, Beer .... Adapted To The Use Of Farmers, As Well As Distillers.
Philadelphia: Printed By John Bioren, 1813.
x, 244 p.; fold. front.; 1 pl. LCP; APS

1264. __________.
The Distiller. Containing 1. Full and particular directions for mashing and distilling .... The Second Edition, enlarged and improved.
Philadelphia: Published By The Author; J. Bioren, Printer, 1818.
xii, 338, [1] p.; fold. front.; 1 pl.

1265. HALL, JONATHAN PRESCOTT, 1796-1862.
An Address Delivered at Middletown, in the State of Rhode Island, 21st Day Of September, 1854. Before The Aquidneck Agricultural Society.
Newport, R.I.: Cranston & Norman, 1854.
55 p. LCP

1266. HALL, LEWIS A.
The Farmer's Railroad; Or A Treatise Uniting Mental With Corporeal Efforts, In The Improvement of the farmer and the farm.
Trenton: Printed By Phillips & Boswell, 1856.
[4], 45, [2] p. On "the great and essential benefits resulting to the farming interest of our country, by giving to it the results of chemical science ...." (p. 1). HSP

1267. HALLOWELL, BENJAMIN, 1799-1877.
Address Of Benjamin Hallowell, (Of Alexandria, Virginia,) At The Meeting Of The Agricultural Society Of Montgomery County, Md., Held At Rockville, Montgomery Co., September 9, 1852. Published By The Request Of The Society.
Washington: Printed At The Congressional Globe Office, 1852.
16 p. LCP

1268. HALSEY, ABRAHAM, 1790-1857.
An Address, Delivered Before The New-York Horticultural Society, At The Anniversary, Sept. 8th, 1830.
New York: Published At The Society's Request, 1830.
22 p. Incl. as Appendix (pp. 15-22): Festival Of The New-York Horticultural Society. Report Of The Inspecting Committee. HSP

1269. HAMILTON CO. [OHIO]. AGRICULTURAL SOCIETY.
First Year's Proceedings Of The Hamilton Co. Agricultural Society, And Report Of The Annual Exhibition Since organized under a Law of the State, Held at Mount Pleasant, Eight Miles North Of Cincinnati, On The 19th And 20th Of September; With The Address Of Col. Wm. H.H. Taylor.
Cincinnati: Printed At The Office Of The "Plow Boy," 1844.
32, 2, [2] p.; illus. T.p. headed: Plow Boy--Extra, One Sheet. HSP

__________.
See also DRAKE, DANIEL, 1785-1852. A Discourse On Intemperance; Delivered At Cincinnati ... Before .... 1828.

1270. HAMMOND, JAMES H.
The Farmer's And Mechanic's Practical Architect; And Guide In Rural Economy.
Boston: John P. Jewett And Company; Cleveland, Ohio: Henry P.B. Jewett [Lithotyped By Cowles And Company; Cambridge: Allen & Farnum, Printers (t.p. verso)], 1858.
224 p.; front.; illus. LCP

1271. HAMMOND, JAMES HENRY, 1807-1864.
An Address Delivered Before The South-Carolina Institute, At Its First Annual Fair, On the 20th November, 1849.
Charleston, S.C.: From The Press Of Walker And James, 1849.
54, [2] p. LCP (wrappers)

1272. __________.
Marl. A Letter Addressed To The Agricultural Society Of Jefferson County, Georgia. (Published By Request Of The Society.)
Augusta: Printed By James McCafferty, 1846.
22 p. This work also incl. in: SOUTH CAROLINA. STATE AGRICULTURAL SOCIETY. The Proceedings .... 1846. LCP

1273. HAMPSHIRE CO. [MASS.]. AGRICULTURAL SOCIETY.
Transactions.
Amherst.
LCP: 1855.

HAMPSHIRE, HAMPDEN AND FRANKLIN [MASS.]. AGRICULTURAL SOCIETY.
See also
ALLEN, SAMUEL CLESSON, 1772-1842. An Address Delivered .... 1830.
BATES, ISAAC CHAPMAN, 1780-1845. An Address To ... 1823.
COLMAN, HENRY, 1785-1849. An Address Before .... 1833.
DICKINSON, SAMUEL F. An Address Delivered At .... 1831.
LAWRENCE, MYRON, 1799-1852. Address Delivered Before .... 1832.
NORTON, JOHN PITKIN, 1822-1852. Address Delivered Before .... 1848.
WEBSTER, NOAH, 1758-1843. An Address, Delivered Before 1818

1274. HAND, FARRELL.
A Treatise On The Culture Of The Tobacco Plants, And The Soils best adapted to it; With the Genuine Method of Curing it, as practised in Virginia. Extracted from a Manuscript on that Subject, wrote by a Man who superintended a large Plantation there for many Years, and now addressed to the Friends of Ireland in general. By Farrell Hand, Gardener. Together, With Remarks on it from his own Experience.
Dublin: Printed for P. Byrne, Bookseller and Stationer, 1782.
29 [i.e. 38] p. incl. fold. front. Pp. 37-8 misnumbered 28-29. LCP

1275. HANNAM, JOHN, 1792-1867.
The Economy Of Waste Manures, A Treatise On The Nature And Use Of Neglected Fertilizers. Written For The Yorkshire Agricultural Society, And Published By Permission Of The Council.
Philadelphia: Carey And Hart, 1844.
94, [2 (pub. adv.)] p. LCP

HANSON, H.C., ed.
See The Florist And Horticultural Journal .... [pdcl.].

1276. HANSON, JOSEPH B., Agent, Philadelphia.
To Farmers. Phosphatic Guano, From Sombrero Island, West Indies, The Richest Deposit Of Lime Known To The World, Containing above 80 per cent. of Bone Phosphate of Lime .... Imported And For Sale By Joseph B. Hanson, Sole Agent In Philad'a .... [wrapper title]
Philadelphia: Ducross, Pr., [1859?].
12 p. APS (wrappers)

1277. HARPER, LOUIS.
Preliminary Report On The Geology And Agriculture Of The State Of Mississippi. By Order Of The Legislature Of Mississippi.
Jackson: E. Barksdale, State Printer, 1857.
Pp. [1-4], [7]-8, [i]-vii, [1-3], [9]-350, [1]; col. fold. front. map; 4 col. maps; 3 sections, 1 fold. LCP; APS

HARPER, ROBERT GOODLOE, 1765-1825.
See NORTH AMERICAN LAND COMPANY, Philadelphia. Observations .... To Which Are Added, Remarks .... In Two Letters from .... 1796.

1278. HARRIS, THADDEUS WILLIAM, 1795-1856.
A Discourse Delivered Before The Massachusetts Horticultural Society, On The Celebration Of Its Fourth Anniversary, October 3, 1832.
Cambridge: E.W. Metcalf And Company, 1832.
96 p. Incl. (pp. 55-96) extracts from Society proceedings. LCP (wrappers); APS

1279. [__________].
A Report On The Insects Of Massachusetts, Injurious To Vegetation. Published Agreeably To An Order Of The Legislature, By The Commissioners On The Zoological And Botanical Survey Of The State.
Cambridge: Folsom, Wells, And Thurston, Printers To The University, 1841.
Pp. [1-2], [v]-viii, [1]-459. Author's name p. vi. LCP; APS

1280. __________.
A Treatise On Some Of The Insects Of New England, Which Are Injurious To Vegetation.
Cambridge: Published By John Owen [Printed At The University Press (t.p. verso)], 1842.
[4], 459 p. This work "... does not differ essentially from the Report ... submitted to the Legislature of Massachusetts ... and this impression has been printed at the expense of the Author for more general circulation ...." (Advertisement). LCP; APS; HSP

HARTFORD CO. [CONN.]. AGRICULTURAL SOCIETY.
See
COLMAN, HENRY, 1785-1849. Agricultural Addresses Delivered At .... 1840.
ELLSWORTH, HENRY LEAVITT, 1791-1858. Sketches Of An Address Delivered Before .... 1818.
NILES, JOHN MILTON, 1787-1856. Oration, Pronounced Before .... 1841.
STUART, ISAAC WILLIAM, 1809-1861. Address Before .... 1845.

HARTFORD CO. [CONN.]. SILK SOCIETY.
See The Silk Culturist, And Farmer's Manual. Published By .... [pdcl.].

1281. HASSEL, CHARLES.
The New Experienced Farrier, Containing One Hundred Cures And Receipts, Original And Selected, Known To Be The Most Approved And Best In The Country.
New Berlin, Pa.: S.G. Miller, Printer, 1843.
48 p. HSP

1282. A Hasty Defence Of The Farmers & Distillers Of York County [Pa.], Against the Aspersions of Temperance Societies, &c.
York: Printed by W.C. Smyth, [1833].
16 p. Incl. (pp. 9-16) the resolutions passed at a "Great Public Meeting In Opposition To Proscriptive Temperance Societies" held in Westmanchester Twp. on Nov. 12, 1833, with Henry Wolf in the chair; and a list of citizens present at the meeting. HSP

1283. HAVEN, CHARLES H.
Views On The Vine Growing Resources Of St. Louis And Adjacent Counties Of Missouri, And On The Important Bearing The Subject May Have On The Future Wealth And Commerce Of The State.
St. Louis: George Knapp And Co., Printers, 1858.
23 p. Incl. (pp. 17-23): Appendix. Proposition For The Formation Of A Vine And Fruit Growing Association .... [i.e. the St. Louis Vine & Fruit Growers' Association (q.v.)]. LCP

1284. HAZARD, THOMAS ROBINSON, 1797-1886.
Address Delivered Before The Aquidneck [R.I.] Agricultural Society, At Their Annual Exhibition, 1852.
Newport, R.I.: Published For The Society By Cranston & Norman, [1853].
24 p. LCP

1285. HAZZI, JOSEPH VON, 1768-1845.
Letter From James Mease, Transmitting A Treatise On The Rearing Of Silk-Worms, By Mr. De Hazzi, Of Munich, With Plates, &c. &c. February 2, 1828. Read, and referred to the Committee on Agriculture.
Washington: Printed By Gales & Seaton, 1828.
108 p.; col. fold. front.; fold. tbl.; illus. T.p. headed: 20th Congress, 1st Session. (Doc. No. 226.) Ho. Of Reps. Silk-Worms. Half t.p. p. 5: A Treatise On The Culture Of Silk In Germany, And Especially In Bavaria: Or Complete Instruction For the Plantation and the Management of Mulberry Trees .... Translated From The German. LCP; HSP

1286. __________.
Letter From James Mease, Transmitting A Treatise On The Rearing Of Silkworms; By Mr. De Hazzi, Of Munich. With Plates, &c. &c. April 21, 1828. Printed by order of the Senate of the United States.
Washington: Printed By Duff Green, 1828.
106 p.; col. fold. front.; 1 fold. tbl.; illus. T.p. headed: 20th Congress, 1st Session. (175-Part 2). Half t.p., p. 3, as above. LCP

1287. HEERINGEN, ERNST.
A Discovery Of The True Cause Of The Disease, Called By The People, Trembles, Or Milk-Sickness.
Louisville: Printed At The Office Of The Kentuckian, 1843.
48 p. The disease attacked cattle. HSP

1288. HEERMANCE, HENRY.
The Farmer's Mine Or Source Of Wealth, Being A Compilation, With The Addition Of New And Important Information On The Subject Of Manure, Together With The Most Approved Methods For The Manufacture Of Vegetable Manure .... To Which Is Added, Productive Farming, By Joseph A. Smith. Revised And Corrected By A.B. Allen ....
New York: Published By Henry Heermance, And For Sale By Saxton & Miles [S.W. Benedict & Co., Print. (t.p. verso)], 1843.
Pp. [1-10], [9]-327. LCP (lacking ½ t.p.); HSP

Hemp. [1799].
See BORDLEY, JOHN BEALE, 1727-1804.

HEPBURN, DAVID, jt. auth.
See GARDINER, JOHN, and __________.

1289. HERBERT, HENRY WILLIAM, 1807-1858.
Hints To Horse-Keepers, A Complete Manual For Horsemen; Embracing How To Breed A Horse. How To Buy A Horse. How To Break A Horse. How To Use A Horse. How To Feed A Horse .... And Chapters On Mules And Ponies. By The Late Henry William Herbert (Frank Forester). With Additions, Including "Rarey's Method Of Horse-Taming," And "Baucher's System Of Horsemanship" .... Beautifully Illustrated.
New York: A.O. Moore & Company [Edward O. Jenkins, Printer & Stereotyper (t.p. verso)], 1859.
LCP: 429 p.; 23 pls.; illus.; lacking front. (see BAL 8165)

1290. HIBBERT AND BUIST, Nurserymen, Philadelphia.
The American Flower Garden Directory, Containing Practical Directions For The Culture Of Plants In The Hot-House, Garden-House, Flower Garden, And Rooms Or Parlours, For Every Month In The Year. With A Description Of The Plants Most Desirable In Each .... Instructions For Erecting A Hot-House, Green-House .... With Lists Of Annuals, Biennials, And Ornamental Shrubs ....
Philadelphia: Printed For The Authors By Adam Waldie, 1832.
Pp. [i]-ix, [1-3], [9]-375, [1]; col. front. For subsequent edns. see: BUIST, ROBERT, 1805-1880. The American Flower Garden Directory .... 1839 etc. LCP; HSP

1291. [HICKEY, WILLIAM, 1787?-1875].
The Flower Garden, Or Monthly Calendar Of Practical Directions For The Culture Of Flowers. By Martin Doyle [pseud.] .... First American Edition. Adapted To The Climate Of The United States, With Notes And Observations By L.D. Gale, M.D. ....
New York: Moore & Payne [H. Ludwig, Printer (t.p. verso)], 1835.
180 p. LCP

1292. HIGGINS, JAMES.
Report Of James Higgins, M.D. State Agricultural Chemist, To The House Of Delegates Of Maryland.
[Annapolis: 1850].
92 p. LCP

1293. __________.
Report Of ... State Agricultural Chemist. To The House Of Delegates Of Maryland.
[Annapolis:] Wm. McNeir, Pr., [1850].
92 p. T.p. headed: (Twenty-five hundred Copies ordered by the Senate.) LCP

1294. __________.
Report Of ... State Agricultural Chemist, To The House Of Delegates Of Maryland.
Annapolis: Riley And Davis, Printers, 1850.
92 p.; col. fold. front. map. T.p. headed: (Document G.) By The House Of Delegates, January 22nd, 1850. Read, and three thousand copies ordered to be printed .... LCP (wrapper)

1295. __________.
The Second Report Of ... State Agricultural Chemist, To The House Of Delegates Of Maryland.
Annapolis: Printed By Thomas E. Martin, 1852.
126 p. LCP (wrapper)

1296. __________.
The Second Report Of ... State Agricultural Chemist, To The House Of Delegates Of Maryland.
Annapolis: Richardson, Cloud And Hope, Printers, 1852.
118 p. LCP

1297. __________.
The Third Report Of ... State Agricultural Chemist, To The House Of Delegates Of Maryland.
Baltimore: From The Press Of Sands & Mills, Office Oe [sic] The "American Farmer," 1853.
160 p. LCP (wrapper); HSP (wrapper)

1298. __________.
The Fourth Annual Report Of ... State Agricultural Chemist, To The House Of Delegates Of The State Of Maryland. By the House of Delegates, February 11th, 1854, read, and 10,000 extra copies ordered to be printed.
Baltimore: Printed By Sherwood & Co., 1854.
92 p. LCP (wrapper)

1299. __________.
Fifth Agricultural Report of ... State Chemist, To The House Of Delegates Of The State Of Maryland, Jan. 8th, 1856. By the House of Delegates, read and thirteen thousand copies ordered to be printed. Jan. 25th, 1856. By the Senate, read and three thousand copies ordered to be printed.
Annapolis: Requa And Wilson, Printers, 1856.
Pp. [1]-14, 19-90, [1-3] (no pp. numbered 15-18, as noted in Erratum at back). LCP (wrapper)

1300. __________.
The Sixth Agricultural Report Of ... State Chemist, To The House Of Delegates Of The State Of Maryland. Jan. 8th, 1858. By the House of Delegates, read and twenty thousand copies ordered to be printed. Jan. 27th, 1858. By the Senate, read and two thousand copies ordered to be printed.
Annapolis: Thomas J. Wilson, 1858.
96, xxii p. With sep.-paginated Appendix: An Essay On The Analysis Of Soils .... By Charles Bickell. LCP (wrapper)

1301. [__________ and CHARLES BICKELL].
Report. [Cap. title; headed:] Office Of State Chemistry ... Baltimore, Nov. 7, 1854. [Text begins:] The following Report on an examination and analysis of Sea Island Cotton Fibre and Seed, and the soil on which it grew ... were made for an intimate friend ... in order to recommend a manure best adapted to the growth of this national staple ....
[Baltimore: 1854].
LCP: Pp. [3]-11, 13-14. Proof sheets, printed on one side only of leaf. Lacking p. 12. Authors' names at end.

1302. HILGARD, EUGENE WOLDEMAR, 1833-1916.
Report On The Geological & Agricultural Survey Of The State Of Mississippi.
Jackson: Mississippi Steam Power Press Print, 1858.
22 p.; errata slip tipped in at back. LCP

1303. __________.
Report On The Geology And Agriculture Of The State Of Mississippi. Printed By Order Of The Legislature.
Jackson, Mississippi: E. Barksdale, State Printer, 1860.
xxiv, 391, [2] p.; fold. front. section; col. fold. map.; fold. section; fold tbl. LCP; APS (lacking fold. front. section) 1304. HILL, D & C. A., Auctioneers, Philadelphia.
Catalogue Of Splendid And Rare Green House And Hot House Plants, To Be Sold By Auction, At Lemon Hill, Formerly The Seat Of The Late Henry Pratt, Deceased, On Tuesday, The 5th Day Of June, 1838 ....
[Philadelphia: 1838].
59 p. LCP; HSP

1304. HILL, D & C. A., Auctioneers, Philadelphia.
Catalogue Of Splendid And Rare Green House And Hot House Plants, To Be Sold By Auction, At Lemon Hill, Formerly The Seat Of The Late Henry Pratt, Deceased, On Tuesday, The 5th Day Of June, 1838 ....
[Philadelphia: 1838].
59 p. LCP; HSP

1305. __________.
The Entire Collection Of Splendid And Rare Green House And Hot House Plants, At Lemon Hill, Formerly The Seat Of The Late Henry Pratt, Deceased, Will Be Sold By Auction ... Under the Superintendance of Peter Mackenzie, for many years Gardener in this Establishment ....
[Philadelphia: 1838].
11 p. HSP

1306. HILLARD, GEORGE STILLMAN, 1808-1879.
Address Before The Norfolk [Mass.] Agricultural Society, At Dedham, Sept. 26, 1860. [wrapper title]
[Boston?: 1860].
15 p. LCP (wrapper)

HILLSBOROUGH [N.H.]. AGRICULTURAL SOCIETY.
See
EASTMAN, MOSES. A Discourse Delivered Before .... 1822.
FOSTER, EDEN BURROUGHS, 1813-1882. An Address On Agriculture: Delivered Before .... [1851].

HINDS, JOHN, pseud.
See BADCOCK, JOHN.

1307. HINGHAM [MASS.]. AGRICULTURAL AND HORTICULTURAL SOCIETY.
Transactions.
Boston.
LCP: Oct. 1858-Mar. 1861 in 1 vol.

1308. [HIRZEL, HANS KASPAR, 1725-1803].
The Rural Socrates; Or An Account Of A Celebrated Philosophical Farmer, Lately Living In Switzerland, And Known By The Name Of Kliyogg.
Hallowell (District Of Maine): Printed by Peter Edes, 1800.
xii, [4], 203, xiii p. Evans 38923 lists under the American editor, Benjamin Vaughan. LCP; APS

__________. The Rural Socrates ....
See also YOUNG, ARTHUR, 1741-1820. Rural Oeconomy .... To Which Is Added The Rural Socrates .... 1776 etc.

1309. An Historical Review Of The Rise, Progress, Present State, And Prospects, Of The Silk Culture, Manufacture, And Trade, In Europe & America. Being An Article Extracted From The American Quarterly Review For December, 1831.
Philadelphia: Printed By Lydia R. Bailey, 1831.
34 p. Discusses Essays On American Silk .... 1830, by Jean d'Homergue and P.S. Du Ponceau (q.v.) LCP; APS

The History of Silk, Cotton, Linen .... 1845.
See GILROY, CLINTON G.

1310. HOARE, CLEMENT, 1789-1849.
A Practical Treatise On The Cultivation Of The Grape Vine On Open Walls. Second American Edition.
Boston: Published By William D. Ticknor [Marden & Kimball, Printers (t.p. verso)], 1840.
144 p.; illus. LCP

1311. __________.
A Practical Treatise On The Cultivation Of The Grape Vine .... Fourth American Edition.
Boston: William D. Ticknor & Company [Printed By John Putnam (t.p. verso)], 1848.
Pp. [1-4], [ix]-x, [11]-180. LCP

1312. Hoffy's Orchardist's Companion. A Quarterly Journal, Devoted To The History, Character, Properties, Modes Of Cultivation, And All Other Matters Appertaining To The Fruits Of The United States, Embellished With Richly Colored Designs ....
Philadelphia.
HSP in LCP: Vol. 1., nos. 1-4; vol. 2, no. 1 (April 1841-March 1842; 1842-3.) All pub. Wrappers to each number.
LCP: Vol. 1, nos. 1-4.
Both sets "Best [i.e. with colored pls.] Edition."

1313. [HOHMAN, JOHANN GEORG, d. ca. 1846].
Albertus Magnus, oder Der lange verborgene Schatz und Haus-Freund, und Getreuer und Christlicher Unterricht für Jedermann. Enthaltend wunderbare und erprobte Mittel und Künste für Gebrechen der Menschen und am Vieh. Aus den Arabischen Schriften des weisen Alchemisten Omar Arey, Emir Chemir Tschasmir, ins Deutsche übersetzt ....
Pennsylvania: Gedruckt für den Käufer, 1839.
100 p. HSP in LCP; LCP

1314. [__________].
Albertus Magnus, oder Der lange verborgene Schatz und Haus-Freund, und Getreuer und Christlicher Unterricht für Jedermann ....
Pennsylvania: Gedruckt für den Käufer, [ca. 1845].
95 p. HSP in LCP; LCP

1315. __________.
Die Land-und Haus-Apotheke, oder getreuer und gründlicher Unterricht für den Bauer und Stadtmann, Enthaltend die allerbesten Mittel, sowohl für die Menschen als für das Vieh besonders für die Pferde. Nebst einem grossen Anhang von der Aechten Färberey ... Erste americanische Auflage.
Reading: Gedruckt bey Carl A. Bruckman, 1818.
[12], 169, [11] p. HSP in LCP; LCP

1316. __________.
Der lange Verborgene Freund, oder: Getreuer und Christlicher Unterricht für jedermann, enthaltend: Wunderbare und probmässige Mittel und Künste, sowohl für die Menschen als das Vieh. Mit Vielen Zeugen bewiesen in diesem Buch ... und zum allerersten Mal in America in Druck erschient.
Reading [Pa.]: Gedruckt für den Verfasser, 1820.
100 p. HSP in LCP; LCP

1317. __________.
Der lange Verborgene Freund, oder Getreuer und Christlicher Unterricht .... nun auf Begehren zum zweytenmal gedruckt.
Ephrata [Pa.]: Gedruckt bey J.B., 1828.
94 p. HSP in LCP; LCP

1318. __________.
Der lange Verborgene Freund, oder Getreuer und Christlicher Unterricht .... Mit vielen Zeugen bewiesen in diesem Buch ... und zum allerersten Mal in Amerika im Druck erscheint.
Dritte durchaus verbesserte Auflage, Gedruckt in diesem Jahre. [Reading: Arnold Puwelle, ca. 1840].
72 p. LCP

1319. __________.
Der Lang Verborgene Freund, oder Getreuer und Christlicher Unterricht .... Mit vielen Zeugen bewiesen in diesem Buch ... und zum allererstenmal in America im Jahre 1820 im Druck erschienen .... nun auf Begehren zum drittenmal gedruckt.
Harrisburg, Pa.: 1840.
84 p. LCP; HSP

1320. __________.
Der Lang verborgene Freund, oder getreuer und christlicher Unterricht .... Zum erstenmal in Amerika im Jahr 1820 im Druck erschienen.
Harrisburg, Pa.: ... nun auf Begehren zum drittenmal gedruckt, 1843.
66, [6] p. LCP

1321. __________.
Der Lang Verborgene Freund, oder Getreuer und Christlicher Unterricht .... zum allererstenmal in Amerika im Jahr 1820 im Druck erschienen .... num auf Begehren zum fünftenmal gedruckt.
New Berlin, Pa.: 1843.
80 p. HSP in LCP; LCP

1322. __________.
Der Lang verborgene Freund. Oder Getreuer und Christlicher Unterricht ... zum allerersten Mal in Amerika im Jahr 1820 im Druck erschienen .... nun auf Begehren zum sechstenmal gedruckt.
Harrisburg, Pa.: 1846.
71 p. HSP in LCP; LCP

1323. __________.
Der Lang Verborgene Freund, enthaltend Wunderbare und Probmässige Mittel und Künste für Menschen und Vieh. Demselben ist beigefügt Dr. G[eorg] F[riedrich] Helfenstein's vielfältig erprobter Hauschatz der Sympthie.
Harrisburg, Pa.: Gedruckt bei Scheffer und Beck, 1853.
108 p. LCP

1324. __________.
Hohmann's Lang Verborgener Freund, enthaltend Wunderbare und Erprobte Heil-Mittel und Künste für Menschen und Vieh.
Harrisburg: Gedruckt bei Theo. F. Scheffer, [1855].
110 p. LCP

1325. [__________].
Der lange verborgene Schatz und Haus-Freund, oder Getreuer und Christlicher Unterricht .... Enthaltend Wunderbare und erprobte Mittel und Künste .... Aus dem arabischen Schriften ... zum Erstenmale in Amerika im Druck erscheint. Herausgegeben von J.H.......s.
Skippacksville, Pa.: Gedruckt bei A. Puwelle, 1837.
123 p. LCP

1326. [__________].
Der lange verborgene Schatz und Haus-Freund, oder getreuer und christlicher Unterricht .... Aus den arabischen Schriften ... ins Deutsche und mit noch vielen andern Kunsten vermehrt. Zweite vermehrte und verbesserte amerikan. Auflage. Herausgegeben von J. H.......s.
Pennsylvania: Gedruckt, 1847.
127 p. LCP

1327. __________.
The Long Lost Friend. A Collection Of Mysterious & Invaluable Arts & Remedies, For Man As Well As Animals. With Many Proofs Of their virtue and efficacy in healing diseases, &c., the greater part of which was never published until they appeared in print for the first time in the U.S. in the year 1820.
Harrisburg, Pa.: T.F. Scheffer, Printer, 1856.
72 p. LCP

1328. HOLCOMB, CHAUNCEY P.
An Address Delivered Before The Md. State Agricultural Society, At Its Annual Exhibition At Baltimore, October 28th, 1853.
Baltimore: From The Press Of Sands & Mills, Office of the "American Farmer," 1853.
24 p. LCP

1329. __________.
Address Delivered Before The Montgomery County, Maryland, Agricultural Society, At Its Annual Exhibition, At Rockville, September 14, 1854. Published By Request Of The Society.
Washington: Printed At The Globe Office, 1854.
14 p. HSP (wrappers)

1330. HOLLAND LAND CO.
230,000 Acres of Land, For Settlers, in the counties of M'Kean and Jefferson, Pennsylvania.
Philadelphia: Hall & Pierie, Printers, [1815?].
Brs. Headed: Interest the ascendant Principle. Signed at end: Benjamin B. Cooper, Attorney in Fact. LCP

1331. HOLLEY, MYRON, 1779-1841.
An Initiatory Discourse, Delivered At Geneva, 27th November, 1828, Before An Assembly, From Which, On That Day, Was Formed The Domestic Horticultural Society Of The Western Parts Of The State Of New-York.
Geneva: Printed By James Bogert, 1828.
27 p. LCP

1332. HOLMES, J.N.
The Annual Address, Before The Oswego County Agriculture Society, At Central Square, Sept. 28th, 1854.
Fulton [N.Y.]: Printed By T.S. Brigham, 1855.
32 p. Incl. (pp. 13-32): Horticulture. An Address ... By Rev. Edward Lord, At Central Square, Sept. 27, 1854. LCP (wrappers)

1333. HOMERGUE, JEAN D'.
The Silk Culturist's Manual: Or A Popular Treatise On The Planting And Cultivation Of Mulberry Trees, The Rearing And Propagation Of Silk Worms, And The Preparation Of The Raw Material For Exportation: Addressed To The Farmers And Planters Of The United States.
Philadelphia: Published And For Sale By Hogan & Thompson [J. Thompson, Printer (t.p. verso)], 1839.
xxxvi, 406 p.; fold. pl. APS

1334. __________, and PETER STEPHEN DU PONCEAU, 1760-1844.
Essays On American Silk, And The Best Means Of Rendering It A Source Of Individual And National Wealth. With Directions to Farmers for Raising Silk Worms.
Philadelphia: John Grigg [Printed by James Kay, Jun. &. Co. Printers to the American Philosophical Society (t.p. verso)], 1830.
xxii, 120, [1] p. LCP; APS

__________. Essays On American Silk ....
See also An Historical Review Of The Rise, Progress, Present Manufacture, And Prospects, Of The Silk Culture .... 1831. Discusses the above essay.

1335. __________.
Report Of The Committee On Agriculture On The Growth And Manufacture Of Silk, To Which Is Annexed, Essays On American Silk, With Directions To Farmers For Raising Silk Worms, May 24, 1830. Printed by order of the House of Representatives.
Washington: Printed By Duff Green, 1830.
88 p.; illus. T.p. headed: 21st Congress, 1st Session. (Doc. No. 126.) Ho. Of Reps. Incl. (pp. 5-16): Report Of The Committee On Agriculture ....; and (pp. 17-88): d'Homergue and Du Ponceau's Essays. LCP

1336. HOMESPUN, PRISCILLA, pseud.
The Universal Receipt Book; Being A Compendious Repository Of Practical Information In Cookery, Preserving, Pickling, Distilling, And All Branches Of Domestic Economy. To which is added, Some Advice To Farmers. Second edition with great additions.
Philadelphia: Published By Isaac Riley; J. Maxwell, printer, 1818.
378 p. LCP

1337. Honig und dessen Nutzen, u.s.w. [cap. title]
[N.p.: ca. 1830].
Brs. Incl. section cap.-titled: Vom Bienensuchen. LCP

1338. HOOPER, EDWARD JAMES, 1803-1882.
Hooper's Western Fruit Book: A Compendious Collection Of Facts, From The Notes And Experience Of Successful Fruit Culturists, Arranged For Practical Use In The Orchard And Garden.
Cincinnati: Moore, Wilstach, Keys & Co. [Stereotyped and Printed by Moore, Wilstach, Keys & Co. (t.p. verso)], 1857.
[8 (adv.)], 333, [3 (pub. adv.)] p.; front.; 4 col. pls. LCP

HOOPES, JOSIAH, Nurseryman.
HOOPES & BROTHER, Nurserymen.
See CHERRY HILL NURSERY, West Chester, Pa.

HOPKINS [CO., KY.] MASTODON COAL AND IRON MINING AND MANUFACTURING COMPANY.
See SAFFORD, JAMES MERRILL, 1822-1907, and RICHARD OWEN, 1810-1890. Report Upon The Mineral And Agricultural Resources .... 1857.

1339. HORRY, ELIAS, 1743-1834.
An Address Delivered in Charleston, Before The Agricultural Society Of South-Carolina, At Its Anniversary Meeting, On Tuesday, The 19th August, 1828. Published by Order of the Society.
Charleston: Printed By A.E. Miller, 1828.
40 p. HSP in LCP; APS

1340. __________.
The Horse: Its Habits, Diseases And Management, In The Stable, And On The Road; With Advice To Purchasers.
New York: Published By Wm. H. Graham, 1847.
126 p. Cover title: The Complete Horse Doctor; Containing The Habits, Diseases And Management Of The Horse .... LCP

1341. The Horticulturist And Journal Of Rural Art And Rural Taste .... Edited [later Established] By A.J. Downing.
Albany, etc.
LCP: Vols. 1-30 (July 1846-Dec. 1875). All pub.

1342. HOSACK, DAVID, 1769-1835.
An Inaugural Discourse, Delivered Before The New-York Horticultural Society At Their Anniversary Meeting, On The 31st Of August, 1824.
New York: Printed By J. Seymour, 1824.
46 p. LCP (wrappers); APS; HSP

1343. [HOSKYNS, CHANDOS WREN, 1812-1876.]
Talpa: Or The Chronicles Of A Clay Farm. An Agricultural Fragment. By C.W.H. With An Introduction And Notes By Lewis F. Allen. To Which Are Added Two Prize Essays On Tile Drainage.
Buffalo: Danforth, Hawley & Co. [Stereotyped by Erastus F. Beadle (t.p. verso)], 1854.
294, [2 (pub. adv.)] p.; 2 pls.; illus. LCP

1344. HOUGH, FRANKLIN BENJAMIN, 1822-1885.
The Comprehensive Farm Record With Directions For Its Use Arranged By Franklin B. Hough ....
New York: C.M. Saxton, Barker & Co. [Albany: Munsell & Rowland, Stereotypers And Printers (t.p. verso)], 1860.
Pp. [1-2], 1-16, 19-20, [1-4], 23-147, [1], [1-10], [iii]-v. Chiefly blank ledger forms. LCP

1345. HOVEY, CHARLES MASON, 1810-1887.
The Fruits Of America, Containing Richly Colored Figures, And Full Descriptions of all the choicest Varieties cultivated In The United States. Volume I.
Boston: C.C. Little & Jas. Brown, And Hovey & Co.; New York: D. Appleton & Co. [Dutton And Wentworth, Printers; c1851 (t.p. verso)].
xii, 100 p.; front.; 48 col. pls.; illus. PHS in APS: Vol. I only (of 3, with 3rd vol. not completed). Nissen 941, LC, and Pritzel 4292 all have 1847 for this vol., with following vols. 1852-6 and [1858?]; this copy, however, has copyright date 1851. See note to Sabin 33206.

1346. __________.
The Fruits Of America Containing Richly Colored Figures, And Full Descriptions Of All The Choicest Varieties Cultivated In The United States. Forty-Eight Richly Colored Plates.
New York: D. Appleton & Company, 1853.
viii, 100 p.; front; 48 col. pls.; illus. LCP

1347. HOWARD, SANFORD.
To The Breeders Of Devon Cattle. [Text begins:] The subscriber is now collecting materials for the Third Volume of the Devon Herd-Book ....
[Boston: 1856].
Brs. Dated at end: Office of the Boston Cultivator, Boston, Mass., June 1, 1856. Refers to: DAVY, JOHN TANNER. Davy's Devon Herd Book .... American Edition, Edited By Sanford Howard .... (q.v.). LCP

1348. HOWES, FREDERICK.
An Address To The Essex Agricultural Society, In Massachusetts, At Topsfield, Oct. 6, 1823.
Salem: Printed For The Society By W. & S.B. Ives, 1824.
23 p. LCP

1349. HOWITT, EMANUEL.
Selections From Letters Written During A Tour Through The United States, In the Summer and Autumn of 1819; Illustrative Of The Character Of The Native Indians, And Of Their Descent From The Lost Ten Tribes Of Israel; As well as descriptive of The Present Situation And Suffering Of Emigrants, And Of The Soil And State Of Agriculture.
Nottingham: Printed and sold by J. Dunn; London: Sold Also By Baldwin, Cradock & Joy, and Darton, Harvey & Darton; Derby: H. Mozley, Richardson & Handford; Mansfield: Collinson and Langley, and all other Booksellers, [1820].
xxi, [1], 230 p. HSP in LCP; LCP; HSP

1350. HOWITT, WILLIAM, 1792-1879.
The Country Year-Book; Or, The Field, The Forest, And The Fireside.
New York: Harper & Brothers, Publishers, 1850.
Pp. [1-2], [5]-423, [1], 1-4 (pub. adv.). LCP

1351. HUBBARD, M.G., Patentee.
Hubbard's Patented Wrought Iron Reapers And Mowers! Single And Combined.
[Penn Yan, N.Y.: 1859].
Brs. Illus. Dated: Penn Yan, January 1st, 1859. LCP

HUFFINGTON, WILLIAM, ed.
See The Delaware Register .... [pdcl.].

HUGHES, GEORGE WURTZ, 1806-1870.
See PRINCE GEORGE'S CO. [MD.]. AGRICULTURAL SOCIETY. Proceedings Of The .... Also The Address By .... 1851.

1352. HUGHES, WILLIAM CARTER.
The American Miller, And Millwright's Assistant. A New Edition. Revised, With Much Additional Matter.
Philadelphia: Henry Carey Baird, (Successor To E.L. Carey); London: Trübner & Co. [Printed by T.K. & P.G. Collins], 1856.
292, 24 (pub. adv.) p.; front.; 11 pls., 2 fold.; illus. LCP: with MS. note by Henry Carey Baird on t.p.: "You will find one or two plates to differ. Send me down a perfect copy. H.C.B."; and on pl. 3: "(This is our new Ed) H.C.B."

1353. HUGHLETT, WILLIAM.
Being desirous of driving my Mills to the extent, it requires a great deal of Wheat, Corn, and Rye, to keep them in operation: I therefore solicit for the refusal of yours ....
[Wilmington: 1796].
[1] p. Headed: Delaware State Mills. LCP (with MS. note to John Dickinson from Wm. Hughlett, dated Feb. 13, 1796, and Dickinson's answer, Feb. 16)

1354. HUMPHREYS, DAVID, 1752-1818.
A Discourse On The Agriculture Of The State Of Connecticut, And The Means Of Making It More Beneficial To The State: Delivered [before the Society for Promoting Agriculture in the State of Connecticut] At New-Haven, On Thursday, 12th September, 1816.
New Haven: Printed by T.G. Woodward, 1816.
42 (i.e. 44) p. Pp. 42-43 unnumbered and p. 44 misnumbered 42. LCP

HUNTERDON CO. [N.J.]. AGRICULTURAL SOCIETY.
See WALL, JAMES WALTER, 1820-1872. An Address, Delivered Before The Agricultural Fair .... 1860.
__________. Speech ... Before The Agricultural Fair .... 1860.

1355. HYATT, JAMES, d. 1904.
Lime And Marl: Their Agricultural Uses. With Explanations Of Their Properties And Management, The Soils To Which They Are Applicable, And The Precautions To Be Observed In Their Use. Especially Adapted To The Wants Of Practical Farmers.
[Philadelphia:] Printed For The [Mount Airy Agricultural] Institute, And For sale by J.W. Moore; New York: Clark & Austin; and at Bookstores and Agricultural Warehouses generally [T.K. And P.G. Collins, Printers (t.p. verso)], 1848.
32 p. incl. front. LCP (wrappers)

1356. HYDE, JAMES F.C.
The Chinese Sugar-Cane; Its History, Mode of Culture, Manufacture Of The Sugar, Etc. With Reports Of Its Success In Different Portions Of The United States, And Letters From Distinguished Men.
Boston: Published By John P. Jewett & Company; Cleveland, Ohio: Henry P.B. Jewett [Stereotyped By Hobart & Robbins, New England Type and Stereotype Foundery (t.p. verso)], 1857.
106 p.; front. LCP

1357. ILLINOIS CENTRAL RAIL ROAD COMPANY.
The Illinois Central Rail Road Company Offer For Sale Over 2,500,000 Acres Selected Prairie, Farm And Wood Lands, In Tracts Of Any Size, To Suit Purchasers, On Long Credits, And At Low Rates Of Interest, Situated On Each Side Of Their Rail Road, Extending All The Way From The Extreme North To The South Of The State Of Illinois.
Chicago: Scott & Fulton, Steam Book And Job Printers, 1855.
32 p.; fold. front. map. The lands lay between Dunleith and Cairo, and between Chicago and Centralia. LCP

1358. __________.
The Illinois Central Railroad Company Offers For Sale Over 1,500,000 Acres Selected Farming And Wood Lands, In Tracts Of Forty Acres And Upwards, To Suit Purchasers, On Long Credits And At Low Rates Of Interest, Situated On Each Side Of Their Railroad ....
Chicago: Illinois Central Rail Road Office [Chicago Daily Press, Print. (wrapper)], 1857.
80 p. incl. front.; 2 maps, 1 fold.; illus. LCP (wrappers)

1359. ILLINOIS. GENERAL ASSEMBLY. HOUSE. COMMITTEE ON AGRICULTURE AND MANUFACTURES.
Report From The Committee ... In Reply To A Resolution Of Inquiry, On Agricultural Statistics. [cap. title]
[Springfield: 1841].
3 p. Title headed: Illinois Legis. H.R. 12th Assem. 2d Session. February 29, 1841. Read and laid on the table. The resolution inquired into "the expediency of adopting some system of collecting the agricultural statistics of the State ...." LCP

1360. Illinois In 1837; A Sketch Descriptive Of The Situation, Boundaries, Face Of The Country, Prominent Districts, Prairies, Rivers, Minerals, Animals, Agricultural Productions, Public Lands ... Also, Suggestions To Emigrants ... Together With A Letter On The Cultivation Of The Prairies, By The Hon. H.L. Ellsworth ....
Philadelphia: Published By S. Augustus Mitchell, And By Grigg & Elliot [Stereotyped By J. Fagan (t.p. verso)], 1837.
143, [1 (adv.)] p.; fold. map. Sabin 34260 gives the publisher as probable author. APS

1361. ILLINOIS LAND COMPANY, Philadelphia.
The Illinois Land Company, Offer For Sale 30,000 Acres, Rich Selected Prairie Farm Lands, In Tracts Of Any Size To Suit Purchasers, On Long Credits, And At Low Prices; Four Important Railroads, Running Through And Around The Lands ....
[Philadelphia:] 1855.
26 p. Wrapper imprint: State Of Illinois: 1855. The lands lay in Champaign, Piatt, and Macon Cos. LCP (wrappers)

1362. ILLINOIS. STATE AGRICULTURAL SOCIETY.
Illinois State Fair, 1859, At Freeport, Stephenson County. Regulations And Premiums For The Seventh Annual Fair Of The Illinois State Agricultural Society, To Be Held At Freeport, On Monday, Tuesday, Wednesday, Thursday And Friday, Sept. 5th, 6th, 7th, 8th and 9th, 1859. $16,000 Offered In Premiums.
Springfield: Bailhache & Baker, Printers, 1859.
29, [3] p. LCP

1363. __________.
Transactions. With The Proceedings Of The County Societies, And Kindred Associations.
Springfield, Ill.
LCP: Vol. 1 (1853-54); vols. 4-7 (1859-68). Subtitle varies slightly.

__________.
See also OVERMAN, C.R. Premium Essay .... 1855.

The Importance Of The British Plantations In America .... 1731.
See HALL, FAYRER.

INDIANA. STATE AGRICULTURAL SOCIETY.
See GREELEY, HORACE, 1811-1872. What The Sister Arts Teach .... An Address Before .... 1853.

1364. INDIANA. STATE BOARD OF AGRICULTURE.
First Annual Report Of The Indiana State Board Of Agriculture: For The Year 1852.
Indianapolis: J.P. Chapman, State Printer, 1852.
322 p. LCP

1365. INGERSOLL, JOSEPH REED, 1786-1868.
Address Delivered Before The Philadelphia Society For Promoting Agriculture, At Their Annual Exhibition, Held at the Lamb Tavern, on the Lancaster Turnpike, Oct. 5, 1849.
Philadelphia: John C. Clark, Printer, 1849.
32 p. LCP (wrappers); HSP (wrappers)

Intending to retire from the bustle of a town life .... [1797; 2 edns.].
See BORDLEY, JOHN BEALE, 1727-1804.

Intimations; On Agriculture And Manufactures .... [1793?].
Intimations; On Manufactures; -- on the Fruits of Agriculture .... [1794].
See BORDLEY, JOHN BEALE, 1727-1804.

1366. IOWA. STATE AGRICULTURAL SOCIETY.
Fourth Annual Report Of The Iowa State Agricultural Society, To The General Assembly, For the Year 1857.
Des Moines: Printed By J. Teesdale, State Printer, Citizen Office, 1858.
vii, [1], 458 p. LCP

1367. __________.
Sixth Annual Report Of The ... To The General Assembly, For The Year 1859.
Des Moines, Iowa: John Teesdale, State Printer, 1860.
480 p. LCP

1368. ISMAR, F.A.
Emanuel Fellenberg's Institution, At Hofwyl, In Switzerland. Two Lectures, Delivered in Georgetown, D.C.
Georgetown, D.C.: Printed At The Columbian Gazette Office, 1831.
52 p. The 1st lecture describes Fellenberg's farm labor educational system; the 2nd proposes the establishment of such an institution in Georgetown. APS

__________.
See also FELLENBERG INSTITUTION, Bolton Farm, Bucks Co., Pa. Fellenberg Institution .... [1831]. Prospectus with Ismar as co-signer.

IVES, JOHN M., ed.
See [MANNING, ROBERT, 1784-1842]. New England Book Of Fruit .... Third Edition .... By John M. Ives.

1369. JACKSON, CHARLES THOMAS, 1805-1880.
An Address Delivered Before The Plymouth County [Mass.] Agricultural Society, At Their Annual Exhibition, In The Town Of Bridgewater, Sept. 25, 1850.
Boston: Printed By Charles C.P. Moody, 1850.
29 p. LCP

1370. __________.
Report On The Geological And Agricultural Survey Of The State Of Rhode- Island, Made Under A Resolve Of Legislature In The Year 1839.
Providence: B. Cranston & Co., 1840.
312 p.; col. fold. map; col. fold. section; illus. LCP; APS

__________.
See also RIDGWAY FARM AND COAL CO., Philadelphia. Report .... Together With The Geological Report Of .... 1856.

JAMES I, King of Gt. Britain, 1566-1625.
See BONOEIL, JOHN. His Maiesties Graciovs Letter .... 1622.

1371. JAMES, CHARLES TILLINGHAST, 1804-1862.
Letters On The Culture And Manufacture Of Cotton: Addressed to Freeman Hunt, Esq., Editor of Hunt's Merchants' Magazine. And Published In The Numbers Of That Journal For February And March, 1850, In Reply To The Communications Of A.A. Lawrence, Esq., Originally published in the Merchants' Magazine for Dec. 1849 and January, 1850.
New York: Printed By Geo. W. Wood, 1850.
35 p. LCP; HSP (wrappers)

1372. JAQUES, GEORGE, 1816-1872.
A Practical Treatise On The Management Of Fruit Trees; With Descriptive Lists Of The Most Valuable Fruits For General Cultivation; Adapted To The Interior Of New England.
Worcester: Erastus N. Tucker [Benj. J. Dodge, Printer (t.p. verso)], 1849.
256 p. LCP

1373. JAY, JOHN, 1817-1894.
A Statistical View of American Agriculture, Its Home Resources and Foreign Markets, With Suggestions for The Schedules of The Federal Census In 1860. An Address Delivered at New York, before the American Geographical and Statistical Society, on the Organization of the Agricultural Section.
New York: D. Appleton & Company; London: Trubner And Company, 1859.
81 p. LCP (wrapper); HSP (wrappers)

JEFFERIES, GEORGE, jt. auth.
See MARKHAM, GERVASE, 1568?-1637.

JEFFERSON CO. [GA.]. AGRICULTURAL SOCIETY.
See HAMMOND, JAMES HENRY, 1807-1864. Marl. A Letter Addressed To .... 1846.

1374. JEFFERSON CO. [N.Y.]. AGRICULTURAL SOCIETY.
Proceedings Of The Second Cattle Show And Fair, Of The Agricultural Society Of Jefferson County; September, 1819.
Watertown: Printed For The Society; Seth Alden Abbey, Printer, [1819].
32 p. HSP

______.
See also
BURR, RICHARD. Authentic Documents ... Being Extracts From Addresses, Reports, &c. Made To The .... 1820.
KIRBY, EDMUND. Address Delivered Before .... 1831.
LE RAY DE CHAUMONT, JAMES DONATIEN, 1760-1840. Addresses .... 1818 etc.
LE RAY DE CHAUMONT, VINCENT. Address Delivered Before .... 1830.
MAULE, JONATHAN, et al. [No title; begins:] Esteemed Friend .... 1820.

1375. [JEFFERSON, THOMAS, 1743-1826].
Observations On The Whale-Fishery. [cap. title]
[New York: 1788].
18 p. Attributed to Jefferson by Tompkins 120. LCP; APS

1376. ______.
A Supplementary note on the mould board described in a letter to Sir John Sinclair, of March 23, 1798, inserted in the American Philosophical transactions, vol. 4, and in Maese's [sic] Domestic Encyclopaedia voce Plough. [cap. title]
[Philadelphia: ca. 1805].
[1] p.; illus. APS

______.
See also
UNITED STATES. SECRETARY OF STATE. Report ... On ... The Cod And Whale Fisheries .... 1791 etc. Jefferson was the Secretary.
WASHINGTON, GEORGE, 1732-1799. Letters On Agriculture .... With Statistical Tables And Remarks, By Thomas Jefferson .... 1847.

1377. JENIFER, DANIEL, 1791-1855.
Remarks Of The Hon. Daniel Jenifer, At The Opening Of The Convention Of Tobacco Planters Of The United States, Held At Washington City, December 15, 1840.
Washington: National Intelligencer Office, 1840.
16 p. LCP

1378. JENKINS, JOHN STILWELL, 1818-1852.
The Life Of Silas Wright, Late Governor Of The State Of New York. With An Appendix, Containing A Selection From His Speeches In The Senate Of The United States, And His Address Read Before The New York State Agricultural Society.
Auburn and Buffalo: John E. Beardsley [Thomas B. Smith, Stereotyper (t.p. verso)], [1847].
378, [2 (pub. adv.)] p. The agricultural address is on pp. 360-78. LCP

1379. JENKS, FRANCIS H., Produce commission merchant, Baltimore.
Circular To The Merchants And Farmers Of The Western Reserve, Ohio, From Francis H. Jenks, (Late Weld & Jenks,) Produce Commission Merchant, No. 8 Bowly's Wharf, Baltimore, Maryland, June, 1851.
[Baltimore: 1851].
52 p. On cheese production. LCP

1380. JEWETT, PAUL.
The New-England Farrier; Or, A Compendium Of Farriery, In Four Parts: Wherein most of the Diseases to which Horses, Neat Cattle, Sheep and Swine are incident, are treated of ... first production of the kind in New-England. Intended For The Use Of Private Gentlemen and Farmers.
Newburyport [Mass.]: Printed By William Barrett, 1795.
46, [2] p. LCP

1381. ______.
The New-England Farrier; Being A Compendium Of Farriery. In Four Parts .... The Second Edition.
Salem: Printed By Joshua Cushing; Providence: Sold By Henry Cushing; Salem: By Cushing & Appleton; Bath: And By Charles Metcalf, 1807.
45, [3] p. LCP

1382. ______.
The New-England Farrier; Or A Compendium Of Farriery, In Four Parts .... To Which Is Added An Appendix, Never before published, comprising many valuable recipes .... Second Edition Enlarged.
Exeter [N.H.]: Printed By John J. Williams, 1822.
105, [3] p. LCP

1383. [______].
The New England Farrier; In Four Parts .... To which is added, An Appendix .... Third Edition.
Woodstock [Vt.]: Printed By David Watson, 1824.
Pp. [1]-84, 73-106. LCP

1384. ______.
The New-England Farrier; Or A Compendium Of Farriery, In Four Parts .... New Edition, With Valuable Additions.
Exeter [N.H.]: Published By Josiah Richardson, 1826.
112 p. LCP

______. New-England Farrier ....
See also RICHARDSON, JOSIAH, comp. The New-England Farrier, And Family Physician; Containing, Firstly .... 1828.

1385. JOHNSON, BENJAMIN PIERCE, 1793-1869.
An Address Before The Cayuga County [N.Y.] Agricultural Society, Delivered September, 1847. Published By The Society.
Auburn, N.Y.: Merrell & How, Printers, 1848.
22 p. LCP

1386. JOHNSON, CUTHBERT WILLIAM, 1799-1878.
The Farmer's And Planter's Encyclopaedia Of Rural Affairs; Embracing All The Most Recent Discoveries In Agricultural Chemistry. Adapted To The Comprehension Of Unscientific Readers. Illustrated By Numerous Engravings Of Animals, Implements, And Other Subjects Interesting To The Agriculturist. Adapted to the United States, By Gouverneur Emerson.
New York: C.M. Saxton, Agricultural Book Publisher; Philadelphia: Lippincott, Grambo & Co.; Boston: B.B. Mussey & Co.; Cincinnati: H.W. Derby & Co. [Philadelphia: Stereotyped By L. Johnson (colop.)], 1855.
vii, [1], 1179 p.; col. front.; 17 pls.; illus. LCP

1387. ________.
The Farmer's Encyclopaedia, And Dictionary Of Rural Affairs; Embracing All The Most Recent Discoveries In Agricultural Chemistry. Adapted To The Comprehension Of Unscientific Readers .... Adapted to the United States.
Philadelphia: Carey And Hart [Stereotyped By L. Johnson; Printed By T.K. & P.G. Collins (t.p. verso)], 1844.
vii, [1], 1165 p.; front.; 16 pls.; illus. LCP

1388. JOHNSON, GEORGE WILLIAM, 1802-1886.
A Dictionary Of Modern Gardening. With One Hundred And Eighty Wood Cuts. Edited, With Numerous Additions, By David Landreth, Of Philadelphia.
Philadelphia: Lea And Blanchard [T.K. And P.G. Collins, Printers (t.p. verso)], 1847.
Pp. [1-4], [17]-635, [1], [1-32 (pub. adv.)]; illus. LCP

1389. JOHNSON, LOUISA.
Every Lady Her Own Flower Gardener. Addressed To The Industrious And Economical. Containing Simple And Practical Directions For Cultivating Plants And Flowers In The Garden And In Rooms. Revised from the Fourteenth London Edition, and Adapted to the Use Of American Ladies.
New York: C.M. Saxton, Agricultural Book Publisher [S.W. Benedict, Stereotyper And Printer (t.p. verso)], 1852.
119 p.; illus. LCP

1390. ________.
Every Lady Her Own Flower Gardener .... Revised from the Fourteenth London Edition ....
New York: C.M. Saxton, Barker & Co.; San Francisco: H.H. Bancroft & Co., 1860.
119 p.; illlus. LCP

1391. JOHNSON, SAMUEL WILLIAM, 1830-1909.
Essays On Peat, Muck, And Commercial Manures, By Samuel W. Johnson, Chemist To The Connecticut State Agricultural Society And Professor Of Analytical And Agricultural Chemistry In Yale College.
Hartford: Published By Brown & Gross [Press Of Williams & Wiley (t.p. verso)], 1859.
Pp. [1]-4, [1-2], i-ii, [7]-58, [1-2], i-ii, [61]-178. Incl. 2 series, Essays on Manures 1857 and Essays on Manures 1858, each with sep. ½ t.p. LCP

1392. ________.
First Annual Report Of Prof. S.W. Johnson, Chemist To The Connecticut State Agricultural Society, And Professor Of Analytical & Agricultural Chemistry in Yale College.
Hartford: Press Of Case, Lockwood And Company, 1858.
58 p. LCP

1393. JOHNSON, STEPHEN WILLIAM.
Rural Economy: Containing A Treatise On Pisé Building; As recommended by the Board of Agriculture in Great Britain, with Improvements by the Author; On Buildings in general; Particularly on the Arrangement of those belonging to Farms: On the Culture of the Vine; and On Turnpike Roads. With Plates.
New Brunswick, N.J.: Printed by William Elliot, For I. Riley & Co., New York, 1806.
Pp. [1-8], [vii]-viii, [1]-246, [1-5]; 8 pls. LCP; APS

1394. ________.
Utunque Quadratus. [Text begins:] Buildings whose walls consist of nothing but common soil were erected by the antients .... Proposals For Publishing By Subscription, Rural Practices; Or, A Treatise On the Mode of erecting Dwelling-houses ... in Pisé .... By S.W. Johnson ....
[New Brunswick: 1806].
Brs. Illus. APS

1395. JOHNSTON, JAMES FINLEY WEIR, 1796-1855.
Catechism Of Agricultural Chemistry And Geology. With An Introduction, By John Pitkin Norton, Of Farmington, Connecticut. From The Eighth English Edition, With Notes and Additions by the Author, prepared expressly for this Edition.
Albany: Erastus H. Pease [Stereotyped And Printed By C. Van Benthuysen And Co. (colop.)], 1845.
3 (pub. adv.), [1], 74 p.; illus. LCP

1396. ________.
Elements Of Agricultural Chemistry And Geology.
New York: Wiley And Putnam [J.P. Wright, Printer (t.p. verso)], 1842.
Pp. [1-4], [7]-249. LCP

1397. ________.
Lectures On The Applications Of Chemistry And Geology To Agriculture. New Edition, With An Appendix, Containing Suggestions For Experiments In Practical Agriculture.
New York: C.M. Saxton, Agricultural Book Publisher, 1850.
Pp. [1-2], [v]-vi, [1]-169, [1-3], [177]-298, [1-2], [303]-481, [1-3], [487]-619, [1-3], [1]-89, [1-13 (pub. adv.)]; front.; illus. 4 parts + Appendix with sep. ½ t.pp. LCP

1398. ________.
Lectures On The Applications Of Chemistry And Geology To Agriculture. New Edition, With An Appendix ....
New York: A.O. Moore, & Co., Agricultural Book Publishers, 1859.
Pp. [1-2], [v]-vi, [1]-169, [1-3], [177]-298, [1-2], [303]-481, [1-3], [487]-619, [1-3], [1]-89, [1], 1-12 (pub. adv.); illus. 4 parts + Appendix with sep. ½ t.pp. LCP

1399. _________.
Notes On North America Agricultural, Economical, And Social. In Two Volumes. Vol. I [-II].
Boston: Charles C. Little And James Brown; Edinburgh And London: William Blackwood And Sons [Edinburgh: Printed By William Blackwood And Sons (colop.)], 1851.
2 vols. LCP: Vol. I: xvi, 415 p.; fold. front. map. II: xii, 512 p. Without ½ t.p. to Vol. I, present in Edinburgh edn. (below)

1400. _________.
Notes On North America Agricultural, Economical, And Social. In Two Volumes. Vol. I [-II].
Edinburgh And London: William Blackwood And Sons [Edinburgh: Printed By William Blackwood And Sons (colop.)], 1851.
2 Vols. Vol. I: pp. [1-6], [v]-xvi, [1]-415; fold. front. map. II: xii, 512 p. LCP

1401. JONES, E., comp.
Review Of Facts And Observations, Made By Naturalists, Botanists, Historians, And Travellers, On The Properties and Productions Of The Sugar Maple Tree, The Process Of Extracting Its Sap, Converting It Into Sugar, &c. &c.
London: Printed By E. Colyer, 1832.
72 p. Half t.p.: The Acer Saccharinum Of Linnaeus. Or, The Sugar Maple Tree. Incl. extracts from Rush, Imlay, Michaux, etc. LCP

1402. JONES, J. RICHTER
Address Delivered Before The Agricultural Society of Susquehanna County [Pa.], October 13, 1853. Published By Order Of The Society.
Harrisburg: Printed At The Office Of The "Keystone," 1854.
15 p. APS

1403. JONES, JOHN O.
Jones' Patent "Improved Elastic Horse-Shoe," Important To Horse-Owners and Horse-Shoers.
[Philadelphia?: 1856].
LCP: pp. [1]-10, 13-28; lacking pp. 11-12. Wrappers.

1404. JONES, THOMAS W., and SIMON JAMES.
Encouragement To Farmers. [Text begins:] Several Citizens of Philadelphia and other parts of Pennsylvania, possessing Land on Conemaugh, Blacklick, Chest, Clearfield, and Paint Creeks ... offer it to settlers on the following terms ....
[Philadelphia: 1801].
Brs. Last paragraph begins: Application may be made to Thomas W. Jones or Simon James, at Beula ....; dated at end: Philadelphia, June 20th, 1801. HSP

1405. JUDD, THOMAS.
An Essay On The Improvement In The Manufacture Of Sugar, Adapted For Louisiana.
Boston: Tuttle, Weeks And Dennett, Printers, 1836.
36 p. APS

1406. [JULIEN, STANISLAS, 1797-1873].
Summary Of The Principal Chinese Treatises Upon The Culture Of The Mulberry And The Rearing Of Silk Worms. Translated From The Chinese.
Washington: Published By Peter Force [Peter Force, Printer (t.p. verso)], 1838.
[4], 198 p.; 10 pls. APS

JUNKIN, GEORGE, 1790-1868, ed.
See The Religious Farmer [pdcl.].

KALAMAZOO CO. [MICH.]. AGRICULTURAL SOCIETY.
See CASS, LEWIS, 1782-1866. Address Of ... Delivered Before .... [1850].

1407. KEAGY, JOHN MILLER, 1792-1837.
A Prospectus Of Woodside Institute, For Practical Education, And Agriculture; Including some of the prominent points of Fellemberg's [sic] Institution in Switzerland.
Harrisburg: Printed By John S. Wiestling, 1828.
8 p.; fold. front. pl. The Institute was 6 miles outside of Philadelphia. LCP

1408. KECHT, J.S.
Der verbesserte praktische Weinbau in Gärten und vorzüglich auf Weinbergen. Mit einer Anweisung den Wein ohne Presse zu keltern. Den amerikanische Weinbauern gewidmet von Heinrich B. Sage. Mit 2 Kupfertafeln.
Reading: gedruckt bey G. Adolph Sage, 1828.
viii, 84 p.; fold. front.; 1 fold. pl. HSP in LCP; LCP

1409. KEIM, GEORGE MAY
Address Before The Schuylkill County [Pa.] Agricultural Society. Delivered October 19, 1854.
Pottsville: Printed For The Society, 1854.
26 p. HSP (wrappers)

1410. KELLEY, E.G.
An Address on Home And Its Embellishments, Delivered Before The Essex Agricultural Society, Of Massachusetts, Oct. 1, 1857.
Newburyport: Herald Press--Wm. H. Huse & Co., Printers, 1857.
45 p. LCP (wrappers)

1411. KELSEY, WILLIAM R.
The Apiarian's Guide, Being A Practical Treatise On The Culture And Management Of Bees. (Fourth Edition--Five Thousand Copies.)
Philadelphia: Printed By William F. Geddes, 1849.
32 p. HSP

1412. KEMP, EDWARD, 1817-1891.
How To Lay Out A Garden: Intended As A General Guide In Choosing, Forming, Or Improving An Estate, (From a Quarter of an Acre to a Hundred Acres in extent,) With Reference To Both Design And Execution. From The Second London Edition, Greatly Enlarged, And Illustrated With Numerous Plans, Sections, And Sketches Of Gardens And Garden Objects.
New York: Wiley & Halsted [R. Craighead, Printer, Stereotyper, and Electrotyper (t.p. verso)], 1858.
xxxi, [1], 403, [1], 4 (pub. adv.) p.; illus. LCP

1413. KENNEDY, JOHN PENDLETON, 1795-1870.
An Address Delivered Before The American Institute, At Chatham Street Chapel, In The City Of New-York, October 17th, 1833. Together with a Detailed Account of the Articles exhibited at The Sixth Annual Fair.
New York: D.K. Minor, 1833.
24 p.; illus. APS; HSP (wrapper)

1414. __________.
Address Delivered Before The Horticultural Society Of Maryland, At Its First Annual Exhibition, June 12, 1833.
Baltimore: John D. Toy, Printer, 1833.
33 p. APS

1415. KENNICOTT, JOHN A. & CHARLES, Nurserymen, W. Northfield, Ill.
Descriptive Catalogue Of Fruit And Ornamental Trees, Shrubs and Plants, Cultivated And For Sale At the Grove Nursery & Garden, West Northfield, Cook Co. Ill.
Chicago: Democrat Print, 1856.
24 p. T.p. headed: 1856-'57. LCP

1416. KENRICK, WILLIAM, 1789-1872.
The American Silk Grower's Guide; Or The Art Of Raising The Mulberry And Silk On The System Of Successive Crops In Each Season.
Boston: George C. Barrett And Russell, Odiorne & Co. [Tuttle And Weeks, Printers (t.p. verso)], 1835.
111 p.; last 7 p. adv. LCP

1417. __________.
The American Silk Grower's Guide .... Second Edition, Enlarged And Improved.
Boston: Weeks, Jordon & Co. [D.H. Ela, Printer (t.p. verso)], 1839.
167, [1] p.; illus. LCP; APS

1418. __________.
Das Americanischen Seidebauers Anweisung, oder die Kunst den Maulbeer-Baum und die Seide zu ziehen. Nach dem System von Mehreren aufeinander folgenden Ernten während einer jeden Jahrzeit. Aus dem Englischen ins Deutsche übersetzt mit hinzugefügten Bemerkungen. Und das Ganze mit einem interesanten Anhang versehen. Durch Joel Schelly.
Philadelphia: gedruckt bey Edmund Y. Schelly, 1838.
viii, 103 p.; illus. HSP in LCP; LCP

1419. __________.
The New American Orchardist, Or An Account Of The Most Valuable Varieties Of Fruit, Adapted To Cultivation In The Climate Of The United States, From The Latitude Of 25° To 54°, With Their Uses, Modes Of Culture, And Management .... Also A Brief Description Of The Most Ornamental Forest Trees, Shrubs, Flowers, &c.
Boston: Carter, Hendee, And Co. And Russell, Odiorne And Co., 1833.
Pp. [i]-xxxvi, [25]-423, [1 (adv.)]. LCP

1420. __________.
The New American Orchardist .... With An Appendix On Vegetables, Ornamental Trees, Shrubs And Flowers. Second Edition, Enlarged And Improved.
Boston: Russell, Odiorne, And Metcalf, And George C. Barrett, 1835.
418, [1] p. LCP

1421. KENT, J. EMERSON.
The Farmer's Hand-Book, Or, Chemistry Applied To Agriculture, The Only Profitable Mode Of Tilling The Soil.
New York: H. Dayton, Publisher; Indianapolis, Ind.: Asher & Company, 1859.
Pp. [1-2], [i]-viii, [13]-108; illus. LCP; HSP

KENT COUNTY [DEL.]. AGRICULTURAL SOCIETY.
See
EMERSON, GOUVERNEUR, 1796-1874. Address Delivered Before .... 1857.
__________. Results Of Extensive Experience ... Communicated To .... 1859 [2 edns.].
RIDGELY, HENRY. Address Delivered By ... Before .... 1859.

1422. KENTUCKY. STATE AGRICULTURAL SOCIETY.
Fourth Exhibition Of The Kentucky State Agricultural Society, To Be Held On The Fair Grounds Of The Kentucky Agricultural And Mechanical Association, At Lexington, Tuesday, Wednesday, Thursday, Friday, And Saturday, September 13, 14, 15, 16, & 17th, 1859. Premiums, Regulations, and Awarding Committees.
Frankfort: Printed At The Kentucky Yeoman Office; S.I.M. Major & Company, 1859.
26, [2] p. LCP

__________.
See also TODD, CHARLES STEWART, 1791-1871. Annual Address Before .... 1839.

1423. KERN, G.M.
Practical Landscape Gardening, With Reference To The Improvement Of Rural Residences, Giving The General Principles Of The Art; With Full Directions For Planting Shade Trees, Shrubbery And Flowers, And Laying Out Grounds.
Cincinnati: Moore, Wilstach, Keys & Co. [Stereotyped and Printed by Moore, Wilstach, Keys & Overend (t.p. verso)], 1855.
328 p.; tinted front.; 5 tinted pls.; illus. LCP

1424. KERSEY, JESSE, 1768-1846.
Lectures On Agriculture, Delivered Before The Downingtown Society For The Acquisition And Promotion Of Natural Knowledge.
West Chester, Penn.: Printed By Simeon Siegfried, 1828.
[4], 80 p. LCP

1425. KING, DANIEL PUTNAM, 1801-1850.
An Address To The Essex [Mass.] Agricultural Society, At Danvers, September 30, 1835, At Their Annual Cattle Show. Published By Order Of The Society.
Salem: Printed At The Gazette Office, 1836.
Pp. [1-2], [5]-22. LCP

1426. KING, WILLIAM STERLING, 1818-1882.
Address Delivered Before The New Hampshire Agricultural Society, At Its Third Annual Exhibition, In Meredith-Bridge, Oct. 7, 1852. With The Remarks Of Hon. Franklin Pierce.
Boston: Bazin & Chandler, Printers, 1853.
40 p. LCP (wrappers)

1427. KIRBY, EDMUND.
Address Delivered Before The Jefferson County [N.Y.] Agricultural Society, At The Annual Cattle Show And Fair, At Watertown, On The 27th Sept. 1831. Published By The Executive Committee.
Watertown: Published by Woodward & Calhoun, 1831.
24 p. Incl. (pp. 17-24): Report of the Viewing Committee and account of the Farmers Jubilee. LCP

1428. KIRK, CALEB.
The following Tracts of Land Are offered for Sale.
York, (P.): Printed By M'Clellan & King, [1805].
Brs. Incl. description of a farm in Manchester Twp., York Co. With German version of text following English. Dated at end: June 11th, 1805. HSP

1429. KIRWAN, RICHARD.
The Manures Most Advantageously Applicable To The Various Kinds Of Soils, And The Causes Of Their Beneficial Effect In Each Particular Instance. From The Sixth London Edition.
Philadelphia: Printed By Kimber, Conrad, And Co., 1807.
94 p. LCP

1430. The Kitchen And Fruit Gardener. A Select Manual Of Kitchen Gardening And Culture Of Fruits, Containing Familiar Directions For The Most Approved Practice In Each Department, Description Of Many Valuable Fruits, And A Calendar Of Work To Be Performed Each Month In The Year. The Whole Adapted To The Climate Of The United States.
Philadelphia: Lea And Blanchard [Stereotyped By J. Fagan; Printed By T.K. And P.G. Collins (t.p. verso)], 1847.
118 p. LCP

1431. KLIPPART, JOHN HANCOCK, 1823-1878.
The Wheat Plant: Its Origin, Culture, Growth, Development, Composition, Varieties, Diseases, Etc., Etc., Together With A Few Remarks On Indian Corn, Its Culture, Etc. One Hundred Illustrations.
Cincinnati: Moore, Wilstach, Keys & Co., 1860.
706, [6 (pub. adv.)] p.; illus. LCP

1432. KNOWLSON, JOHN C.
The Complete Farrier, Or Horse Doctor; A Guide For The Treatment Of Diseases Of Horses: To Which Are Added, Valuable Hints For Choosing A Good Horse, And Directions For Their Training; The Whole Being The Result Of Fifty Years' Extensive Practice.
New York: Published By Ensign, Bridgman, & Fanning, 1858.
64 p.; illus. LCP

1433. KOOGLE, J.D.
The Farmer's Own Book: A Treatise On The Numerous Diseases of the Horse, With An Explanation Of Their Symptoms, And The Course Of Treatment To Be Pursued; Also Other Useful Information.
Baltimore: Printed By McCoull & Slater, 1857.
206, [3], iii p. LCP

1434. KRAFFT, MICHAEL AUGUST.
The American Distiller, Or, The Theory And Practice Of Distilling, According To The Latest Discoveries And Improvements, Including The Most Improved Methods Of Constructing Stills, And Of Rectification.
Philadelphia: Printed For Thomas Dobson; Archibald Bartram, Printer, 1804.
Pp. [1-12], [9]-219; 2 fold. pls. HSP in LCP (lacking 1 pl.); LCP

1435. KRAUSS, JOHANN, 1770-1819.
Oeconomisches Haus-und Kunst-Buch, oder Sammlung ausgesuchter Vorschriften, Zum Nutzen und Gebrauch fur Land-und Hauswirthe, Handwerker, Kunstler und Kunst-Liebhaber. Zusammengetragen Aus den besten Englischen und Deutschen Schriften.
Allentown [Pa.]: Gedruckt bey Henrich Ebner, 1819.
452 p. HSP in LCP; LCP

1436. Kurzgefasstes Arzney-Büchlein, für Menschen und Vieh, Darinnen 128 auserlesene Recepte, nebst einer Prognostischen Tafel.
Vienna: gedruckt; Ephrata [Pa.]: nachgedruckt, 1791.
24 p. Copy with reversed brackets around page numbers 11, 13, 23, 24. HSP in LCP; LCP

1437. Kurzgefasstes Arzney-Büchlein, für Menschen und Vieh, Darinnen CXXVIII. auserlesene Recepten ....
Vienna: gedruckt; Ephrata [Pa.]: nachgedruckt, 1791.
24 p. Copy with Roman numeral in title and reversed brackets around p. numbers 10, 11, 23, 24. LCP

1438. Kurzgefasstes Arzney-Büchlein, für Menschen und Vieh, Darinnen CXXVII auserlesne Recepten ....
Vienna: gedruckt; Ephrata [Pa.]: nachgedruckt, 1791.
LCP: pp. [1]-22 only (of 24; lacking last leaf). Copy with Roman numeral in title and reversed brackets around p. numbers 5, 10, 11, 17-22.

1439. Kurzgefasstes Arzney-Büchlein, für Menschen und Vieh, darinnen ein hundert und acht und zwanzig auserlesene Recepte enthalten sind; Nebst einer prognostischen Tafel.
Reading [Pa.]: Gedruckt bey Jungmann und Gruber, 1794.
24 p. LCP

1440. Kurzgefasstes Arzney-Büchlein für Menschen und Vieh, Darinnen CXXX auserlesene Recepten.
Ephrata [Pa.]: zum sechsten mal gedruckt, 1795.
24 p. LCP

1441. Kurzgefasstes Arzney-Büchlein, für Menschen und Vieh, Darinnen CXXX auserlesene Recepten. Die achte Auflage.
Reading [Pa.]: bey Benjamin Mayer, 1797.
LCP: pp. [3]-22 only (of 24; lacking t.p. and last leaf. Title and imprint transcribed from Early American Imprints microfilm of MiU-C copy)

1442. Kurz gefasstes, Ross-Arzney Büchlein.
Ephrata [Pa.]: zum zweiten mal gedruckt, bey J. Baumann, 1803.
24 p. LCP

1443. Kurz gefasstes Ross Arzeney Buchlein.
[Reading?:] Gedruckt in diesem Jahr, [1815?].
12 p. LCP

1444. [LAMBERT, JOHN].
A Short And Practical Essay, On Farming: Being the Experience of a farmer, of about sixty years of age .... Shewing the means whereby these worn out lands may be improved ....
Philadelphia: [1798].
8 p. LCP

LANCASTER CO. [PA.]. AGRICULTURAL SOCIETY.
See GOWEN, JAMES. Address Delivered Before The .... 1852 [2 edns.].

1445. [LANDIS, CHARLES KLINE, 1835-1900].
Practical Hints At A Glance Upon Farming Written For The Weymouth Stockholders.
Philadelphia: 1855.
20 p. LCP (wrappers; with fold. leaf, Second Annual Statement Of The Weymouth Farm & Agricultural Company, at front)

1446. LANDRETH, DAVID, 1802-1880.
Letter From David Landreth, Of Philadelphia, To The Commissioner Of Patents.
[Philadelphia: 1858].
11 p. Criticizes the Agricultural Division of the Patent Office. Dated at end: June 21, 1858. HSP

__________.
See also JOHNSON, GEORGE WILLIAM, 1802-1886. A Dictionary Of Modern Gardening .... Edited, With Numerous Additions, By .... 1847.

LANDRETH FAMILY, Nurserymen, Philadelphia and Bristol.
Nursery established in 1784 in Philadelphia by David Landreth the elder (1752-1836), who was assisted by his brother Cuthbert. David the younger (1802-1880) took control in 1828, and moved the nursery to Bristol in 1847 while maintaining a warehouse in Philadelphia. See DAB and Bailey.

1447. __________.
A Catalogue Of Greenhouse Plants, Hardy Trees, Evergreen Shrubs, Flowering Shrubs, Bulbous Rooted, And Herbaceous Plants, Arranged By Their Botanic And English Names. With a collection of the most esteemed varieties of Fruit Trees Sold By David And Cuthbert Lardreth ....
Philadelphia: Printed For The Proprietors; Fry and Kammerer, Printers, 1811.
47 p. APS

1448. __________.
A Catalogue Of Greenhouse Plants, Hardy Trees, Evergreen Shrubs .... Cultivated At The Nurseries And Gardens Of David And Cuthbert Landreth ....
Philadelphia: Printed for D. & C. Landreth; I. Ashmead & Co. Printers, 1824.
48 p. HSP

1449. __________.
Catalogue Of Green-House Plants, Hardy Trees, Evergreen Shrubs .... Cultivated And For Sale At The Nurseries And Gardens Of David And Cuthbert Landreth ....
Philadelphia: Printed for D. & C. Landreth; I. Ashmead, Printer, 1826.
56 p. LCP

1450. __________.
Catalogue Of Select Annual, Biennial, And Perennial Flower Seeds, Cultivated By David Landreth .... With observations on the preparation of the soil adapted to them, and on their general culture. [cap. title]
[Philadelphia:] [Stavely & M'Calla, Prs. (colop.)], [1855].
7, [1] p. Dated according to addresses given in the Philadelphia directories. LCP

1451. __________.
Catalogue Of The Principal Garden Seeds, Cultivated At The Horticultural Grounds Of David Landreth & Co. (Late D. & C. Landreth,) Near Philadelphia, And Sold At Their Warehouse, No. 63 Chestnut Street.
Philadelphia: Printed By William Stavely, 1836.
16 p. APS

1452. __________.
Landreths' Agricultural Warehouse, Nos. 21 & 23 South Sixth Street, Near The State House, Philadelphia. [cap. title]
[Philadelphia: Stavely & M'Calla, Printers (colop.)], [1856?].
8 p.; illus. Describes agricultural machinery. LCP

__________.
See also
ALMANACS. PHILADELPHIA, 1856-58. Landreths' Rural Register And Almanac ....
The Floral Magazine .... Published By .... [pdcl.].

Der lange verborgene Schatz und Haus-Freund .... 1837.
See HOHMAN, JOHANN GEORG, d. ca. 1846.

1453. LANGSTROTH, LORENZO LORRAINE, 1810-1895.
Langstroth On The Hive And The Honey-Bee, A Bee Keeper's Manual.
Northampton: Hopkins, Bridgman & Company [Greenfield: C.A. Mirick, Printer (t.p. verso)], 1853.
Pp. [i]-xvi, [13]-384. LC calls for 2 fronts. APS

1454. __________.
A Practical Treatise On The Hive And Honey-Bee. With An Introduction. By Rev. Robert Baird, D.D. Second Edition, Enlarged, And Illustrated With Numerous Engravings.
New York: C.M. Saxton & Co., Agricultural Booksellers [Greenfield, Ms.: C.A. Mirick, Printer (t.p. verso)], 1857.
Pp. [1]-492, [511]-534; 18 pls. following p. 492. LCP

__________.
See also
COLVIN, RICHARD. A Small Treatise ... Compiled By Permission From Langstroth On The Honey Bee. 1859.
MAHAN, PHINEAS JENKS. A Small Treatise .... Abbreviated by permission From Rev. L.L. Langstroth's Bee-Keeper's Manual. 1858.

1455. [LASTEYRIE DU SAILLANT, CHARLES PHILIBERT, comte, 1759-1849.
Questions Relating To The Culture Of The Cotton-Tree. [cap. title]
[N.p.: 1808?].
3 p. Consists of 44 questions. APS

1456. __________.
A Treatise On The Culture, Preparation, History, And Analysis Of Pastel, Or Woad: The Different Methods Of Extracting The Coloring Matter, And The Manner Of Using It, And Indigo, In Dyeing. To Which Is Added, Information Upon The Art Of Extracting Indigo .... Translated From The French, By H.A.S. Dearborn.
Boston: Printed By Rowe & Hooper, 1816.
140 p. LCP

1457. LATHROP, LEONARD E.
The Farmer's Library, Or Essays Designed To Encourage The Pursuits And Promote The Science Of Agriculture.
Rutland [Vt.]: Printed By William Fry, 1825.
215, [1] p. LCP

1458. __________.
The Farmer's Library. Or Essays Designed To Encourage The Pursuits .... Second Edition, Corrected & Enlarged.
Windsor [Vt.]: Published By P. Merrifield; W. Spooner, Printer, 1826.
Pp. [1-2], [v]-x, [13]-300. LCP

1459. __________.
The Farmer's Library: Or Essays Designed To Encourage The Pursuits .... Third Edition, Corrected And Enlarged.
Rochester, N.Y.: Printed By Marshall & Dean, 1828.
344 p. LCP; APS; HSP

1460. LATROBE, JOHN HAZLEHURST BONEVAL, 1803-1891.
Address Delivered Before The Horticultural Society, Of Maryland, At Its Second Anniversary, May 15, 1834.
Baltimore: John D. Toy, Printer, 1834.
24 p. LCP (wrappers); HSP (wrappers)

1461. __________.
Manual Labour School. Address On The Subject Of A Manual Labour School. And An Address To The Citizens Of Baltimore.
Baltimore: Printed By John D. Toy, 1840.
20 p. Proposes a farm-school for indigent boys. HSP (wrappers)

1462. LAVENIR, Auctioneer, Germantown, Pa.
Elegant Seat And valuable Lots for Sale. On Tuesday, the 24th September, at 2 o'clock in the afternoon, will be exposed to Public Sale, on the Premises ....
Germantown: Printed by Michael Billmeyer, [1799].
Brs. Describes a farm in Chestnut Hill outside of Philadelphia. Signed and dated at end: Lavenir. September 3d, 1799. HSP

LAWES, JOHN BENNETT, 1814-1900, jt. auth.
See DUMAS, JEAN BAPTISTE ANDRE, 1800-1884. Manures .... 1848. Incl. The Action Of Salt .... by A.C. Becquerel and __________.

1463. LAWRENCE, B.
The Complete Cattle-Keeper, Or, Farmer's And Grazier's Guide In the Choice and Management of Neat Cattle And Sheep; Including Useful Observations And Suggestions Relative To The Comparative Value Of The Various Breeds .... Hints To Dairymen .... A Description Of The Internal Structure of neat cattle ....
Philadelphia: E.L. Carey And A. Hart; Boston: Carter & Hendee; New York: J & C. & H. Carvill, And Collins & Hannay; Baltimore: W. & J. Neal; Washington: Thompson & Homans, 1832.
201 p. LCP

1464. LAWRENCE, JOHN, 1753-1839.
The Horse In All His Varieties And Uses; His Breeding, Rearing And Management, Whether In Labor Or In Rest; With Rules, Occasionally Interspersed, For His Preservation From Disease.
Philadelphia: E.L. Carey And A. Hart; Boston: For Sale By Carter And Hendee; New York: G. And C. And H. Carvill; Baltimore: W. And J. Neal; Washington: Thompson And Homans, 1830.
Pp. [1-4], [3]-328, 1-12 (pub. adv.). LCP

1465. [__________].
A Treatise On Breeding, Rearing, And Fattening, All Kinds Of Poultry, Cows, Swine, And Other Domestic Animals. By B. Moubray, Esq. [pseud.]. Reprinted from the sixth London Edition. With Such Abridgments, And Additions, As, it was conceived, would render it best adapted to ... the United States. By Thomas G. Fessenden ....
Boston: Published By Lilly & Wait; And Carter & Hendee, 1832.
266 p. incl. front.; illus. LCP

1466. LAWRENCE, MYRON, 1799-1852.
Address Delivered Before The Hampshire, Franklin And Hampden [Mass.] Agricultural Society, At Their Annual Fair, Holden At Northampton, October 24, 1832. Published By Request.
Northampton: John Metcalf, Printer, 1832.
16 p. LCP

1467. LAWSON, WILLIAM.
A New Orchard and Garden: Or The best way for planting, grafting, and to make any ground good .... With the Country Housewifes Garden for herbs .... As Also The Husbandry of Bees ... all being the experience of 48. yeeres labour, and now the third time corrected and much enlarged .... [wrapper title]
London: Printed by J.H. for Francis William, 1626; Philadelphia: Reprinted by Robert Pearsall Smith, 1858. [wrapper imprint]
39 p.; illus. This reprint incl. only the first-named work. LCP

1468. LEBANON CO. [PA.]. HORSE COMPANY.
Charter Of The Lebanon County Horse Company, For The Detection Of Horse Thieves, And The Recovery Of Stolen Horses. Incorporated April 14th, 1853.
Lebanon: Printed By J. Hartman & Son, 1853.
8 p. HSP

1469. __________.
Frey-Brief der Libanon Caunty Pferde Gesellschaft, für die Entdeckung von Pferde-Dieben, und die Wieder-Erhaltung Gestohlener Pferde. Incorporiert den 14th April, 1853.
Lebanon: Gedruckt bey Jos. Hartman und Sohn, 1853.
8 p. HSP

Lectures Upon Agriculture .... [1788?].
See PHILADELPHIA. SOCIETY FOR PROMOTING AGRICULTURE.

1470. LEE, ZACCHEUS COLLINS, 1805-1859.
Address Before The Chester County [Pa.]. Horticultural Society. At Their June Exhibition, In The Borough Of West Chester, 1858.
West Chester, Pa.: Printed At The Village Record Office, 1858.
24 p. LCP; HSP

1471. __________.
Address Delivered Before The Horticultural Society Of Maryland, At Its Annual Exhibition, June 6, 1839.
Baltimore: John D. Toy, Printer, 1839.
32 p. LCP (wrappers)

1472. LEHIGH CO. [PA.]. AGRICULTURAL SOCIETY.
Report Of The Managers Of The Agricultural Society Of Lehigh County. Third Annual Meeting, October 10, 1854.
Allentown: Haines & Diefenderfer, Printers, 1855.
22 p. Incl. (pp. 17-22): Annual Address ... by Robert Emmet Wright. LCP (wrappers)

1473. LEIB, ISAAC.
Wohlerfahrner Pferde-Arzt; enthaltend Mittel für die Heilung aller bekannter und verschiedenartigen Krankheiten und Seuchen der Pferde; welche nach einer fünf und zwanzigjährigen Ausubung der Ross-Heilkunst bewahrt und untrüglich befundedn wurden.
Lebanon: Gedruckt bey Jos. Hartman, 1842.
xii, 184 p. incl. front. HSP in LCP; LCP

1474. [LEIB, MICHAEL, 1761-1822].
Requisites For, And Complete Method Of Hog-Teaching.
Philadelphia: Printed for the Author, [1802].
11 p. HSP in LCP; LCP

1475. LEINAU, GEORGE A., Fertilizer merchant, Philadelphia.
Farmers And Planters Attention. Owners Of Poor & Worn Out Land, Use Leinau's American Fertilizer As A Renovater Of The Soil, It Has No Equal .... For Wheat, Corn, Potatoes, Tobacco, Cotton and and [sic] the Sugar Cane .... George A. Leinau, No. 19 South Front St., Phila. [wrapper title]
[Philadelphia:] Duross Pr., [1856?].
16 p. Chiefly testimonials. LCP (wrappers)

1476. LELIEVRE, J.F.
Nouveau Jardinier De La Louisiane, Contenant Les Instructions Necessaires Aux Personnes qui s'occupent de Jardinage.
New Orleans: Chez J.F. Lelievre, Libraire [Imprimé par Gaux et Co. (½ t.p. verso)], 1838.
200 p. LCP

1477. LE RAY DE CHAUMONT, JAMES DONATIEN, 1760-1840.
An Address, Delivered At the Meeting At [sic] The Agricultural Society Of Jefferson County [N.Y.], December 29, 1817. To Which Is Subjoined, The Articles Of Association And The By-Laws Of The Society, &c. &c. Published By Order Of The Association.
New York: Printed By William A. Mercein, 1818.
20 p. APS; HSP

1478. __________.
An Address, Delivered At The Meeting Of The Agricultural Society Of Jefferson County [N.Y.], December 29, 1817 ....
Watertown: Printed By John H. Lord, Jr., 1818.
16 p. LCP

1479. __________.
An Address Delivered Before The Jefferson County [N.Y.] Agricultural Society, 15th September, 1829, At Watertown, N.Y. To Which Is Subjoined, Notes, and an Abstract of the Report of the Viewing Committee.
[Watertown?:] Published By Order Of The Society, 1829.
21 p. LCP

1480. __________.
Address Delivered On The 29th Of September, 1824, Before The Jefferson County [N.Y.] Agricultural Society.
[Watertown?:] Printed By Order Of The Society, [1824].
24 p. APS

1481. __________.
Remarks Addressed To The Jefferson County [N.Y.] Agricultural Society, At Their Semi-Annual Meeting, In March, 1820.
New York: Printed By William A. Mercein, 1820.
22 p. LCP; APS; HSP

1482. LE RAY DE CHAUMONT, VINCENT.
Address Delivered Before The Jefferson County [N.Y.] Agricultural Society, At The Annual Cattle Show, 28th September, 1830.
[Watertown?:] Published At The Request Of The Society, 1830.
28 p. LCP; APS

1483. Lettre d'un Américan sur les productions de l'Amérique, qui peuvent être des objets de commerce pour la Hollande. [cap. title]
[Amsterdam: 1783].
4 p. LCP

1484. LEUCHARS, ROBERT B.
A Practical Treatise On The Construction, Heating, And Ventilation Of Hot-Houses; Including Conservatories, Green-Houses, Graperies, And Other Kinds Of Horticultural Structures. With Practical Directions For Their Management, In Regard To Light, Heat, And Air. Illustrated With Numerous Engravings.
Boston: John P. Jewett And Company [Stereotyped by Hobart & Robbins; New England Type And Stereotype Foundery (t.p. verso)], 1851.
vi, 366 p.; front.; illus. HSP in LSP

1485. LIEBIG, JUSTUS VON, 1803-1873.
Chemistry In Its Application To Agriculture And Physiology. Edited From The Manuscript Of The Author By Lyon Playfair, Ph.D. From The Last London Edition, Much Improved.
Philadelphia: James M. Campbell & Co; New York: Saxton & Miles; Stereotyped By C.W. Murray & Co., 1843.
135 p. Copy with 4-paragraph Editor's Preface, ending "... to their fertility." LCP; HSP (wrappers; wrapper imprint does not incl. stereotyper)

1486. __________.
Chemistry In Its Application To Agriculture And Physiology ....
Philadelphia: James M. Campbell & Co.; New York: Saxton & Miles, 1843.
135 p. Copy with 5th paragraph, beginning "It is necessary to state that all additions and alterations," added to Editor's Preface. LCP

1487. __________.
Chemistry In Its Application To Agriculture And Physiology. Edited From The Manuscript Of The Author By Lyon Playfair ... And William Gregory .... From The Fourth London Edition, Revised And Enlarged.
New York: Wiley & Putnam [William Osborn, Printer; T.B. Smith, Stereotyper (t.p. verso)], 1847.
xiv, 401 p. LCP

1488. __________.
Chemistry In Its Application To Agriculture And Physiology. Edited From The Manuscript Of The Author By Lyon Playfair, Ph.D. From The Last London Edition, Much Improved.
Philadelphia: James M. Campbell [Printed By King And Baird (t.p. verso)], 1847.
135 p. HSP (wrappers)

1489. __________.
Familiar Letters On Chemistry, And Its Relation To Commerce, Physiology, And Agriculture. Edited By John Gardner, M.D. ....
New York: D. Appleton & Co.; Philadelphia: George S. Appleton, 1843.
Pp. [1-2], [5]-180. LCP (wrappers)

1490. __________.
Familiar Letters On Chemistry, And Its Relation To Commerce, Physiology And Agriculture. Edited By John Gardner, M.D. ....
New York: J. Winchester, New World Press, 1843.
53, [1 (pub. adv.)] p. LCP

1491. __________.
Organic Chemistry In Its Applications To Agriculture And Physiology. Edited From The Manuscript Of The Author By Lyon Playfair, Ph.D. First American Edition, With An Introduction, Notes, And Appendix, By John W. Webster ....
Cambridge: Published By John Owen; Boston: James Munroe And Company, And C.C. Little And J. Brown; New York: Wiley And Putnam, And Geo. C. Thorburn; Philadelphia: Thomas, Cowperthwait, And Company, And Cary [sic] And Hart; Baltimore: Cushing And Brother [Cambridge: Folsom, Wells, And Thurber, Printers To The University (t.p. verso)], 1841.
xx, 435, [1] p. LCP

1492. __________.
Organic Chemistry In Its Applications To Agriculture And Physiology .... Edited From The Manuscript Of The Author By Lyon Playfair, Ph.D. Second American Edition, With An Introduction, Notes, And Appendix, By John W. Webster, M.D. ....
Cambridge: Published By John Owen; Boston: James Munroe And Company, And C.C. Little And J. Brown; New York: Wiley And Putnam, And Geo. C. Thorburn, _et al_. [Thomas G. Wells, Printer To The University (t.p. verso)], 1841.
lxi, [3], 424, 3+[1] (pub. adv.) p. APS

1493. __________.
Principles Of Agricultural Chemistry, With Special Reference To The Late Researches Made In England.
New York: John Wiley [R. Craighead, Printer (t.p. verso)], 1855.
[1 (pub. adv.], vii, [1], 105 p. LCP

1494. __________.
Professor Liebig's Complete Works On Chemistry. Comprising His Agricultural Chemistry ... Animal Chemistry ... And Familiar Letters On Chemistry, And Its Relation To Commerce, Physiology And Agriculture, Etc., Etc., Etc.
Philadelphia: T.B. Peterson, [1856].
LCP: pp. [1]-135, [1], [1]-111, [1], 3-55; lacking sep. t.p. to last section.

1495. LINCOLN, LEVI, 1782-1868.
Address, Delivered Before The Worcester [Mass.] Agricultural Society, October 7, 1819, Being Their First Anniversary Cattle Show and Exhibition of Manufactures. Published By Order Of The Trustees.
Worcester: Printed By Manning & Trumbull, Dec. 1819.
30 p. LCP

1496. LINCOLN, WILLIAM, 1801-1843.
An Address Delivered Before The Massachusetts Horticultural Society At Their Ninth Anniversary, September 20, 1837.
Boston: Dutton And Wentworth, Printers, 1837.
56 p. LCP (wrappers)

1497. LINDLEY, JOHN, 1799-1865.
The Theory Of Horticulture: Or, An Attempt To Explain The Principal Operations Of Gardening Upon Physiological Principles. First American Edition, With Notes, Etc. By A.J. Downing And A. Gray.
New York: Wiley And Putnam; Boston: C.C. Little & Co. [New York: J.P. Wright, Printer (t.p. verso)], 1851.
xxiv, 346, [4 (pub. adv.)] p.; illus. LCP

1498. __________.
The Theory of Horticulture: or, An Attempt To Explain The Principal Operations Of Gardening Upon Physiological Principles. Second American Edition, With Notes, Etc. By A.J. Downing.
New York: John Wiley [R. Craighead, Printer (t.p. verso)], 1852.
xx, 364 p.; illus. LCP

LINNAEAN BOTANIC GARDEN AND NURSERIES, Flushing, N.Y.
See
PRINCE FAMILY, Nurserymen, Flushing, N.Y.
WINTER & CO., Nurserymen, Flushing, N.Y.

1499. LINSLEY, DANIEL CHIPMAN.
Morgan Horses: A Premium Essay On The Origin, History, And Characteristics Of This Remarkable American Breed Of Horses; Tracing The Pedigree From The Original Justin Morgan, Through The Most Noted Of His Progeny, Down To The Present Time. With Numerous Portraits. To Which Are Added Hints For Breeding, Breaking, And General Use And Management Of Horses, With Practical Directions For Training Them For Exhibition At Agricultural Fairs.
New York: C.M. Saxton And Company, Agricultural Book Publishers [E.O. Jenkins, Printer and Stereotyper (t.p. verso)], 1857.
Pp. [i]-viii, [1-2], 9-340, 1-11 + [1-3] (pub. adv.); front.; 12 pls. LCP

1500. LITHEN, JOHN, Nurseryman, Philadelphia.
Catalogue Of Garden-Seeds, Flower-Roots, Fruit and Forest-Trees, Evergreen and Flowering Shrubs, &c. For Sale by John Lithen, Nursery and Seedsman, South-Side of Centre-Square, Philadelphia.
[Philadelphia: ca. 1800].
Brs. Lithen appears in Philadelphia directories for 1800 only. HSP

1501. The Little Eagle, Containing A Short Description Of A Tract of Land, Situated In Clearfield County, Pa. Together with many advantages appertaining thereto, respecting its healthiness, which is offered for sale on moderate terms ....
Harrisburg, Pa.: 1828.
14 p. LCP

1502. LIVINGSTON, ROBERT R., 1746-1813.
Essay On Sheep; Their Varieties - Account Of The Merinoes Of Spain, France, &c. Reflections On The Best Method Of Treating Them, And Raising A Flock In The United States; Together With Miscellaneous Remarks On Sheep And Woollen Manufactures. Published by Order of the Legislature of the State of New-York.
New York: Printed By T. And J. Swords, 1809.
189 p. LCP

1503. __________.
Essay On Sheep; Their Varieties--Account Of The Merinoes Of Spain, France, &c. ....
Trenton: Published By James Oram, 1810.
149, [3] p.; col. front. LCP (lacking front.); HSP

1504. __________.
Essay On Sheep: Their Varieties - Account Of The Merinoes Of Spain, France, &c. ....
Concord, N.H.: Published By Daniel Cooledge; Boston: Sold also by Thomas and Andrews, West and Richardson, Bradford and Read, and Caleb Bingham [I. & W.R. Hill, Printers (colop.)], 1813.
143 p. LCP; APS

1505. LLOYD, DAVID.
Economy Of Agriculture: Being A Series Of Compendious Essays On Different Branches Of Farming.
Germantown, Pa.: P.R. Freas & Co., Printers, 1832.
120 p. LCP; HSP

1506. LOGAN, GEORGE, 1753-1821.
An Address On The Errors Of Husbandry, In The United States. Delivered Before The Philadelphia Society For Promoting Agriculture, At Their Annual Meeting, January 14, 1818.
Philadelphia: Printed By Lydia R. Bailey, 1818.
11 p. LCP; APS; HSP

1507. __________.
Agricultural Experiments, On Gypsum, Or Plaister of Paris: With Some Observations, On the Fertilizing Quality, And Natural History Of That Fossil.
Philadelphia: Printed By Francis And Robert Bailey, 1797.
18 p. LCP

1508. __________.
Fourteen Agricultural Experiments, To Ascertain The Best Rotation Of Crops: Addressed To The "Philadelphia Agricultural Society."
Philadelphia: Printed By Francis And Robert Bailey, 1797.
Pp. [1-4], [3]-41. LCP

1509. __________.
A Letter To The Citizens Of Pennsylvania, On The Necessity Of Promoting Agriculture, Manufactures, and the useful Arts.
Lancaster: W. & R. Dickson, March 14, 1800.
28 p. HSP in LCP

1510. __________.
A Letter To The Citizens Of Pennsylvania .... Second Edition.
Philadelphia: Printed By Patterson & Cochran, May 1, 1800.
30 p. HSP in LCP; LCP; APS

LONGWORTH, NICHOLAS, 1782-1863. Directions For The Cultivation Of The Strawberry.
See BUCHANAN, ROBERT, 1797-1879. The Culture Of The Grape ... With An Appendix Containing .... 1852 etc.

1511. LORAIN, JOHN.
Hints To Emigrants, Or A Comparative Estimate Of The Advantages of Pennsylvania, And Of The Western Territory, &c.
Philadelphia: Published By Littell & Henry; A. Waldie, Printer, 1819.
Pp. [1-4], [3]-144. HSP in LCP; LCP

1512. __________.
Nature And Reason Harmonized In The Practice Of Husbandry. With An Alphabetical Index.
Philadelphia: H.C. Carey & I. Lea, 1825.
x, [2], 563 p. HSP in LCP; LCP; APS

LORD, EDWARD. Horticulture. An Address Before The Oswego County Agricultural Society .... 1854.
See HOLMES, J.N. The Annual Address, Before The Oswego County Agricultural Society .... 1855. Incl. also Lord's address.

1513. LOUBAT, ALPHONSE.
The American Vine Dresser's Guide. Second Edition.
New York: Published By G. & C. Carvill; Printed at Grattan's Office, 1829.
142 p. English and French text on facing pages. LCP

1514. LOUDON, JANE WEBB, 1807-1858.
Gardening For Ladies; And Companion To The Flower-Garden. First American, From The Third London Edition. Edited By A.J. Downing ....
New York: Wiley And Putnam [Jas. P. Wright, Printer (t.p. verso)], 1843.
96, [4], 347 p.; illus. LCP

1515. LOUISIANA. AGRICULTURISTS' AND MECHANICS' ASSOCIATION.
Annual State Fair, January 5, 1846. Proceedings Of The Agriculturists' And Mechanics' Association Of Louisiana, Containing The Address Delivered By T.B. Thorpe, With Notes Appended: Also The Awards of the Various Committees, With The Detailed Reports On Agricultural Products; Of Sugar Culture In Louisiana; Of Cotton, Tobacco, &c., &c. .... The Whole Published by order of the Association.
New Orleans: Published By B.M. Norman, 1846.
28 p. LCP

__________.
See also ROST, PIERRE ADOLPHE, 1797-1868. Oration Delivered Before .... 1845.

1516. LOVERING, JOSEPH S.
A Detailed Account Of Experiments And Observations Upon The Sorghum Saccharatum Or Chinese Sugar Cane, Made With The View Of Determining Its Value As A Sugar Producing Plant, From September 28, To December 20, 1857, At Oakhill, Philadelphia County, Pa.
Philadelphia: Crissy & Markley, Printers, 1858.
27 p. HSP (wrappers)

1517. __________.
A Detailed Account Of Experiments And Observations Upon The Sorghum Saccharatum ....
Philadelphia: H.A. Dreer, Seedsman And Florist, 1858.
24, [24 (adv.)] p. HSP (wrappers)

__________. A Detailed Account of Experiments ....
See also OLCOTT, HENRY STEEL, 1832-1907. Supplement To Sorgho And Imphee .... Containing The American Experiments of 1837, And J.S. Lovering's Pamphlet. 1858.

1518. LOWELL, JOHN, 1769-1840.
Address Delivered Before The Massachusetts Agricultural Society [i.e. Society for Promoting Agriculture] At The Brighton Cattle Show, October 13, 1818. Published by order of the Trustees.
Boston: Printed By Wells And Lilly, 1818.
22, [2 (pub. adv.)] p. LCP

1519. LUMPKIN, JOSEPH HENRY, 1799-1867.
An Address Delivered Before The South-Carolina Institute, At Its Second Annual Fair, On the 19th November, 1850.
Charleston, S.C.: Walker & James, 1851.
52 p. LCP (wrapper)

1520. LUNT, GEORGE, 1803-1885.
An Address Delivered Before The Massachusetts Horticultural Society, On The Dedication Of Horticultural Hall, May 15, 1845.
Boston: Printed By Dutton And Wentworth, 1845.
26 p. LCP (wrappers)

1521. LYFORD, STEPHEN C.
An Address Delivered Before The Strafford [N.H.] Agricultural Society, October 20, 1824, At Wolfborough.
Concord: Printed By Jacob B. Moore, 1825.
16 p. LCP

1522. LYNCH, JAMES.
Address, Delivered Before The American Institute Of The City Of New-York At Their Late Fair, Held At Masonic Hall, October, 1829. Accompanied With A Report On The Fair ....
New York: Printed By John M. Danforth, 1829.
31 p. LCP; HSP

MAC---
See also under
MC---
M'H---

1523. MACKENZIE, GEORGE STEUART, 1780-1848.
A Treatise On The Diseases And Management Of Sheep; With Introductory Remarks On Their Anatomical Structure; And An Appendix Containing Documents Exhibiting The Value Of The Merino Breed Of Sheep, And Their Progress In Scotland.
New York: Printed and Published by I. Riley, 1810.
100, 45 p.; 5 pls. Sep.-paginated Appendix. LCP

1524. MacMAHON, BERNARD, 1775?-1816, Nurseryman, Philadelphia.
The American Gardener's Calendar; Adapted To The Climates And Seasons Of The United States. Containing A Complete Account Of All The Work Necessary To Be Done In The Kitchen-Garden, Fruit-Garden, Orchard, Vineyard, Nursery .... For Every Month In The Year .... Also, General as well as minute instructions, for laying out, or erecting, each and every of the above departments .... To Which Are Annexed, Extensive Catalogues of the different kinds of plants, which may be cultivated ....
Philadelphia: Printed By B. Graves; For The Author, 1806.
v, [1], 648, [18] p.; fold. tbl. LCP; APS; HSP

1525. __________.
The American Gardener's Calendar .... Second Edition, Improved.
Philadelphia: Published By T.P. M'Mahon; William Fry, Printer, 1819.
618 p.; fold. tbl. LCP; HSP

1526. __________.
The American Gardener's Calendar .... Eighth Edition Improved.
Philadelphia: Published By A. M'Mahon; John Young, Printer, 1832.
LCP: 616 p. only (of 618; lacking last leaf; lacking fold. tbl.?)

1527. __________.
The American Gardener's Calendar .... Ninth Edition, Greatly Improved.
Philadelphia: Published By A. M'Mahon [E.G. Dorsey, Printer (t.p. verso)], 1839.
8 (pub. adv.), 613 p.; fold. tbl. LCP

1528. __________.
The American Gardener's Calendar .... Tenth Edition, Greatly Improved.
Philadelphia: Lippincott, Grambo And Co., Successors To Grigg, Elliot & Co., 1851.
613 p.; fold. tbl. LCP

1529. __________.
The American Gardener's Calendar .... Eleventh Edition, With A Memoir Of The Author: Revised And Illustrated Under The Supervision Of J. Jay Smith.
Philadelphia: J.B. Lippincott & Co., 1857.
637 p.; front.; illus. LCP

1530. __________.
A Catalogue Of American Seeds Sold By Bernard M'Mahon, Seedsman, Philadelphia. 180[4].
[Philadelphia:] Printed By Bartholomew Graves, [1804].
30 p. The last digit of the date is added in MS. HSP in LCP; APS

1531. __________.
A Catalogue Of Garden, Herb, Flower, Tree, Shrub, And Grass Seeds; Gardening, Agricultural And Botanical Books, Garden Tools, &c. Sold By Bernard M'Mahon, No. 39, South Second Street, Philadelphia: Who Has Likewise For Sale, Plants Of above fifty varieties of the most superior English, Irish & Scotch Gooseberries.
Philadelphia: Printed By William Duane, [1813?].
36 p. LCP; HSP (with MacMahon's address changed in MS. on t.p. to No. 13, South Second, and with MS. date 1815. MacMahon is in city directories at No. 39 in 1813 and at No. 13 in 1814)

1532. MacNEVEN, WILLIAM JAMES, 1763-1841.
Introductory Discourse To A Few Lectures On The Application Of Chemistry To Agriculture, Delivered Before The New-York Athenaeum In The Winter Of 1825.
New York: G. & C. Carvill; Palmer & Co. Printers to the New-York Athenaeum, 1825.
40 p. APS; HSP

1533. MADISON CO. [N.Y.]. AGRICULTURAL SOCIETY.
Transactions.
Hamilton, N.Y.
LCP: 1845.

1534. MADISON, JAMES, 1751-1836.
An Address Delivered Before The Agricultural Society Of Albemarle [Va.], On Tuesday, May 12, 1818.
Richard: Printed by Shepherd & Pollard, 1818.
31 p. LCP; APS

1535. Magazine Of Gardening And Botany. H.F. Dickehut, Editor.
Baltimore.
LCP: Vol. 1, no. 1 (Jan. 1834). All pub.

1536. MAHAN, JASON M.
Directions For Cultivating The Italian Mulberry, Management Of Silk Worms, &c.
Bethania, Pa.: Reuben Chambers, Printer, 1834.
16 p. HSP

1537. MAHAN, PHINEAS JENKS.
A Small Treatise, Containing A Few Practical Directions For Managing The Bee And Hive. Abbreviated by permission From Rev. L.L. Langstroth's Bee-Keeper's Manual.
Baltimore: Printed By Sherwood & Co., 1858.
47, [1] p.; illus. APS (wrappers)

1538. __________.
A Small Treatise, Containing A Few Practical Directions .... From The Rev. L.L. Langstroth's Bee-Keeper's Manual ....
Philadelphia: Merrihew & Thompson, Printers, [1858].
APS: fragment. 4+4 (adv. on col. paper) p.; t.p., pp. 9-24, 5-8 (adv. on col. paper); illus. Apparently lacking pp. 3-8 and 8 (?) p. at end.

MANLEY, CHARLES D.
See DELAWARE CO. [PA.]. INSTITUTE OF SCIENCE. Report Of ... The Sixth Annual Exhibition Of .... Also, The Address .... 1851.

1539. MANLEY, JOHN, comp.
Cattaraugus County: Embracing Its Agricultural Society, Newspapers, Civil List, (From The Organization Of The County To 1857,) Biographies Of Old Pioneers, (With Portraits,) ... Colonial and State Governors of New York; Names Of Towns And Post Offices, With The Statistics Of Each Town.
Little Valley, N.Y.: Published By John Manley; N.Y.: Hosford & Co., Stationers And Printers, 1857.
140 p.; 8 pls. Incl. Officers, Premiums for the 16th Annual Fair, By-Laws, Abstracts of the Treasurer's Report, etc. of the Cattaraugus County Agricultural and Horticultural Society. With numerous advs. throughout pp. 109-137. HSP

1540. __________.
Cattaraugus County: Embracing Its Agricultural Society, Newspapers, Civil List ....
Little Valley, N.Y.: Published By John Manley; N.Y.: Hosford & Co. Stationers And Printers, 1857.
136 p.; without pls. As above except that advs. from pp. 135-37 appear on back wrapper and verso of front wrapper in this copy. HSP (wrappers)

1541. MANN, JAMES.
The American Bird-Keeper's Manual; Or Direction For The Proper Management Of American And Foreign Singing Birds. With Particular Instructions For The Breeding Of Canary Birds .... Together With Some Remarks Upon The Diseases To Which Birds Are Liable ....
Boston: Published By The Author; And Sold By Little And Brown, And W.D. Ticknor And Co.; New York: C.S. Francis And Co.; Portland: E.C. Andrews [Boston: Printed By Freeman And Bolles (t.p. verso)], 1848.
166 p. LCP

1542. MANNING, ROBERT, 1784-1842.
Book Of Fruits, Being A Descriptive Catalogue Of The Most Valuable Varieties Of The Pear, Apple, Peach, Plum & Cherry, For New-England Culture. To Which Is Added The Gooseberry, Currant, Raspberry, Strawberry, And The Grape; With Modes Of Culture. Also, Hardy Ornamental Trees And Shrubs. With Plates. First Series for 1838.
Salem: Published By Ives & Jewett [Salem Observer Press (t.p. verso)], 1838.
120, [11 (adv.)] p.; col. front.; 3 col. pls.; illus. LCP

1543. [__________].
New England Book Of Fruit; Containing An Abridgement Of Manning's Descriptive Catalogue Of The Most Valuable Varieties Of The Pear, Apple .... Third Edition, Revised, And Enlarged. By John M. Ives.
Salem: W. & S.B. Ives; Boston: W.J. Reynolds & Co., B.B. Mussey & Co.; And For Sale At Most Of The Agricultural And Seed Stores In The United States [Boston: Printed By S.N. Dickinson & Co. (½ t.p. verso)], 1847.
viii, 144 p.; extra col. eng. t.p.; col. front.; illus. LCP

1544. __________.
The New England Fruit Book. Being A Descriptive Catalogue Of The Most Valuable Varieties Of The Pear, Apple, Peach, Plum & Cherry, For New-England Culture. To Which Is Added Other Varieties; Also The Grape, Quince, Gooseberry, Currant, And Strawberry; With Outlines Of Many Of The Finest Pears, Drawn From Nature; With Directions For Pruning, Grafting, Budding, And General Modes Of Culture. Second Edition, Enlarged. By John M. Ives.
Salem, Mass.: Published By W. & S.B. Ives; Boston: B.B. Mussey [Boston: Printed By Freeman And Bolles (½ t.p. verso)], 1844.
[8], 133 p.; col. front.; illus. LCP

1545. MANNY, JOHN H., <u>et al</u>.
Remonstrance of John H. Manny and others, against the extension of Letters Patent of Cyrus H. McCormick, Obed Hussey, and Moore and Haskell. [cap. title]
[Rockford, Ill.?: 1855].
10 p. Title headed: Remonstrance. To the Senate and House of Representatives of the United States, in Congress assembled. Concerns reaping machines. LCP

MANNY'S PATENT ADJUSTABLE REAPING AND MOWING MACHINES, Hoosick Falls, N.Y.
See BALL, LEVI CHANDLER, 1809-1875. Address Delivered At ... The Bennington Co. [Vt.] Agricultural Society .... 1852. Incl. prospectus for the machines.

Manuel-Guide Des Voyageurs Aux Etats-Unis .... 1818.
See FERNAGUS DE GELONE, -----.

1546. MANWARING, CHRISTOPHER, 1774-1832.
Essays, Historical, Moral, Political And Agricultural.
New London: Printed By Samuel Green, For The Author, 1829.
204, [1] p. LCP

1547. MAPES, JAMES JAY, 1806-1866.
Address, Delivered At The Opening Of The 18th Annual Fair Of The American Institute, At Niblo's Garden, Tuesday, Oct. 7, 1845.
New York: James Van Norden & Co., Printers, 1845.
8 p. LCP

1548. MARCANDIER, -----.
An Abstract Of The most useful Parts of a late Treatise on Hemp, translated from the French of M. Marcandier, Magistrate of Bourges, and inscribed by the Editor at London, to the laudable Society for promoting Arts, Manufactures, &c. Being much recommended to the Growers and Manufacturers of that valuable Material .... Together With some Observations upon the Prospect of singular Advantage which may be derived to Great-Britain and her Colonies from their early adopting the Method prescribed. To which is added, some Account of the Use of the Horse-Chesnut; And A Plan of the Pennsylvania Hemp Brake.
Boston: Printed and Sold by Edes & Gill, 1766.
[4], 30, [1] p.; illus. T.p. headed: Published for the general benefit of Great-Britain and the North American Colonies. HSP in LCP

1549. [MARCET, JANE HALDIMAND, 1769-1858].
Conversations On Vegetable Physiology; Comprehending The Elements Of Botany, With Their Application To Agriculture. By The Author Of Conversations On Chemistry .... Adapted To The Use Of Schools By Rev. J.L. Blake, A.M.
Boston: Published By Crocker & Brewster; New York: J. Leavitt, 1830.
372 p.; col. front.; 9 pls., 5 col. LCP

1550. [__________].
Conversations On Vegetable Physiology; Comprehending The Elements Of Botany .... Third American Edition, With Engravings.
Philadelphia: E.L. Carey & A. Hart; Baltimore: Carey, Hart & Co. and William & Joseph Neal; Boston: Crocker & Brewster, Allen & Ticknor, Russell, Odiorne & Co. and Lilly, Wait & Co., et al., 1833.
372 p.; col. front.; 3 col. pls. HSP

MARION CO. [IND.]. AGRICULTURAL SOCIETY.
See SMITH, OLIVER HAMPTON, 1794-1859. Address ... Delivered On The Fair Grounds .... 1856.

1551. MARKHAM, GERVASE, 1568?-1637.
The Citizen And Countryman's Experienced Farrier. Containing, I. The most best approved Method of Ordering, Dieting, Exercising, Purging, Scouring, And Cleansing Of Horses .... II. A Certain Sure Method to know the true State of any Horse's Body .... To all which is added: A valuable and fine collection of the surest and best Receipts ... By J. Markham, G. Jefferies, and Discreet Indians.
London: Printed; Wilmington: Re-printed, and Sold by James Adams, 1764.
364 p. Compiled by John Millis from Markham's works and including "suitable approved Receipts from able Farriers" in the neighborhood (pp. iii-iv), esp. George Jefferies. HSP in LCP (lacking last 4 leaves); LCP

1552. __________.
The Citizen And Countryman's Experienced Farrier. Containing, I. The most best approved method of ordering, dieting, exercising, purging, scowring, and cleansing of horses ... By J. Markham, G. Jefferies, and Discreet Indians.
Baltimore: Printed by Samuel Sower, 1797.
317 p. LCP

1553. MARSH, GEORGE PERKINS, 1801-1882.
Address Delivered Before The Agricultural Society Of Rutland County [Vt.], Sept. 30, 1847. Published By Request Of The Society.
Rutland, Vt.: Printed At The Herald Office, 1848.
24 p. LCP

1554. MARSHALL, CHARLES, d. 1818.
An Introduction To The Knowledge And Practice Of Gardening. First American From The Second London Edition, Considerably Enlarged and Improved. To Which Is Added, An Essay On Quick-Lime, As A Cement And As A Manure, By James Anderson .... Vol. I [- II].
Boston: Printed By Samuel Etheridge, For Joseph Nancrede, 1799.
LCP: 2 vols. Vol. I: x, 276 p. II: [2], 134, 115, [1], [4 (pub. adv.)] p. Sep. t.p. and pagination for Anderson's essay, Vol. II. Copy with catchword "Section" on last p. of Vol. I and with "End of Vol. I" added in MS.

1555. __________.
An Introduction To The Knowledge And Practice Of Gardening .... Vol. I.
Boston: Printed By Samuel Etheridge, For Joseph Nancrede, 1799.
HSP in LCP: Vol. I only (of 2): x, 276 p. Copy without catchword on last p. and with "End Of Vol. I" in letterpress.

1556. MARSHALL, JOSIAH T.
The Farmers And Emigrants Complete Guide, Or A Hand Book, With Copious Hints, Recipes, And Tables, Designed For The Farmer And Emigrant.
Cincinnati: Applegate & Co., 1855.
492, [12 (pub. adv.)] p.; illus. APS

1557. __________.
The Farmer's And Emigrant's Hand-Book: Being A Full And Complete Guide For The Farmer And The Emigrant. Comprising The Clearing Of Forest And Prairie Land---Gardening---Farming Generally---Farriery---Cookery---And The Prevention And Cure Of Diseases, With Copious Hints, Recipes, And Tables. Second Edition, Revised.
New York: D. Appleton & Co; Philadelphia: Geo. S. Appleton, 1845.
492 p.; illus. LCP; APS (with [10]+24 p. pub. adv. at back)

1558. MARTIN, WILLIAM.
A Rough Sketch Of The Renewal System Of Training The Grape Vine On Arbors, Trellises, Or Walls.
Pittsburgh: Printed By W.S. Haven, 1854.
16 p.; 5 pls. LCP

1559. MARYLAND. AGRICULTURAL COLLEGE.
Report Of The Register Of The Maryland Agricultural College, To The Board Of Trustees; Act Of Incorporation, With Amendments Thereto; List Of Officers, And Names Of Stockholders, With Number Of Shares Held By Each. Published By Order Of The Board Of Trustees.
Baltimore: Printed By Samuel S. Mills, 1858.
29 p.; errata slip tipped in at back. LCP

1560. __________.
Second Circular Of The Maryland Agricultural College. 1860.
Washington: Henry Polkinhorn, Book And Job Printer, 1860.
28 p. HSP (wrapper)

MARYLAND. AGRICULTURAL SOCIETY.
See MAXCY, VIRGIL, 1785-1844. Address To The .... 1820.

1561. MARYLAND. GENERAL ASSEMBLY. HOUSE. SELECT COMMITTEE ....
Report Of The Select Committee Appointed To Enquire Into The Expediency Of Establishing An Agricultural School, And A Pattern Farm, In Each Of the Counties of this State.
Annapolis: Jeremiah Hughes, Printer, 1836.
24 p. T.p. headed: (B.) Mr. Teackle, Chairman. HSP

1562. MARYLAND. HORTICULTURAL SOCIETY.
The Constitution And By-Laws Of The Horitucltural Society Of Maryland.
Baltimore: Printed By Joseph Robinson, 1833.
36 p. LCP (wrappers); APS (wrappers)

1563. __________.
Constitution And By-Laws Of The Horticultural Society Of Maryland. (Adopted July 21st, 1851.)
Baltimore: Sands & Mills, Printers, Office "American Farmer," 1851.
8 p. LCP

__________.
See also
DAVIDGE, FRANCIS H. Address Delivered Before .... 1835.
KENNEDY, JOHN PENDLETON, 1795-1870. Address Delivered Before .... 1833.
LATROBE, JOHN HAZLEHURST BONEVAL, 1803-1891. Address Delivered Before .... 1834.
LEE, ZACCHEUS COLLINS, Address Delivered Before .... 1839.

MARYLAND. STATE AGRICULTURAL CHEMIST.
See
HIGGINS, JAMES. Reports .... 1850 etc.
TYSON, PHILIP THOMAS, 1799-1877. Report. 1860.

1564. MARYLAND. STATE AGRICULTURAL SOCIETY.
List Of Premiums And Rules And Regulations For The Tenth Cattle Show And Agricultural & Horticultural Exhibition To Be Held By The Maryland State Agricultural Society, At the Show Grounds in the City of Baltimore, On Tuesday, Wednesday, Thursday and Friday, October 20th, 21st, 22d, 23d, 1857.
Baltimore: Printed By Samuel S. Mills, 1857.
36 p. LCP

__________.
See also HOLCOMB, CHAUNCEY P. An Address Delivered Before .... 1853.

1565. MASON, CYRUS.
The Oration On The Thirteenth Anniversary Of The American Institute, Delivered By Cyrus Mason, At The Broadway Tabernacle, October 15th, 1840. Second Edition.
New York: D. Appleton & Co. [Hopkins & Jennings, Printers (t.p. verso)], 1840.
38 p. HSP (wrappers)

1566. MASON, RICHARD.
The Gentleman's New Pocket Farrier, Comprising A General Description Of The Noble And Useful Animal The Horse; Together With The quickest and simplest mode of fattening; necessary treatment while undergoing excessive fatigue .... Fourth Edition, Enlarged And Improved. To Which Is Added An Appendix, Containing Observations and Receipts for the Cure of most of the common distempers .... Also, Annals Of The Turf, Or Virginia Stud Book.
Richmond [Va.]: Printed By Peter Cottom, 1828.
300 p.; front.; 4 pls. LCP

1567. MASSACHUSETTS. AGRICULTURAL SURVEY OF THE STATE.
Report Of The Agricultural Meeting, Held In Boston, January 13, 1840, Containing The Remarks On That Occasion Of The Hon. Daniel Webster, Of The U.S. Senate, And Of Professor Silliman, M.D., L.L.D., Of Yale College. With Notes By Henry Colman, Commissioner for the Agricultural Survey of the State.
Salem: Printed At The Gazette Office; Salem Power Press, 1840.
36, 7, [1] p. The meeting was called by Colman to discuss the Survey. LCP

__________.
See also COLMAN, HENRY, 1785-1849. Letter .... 1837; Reports .... 1838 etc.

1568. MASSACHUSETTS. GENERAL COURT. SELECT COMMITTEE.
Report And Resolves Providing For A Board Of Agriculture.
[Boston: 1838].
12 p. T.p. headed: Senate. No. 84. LCP

1569. MASSACHUSETTS. GENERAL COURT. SENATE. COMMISSION ON ARTIFICIAL PROPAGATION OF FISH.
Report Of Commissioners Appointed Under Resolve Of 1856, Chap. 58, Concerning The Artificial Propagation of Fish, With Other Documents. [wrapper title]
Boston: William White, Printer To The State, 1857. [wrapper imprint]
54 p. P. 1 headed: Senate. No. 193. Commonwealth of Massachusetts. APS (wrapper)

1570. MASSACHUSETTS. HORTICULTURAL SOCIETY.
Massachusetts Horticultural Society.
Boston: Press Of Isaac R. Butts, 1829.
51, [1] p. Constitution, by-laws, list of members, etc. LCP

1571. __________.
Report Of The Twentieth Annual Exhibition Of The Massachusetts Horticultural Society, And Third Trienniel Festival, Held At Faneuil Hall, September 19, 20, 21, 22, 1848.
Boston: Printed By Tuttle & Dennett, 1848.
61 p. LCP (wrapper)

1572. __________.
Transactions.
Boston.
LCP: 1837-38; 1843-46; 1852; March, 1859; 1860; 1866; 1876, part II.
Also known as: Report Of Transactions; Journal Of Proceedings.

__________.
See also
COOK, ZEBEDEE, JR. Ar Address, Pronounced Before .... 1830.
DEARBORN, HENRY ALEXANDER SCAMMEL, 1783-1851. An Address Delivered Before .... 1829.
HARRIS, THADDEUS WILLIAM, 1795-1856. A Discourse Delivered Before .... 1832.
LINCOLN, WILLIAM, 1801-1843. An Address Delivered Before .... 1837.
LUNT, GEORGE, 1803-1885. An Address Delivered Before .... 1845.
WARD, MALTHUS AUGUSTUS. An Address Pronounced Before .... 1831.
WESTON, EZRA, 1771-1842. An Address Delivered Before .... 1836.

1573. MASSACHUSETTS. SCHOOL OF AGRICULTURE.
Massachusetts School Of Agriculture. List Of Officers ....
Boston: J.H. Eastburn's Press, 1858.
8 p. Solicits donations for the proposed school. APS (wrapper)

MASSACHUSETTS. SECRETARY OF THE COMMONWEALTH. Responsible for state agricultural business until the establishment of the State Board of Agriculture (q.v.) in 1852.

1574. __________.
A Synopsis Of The Several Communications On The Cause And Cure Of The Potato Rot; Received By The Executive Of Massachusetts. Prepared And Published Under The Authority Of The Legislature, By Amasa Walker, Secretary Of The Commonwealth.
Boston: Dutton & Wentworth, State Printers, 1852.
48 p. LCP

1575. __________.
Transactions of the Agricultural Societies in the State of Massachusetts.
Boston.
LCP: 1846-52.
1846 vol. has title: Abstracts From The Returns Of Agricultural Societies; 1852 vol. has subtitle: Also The Proceedings Of The State Board Of Agriculture.

MASSACHUSETTS. SOCIETY FOR PROMOTING AGRICULTURE. Georgick Papers.
See
__________. Papers On Agriculture. 1809.

1576. __________.
Inquiries By The Agricultural Society. [P. 3 continues:] With a view to collect the most accurate Information on the principal Branches of Agriculture, as now practiced, and thus be enabled to propagate the knowledge of whatever shall be found useful ... the following Inquiries are addressed to Mr. [ ] by the Trustees of the Massachusetts Society for Promoting Agriculture.
[Boston: Printed By Young & Minns, Printers To The State, 1800 (colop.)].
29 p. Incl. 50 questions with blanks for answers. LCP; APS

1577. __________.
Laws And Regulations Of The Massachusetts Society .... Containing, The Names of its present Officers and Members, List of Premiums, &c. With Some interesting Extracts from Foreign and Domestick Publications. By The Trustees.
Boston: Printed By Isaiah Thomas And Ebenezer T. Andrews, 1793.
56 p. HSP in LCP; LCP; APS (lacking last leaf)

1578. __________.
The Massachusetts Agricultural Repository, And Journal. Published By The Trustees Of The Massachusetts Society ....
Boston.
APS: Vols. 3-4 (Nov. 1813-July 1817)
HSP: Vol. 8, no. 3 (Jan. 1825)

1579. __________.
On The Culture Of Potatoes: Extracted From Communications Made To The Board Of Agriculture In Great Britain. Published by Order of the Trustees of the Massachusetts Society ....
Boston: Printed By Young & Minns, Printers To The State And To The Agricultural Society, 1798.
25 p. LCP; APS

1580. __________.
Papers on Agriculture.
Boston.
LCP: 1801; 1803; 1804; 1804 [i.e. 1805]; 1807; 1809; 1811.
APS: 1803; 1804; 1804 [i.e. 1805; changed in MS. on t.p.); 1809.
In 1809 known as: Georgick Papers.

1581. __________.
Rules And Regulations Of The Massachusetts Society .... Containing, The Names Of Its Present Members And Officers, Several Important Communications, And Interesting Extracts, From Foreign And Domestic Publications, With A List Of Premiums, Donations, &c. By The Trustees.
Boston: Printed By Thomas Fleet, Jun., 1796.
78 p.; 1 pl. LCP; HSP

1582. __________.
Transactions.
Boston.
APS: New series, vol. 1 (1858)

__________.
See also
ALLEN, STEPHEN MERRIL, 1819-1894. Fibrilia ... Addresses, Before .... 1860.
COLMAN, HENRY, 1785-1849. An Address Delivered Before .... 1822.
DANA, SAMUEL LUTHER, 1795-1868. Manures: A Prize Essay. 1843 etc.
DEARBORN, HENRY ALEXANDER SCAMMEL, 1783-1851. An Address Delivered Before .... 1835.
LOWELL, JOHN, 1769-1840. Address Delivered Before .... 1818.
QUINCY, JOSIAH, 1772-1864. An Address Delivered Before .... [1819].
PECK, WILLIAM DANDRIDGE, 1763-1822. Natural History Of The Slug Worm. By Order Of The .... 1799.

MASSACHUSETTS. STATE BOARD OF AGRICULTURE. Established in 1852 to handle the agricultural duties of the Secretary of the Commonwealth (q.v.).

1583. ________.
Abstract Of Returns Of The Agricultural Societies Of Massachusetts, 1858. Edited By Charles L. Flint, Secretary Of The Board Of Agriculture.
Boston: William White, Printer To The State, 1859.
vvi, [1], 348 p.; front. LCP; HSP

1584. ________.
The Agriculture Of Massachusetts, As Shown In Returns Of The Agricultural Societies, 1853. Prepared By Charles L. Flint, Secretary of the Board of Agriculture.
Boston: William White, Printer To The State, 1854.
vi, [2], 492 p.; front. ; 5 pls. LCP

1585. ________.
First Annual Report Of The Secretary Of The Board Of Agriculture, Together With The Reports Of Committees, Appointed To Visit The County Societies. With An Appendix, Containing An Abstract Of The Finances Of The County Societies.
Boston: William White, Printer To The State, 1854.
[4], 186 p.; front. LCP

1586. ________.
Second Annual Report Of The Secretary Of The ... Together With The Reports Of Committees Appointed To Visit The County Societies, With An Appendix, Containing An Abstract Of The Finances Of The County Societies.
Boston: William White, Printer To The State, 1855.
[4], 202 p. LCP (wrappers)

1587. ________.
Sixth Annual Report Of The Secretary Of The ... Together With Reports Of Committees Appointed To Visit The County Societies, With An Appendix Containing An Abstract Of The Finances Of The County Societies.
Boston: William White, Printer To The State, 1859.
159, [1], cxxxiii, [1] p.; front.; illus. LCP; HSP

________.
See also MASSACHUSETTS. GENERAL COURT. SELECT COMMITTEE. Report And Resolves Providing For .... [1838].

MASSACHUSETTS. ZOOLOGICAL AND BOTANICAL SURVEY. A Report On The Insects .... 1841.
See HARRIS, THADDEUS WILLIAM, 1795-1856.

1588. MATLACK, TIMOTHY, 1736-1829.
An Oration, Delivered March 16, 1780, Before The Patron, Vice-Presidents And Members Of The American Philosophical Society, Held At Philadelphia, For Promoting Useful Knowledge.
Philadelphia: Printed By Styner And Cist, 1780.
27 p. On the "Rise, Progress and Antiquity of Agriculture." APS

1589. MATHER, THOMAS G.
Address Delivered Before The Middlesex County [Conn.] Agricultural Society, At Its Annual Exhibition, October, 13th, 1843.
Middletown, Conn.: Wm. D. Starr, Printer, 1843.
19, [1] p. LCP (wrappers)

1590. MAUCK, ROBERT.
[No heading; text begins:] Whereas on the 7th of December A.D. 1829, I, Robert Mauck, of the county of Gallia in the Staate [sic] of Ohio, obtained from the proper Officers ... a Patent for "purifying or rectifying, Peach, Apple & Cider Brandy & Whiskey by re-distillation and filtration ...."
[Gallipolis: 1830].
Brs. Incl. Instructions For Rectifying. LCP

1591. MAULE, JONATHAN, et al.
[No title; text begins:] Esteemed Friend, As thou art now in Philadelphia, and will, it is altogether probable, have an opportunity of seeing many friends ....
[New York?: 1820].
2 p. Dated at head: Chaumont, 4th Mo. 10, 1820. Signed at end: Jonathan Maule, George Truman, etc.; with Note signed: John Brown & Samuel Brown. Compares agricultural resources of Jefferson Co., N.Y., and Bucks Co., Pa. LCP

1592. MAXCY, VIRGIL, 1785-1844.
Address To The Agricultural Society Of Maryland. Delivered At their Anniversary Meeting held in the Chamber of the House of Delegates, at Annapolis, On Wednesday the 15th December, 1819.
Annapolis: Printed By J. Green, January 1820.
19 p. APS (last leaf impf.); HSP

MC---
See also under
MAC---
M'H---

1593. McCORMICK, CYRUS HALL, 1809-1884.
Please Read! Memoranda On The Case Of C.H. McCormick, For The renewal of his Patent, now before Congress. April 4, 1854. With Additions Made Since A Trial At Law. January 2, 1855. [cap. title]
[N.p.: 1855].
14 p. Incl. remarks on his reaping machine. LCP

________.
See also
MANNY, JOHN H. Remonstrance ... against the extension of Letters Patent of .... [1855].
SEWARD, WILLIAM HENRY, 1801-1872. The Reaper .... 1854. Patent infringement case.

1594. McCOUN, WILLIAM T.
Address, Delivered Before The Queen's County [N.Y.] Agricultural Society, At Its First Anniversary, At Hempstead, October 13th, 1842.
New York: Vinten, Printer, 1843.
16 p. LCP

1595. McCULLOH, RICHARD SEARS.
Reply To The Review By Mr. Avequin, Of New Oreleans, Of M'Culloh's Report On Sugar. Published In De Bow's Commercial Review.
Philadelphia: C. Sherman, Printer, 1848.
41 p. J.B. Avequin's review appeared in the July, 1848 issue of De Bow's Review. HSP

1596. McKILLOP, JOHN & CO., Commercial agents, Baltimore.
Harvest Prospects, July, 1855, Compiled From Authentic and Reliable Sources, Obtained From Local Residents In Nearly Every County In Each State; With Summary and General Remarks. By John McKillop & Co., Of The Commercial Agency .... [wrapper title]
Baltimore: "The Printing Office," 1855. [wrapper imprint]
42 p. + slip tipped in at end; signed: Tappan & McKillop, New-York, August 1, 1855. Imprint covered by label of: Fras. W. Tappan & Co. Commercial & Collecting Agency .... LCP

McMINN, J.M. A Contribution To The Classification Of The Species And Varieties of The Grape Vine.
See [SPANGLER, ANDREW M., pub.]. Both Sides Of The Grape Question .... 1860. Incl. McMinn's essay.

1597. McNAUGHTON, JAMES, 1796-1874.
Address Delivered Before The New-York State Agricultural Society, At The Capitol In The City Of Albany. February 10, 1837.
Albany: Printed By Packard And Van Benthuysen, 1837.
16 p. LCP

1598. MEASE, JAMES, 1771-1846.
An Address On The Progress Of Agriculture, With Hints For Its Improvement In The United States: Delivered Before The Philadelphia Society for Promoting Agriculture, At Their Annual Meeting, January 14th, 1817.
[Philadelphia:] Published By Order Of The Society, 1817.
45 p. LCP; APS; HSP

1599. __________.
An Address On The Progress Of Agriculture, With Hints .... Delivered Before The Philadelphia Society for Promoting Agriculture, At Their Annual Meeting, January 14th, 1817.
[Philadelphia: 1817].
Pp. [1-2], [v]-xl, 41-45. This copy is from the same setting of type as is the above copy except for t.p. and page numbers. Without list of officers (p. 3 in above copy) and with the latter's p. 4 text on t.p. verso instead. HSP

1600. __________.
Address On The Subject Of Establishing A Pattern Farm In The Vicinity Of Philadelphia: Delivered At The Annual Meeting Of The "Philadelphia Society For Promoting Agriculture."
Philadelphia: Printed By Order Of The Society, 1818.
14 p. LCP; APS; HSP

1601. __________.
Introductory Lectures To A Course Of Lectures Upon Comparative Anatomy, And The Diseases Of Domestic Animals. Delivered November 3, 1813.
Philadelphia: Printed By Lydia R. Bailey, 1814.
[4], 51 p. Delivered before the Philadelphia Society For Promoting Agriculture. LCP; APS; HSP

__________. Letter From James Mease Transmitting A Treatise On The Rearing Of Silkworms ....
See HAZZI, JOSEPH VON, 1768-1845.

1602. __________.
On Some Of The Vegetable Materials From Which Cordage, Twine And Thread, Are Made.
[New Haven: 1832].
14 p. Tp. headed: From the American Journal of Science and Arts, No. 1. Vol. XXI. LCP; APS

__________.
See also Archives Of Useful Knowledge .... By .... [pdcl.].

1603. MECHI, JOHN JOSEPH, 1802-1880.
How To Farm Profitably; Or, The Sayings & Doings Of Mr. Alderman Mechi. With Illustrations, From Photographs By Mayall.
London and New York: Routledge, Warnes, & Routledge [London: Cox And Wyman, Printers (colop.)], 1859.
[2 (pub. adv.)], iv, 284, [2 (pub. adv.)] p.; front.; 3 pls. LCP

1604. The Medical and Agricultural Register. Edited by Daniel Adams.
Boston.
LCP; APS: Vol. 1, nos. 1-24 (Jan. 1806-Dec. 1807). All pub.

1605. MEEHAN, THOMAS, 1826-1901.
The American Handbook Of Ornamental Trees.
Philadelphia: Lippincott, Grambo, And Co., 1853.
Pp. [i]-xv, [1], [25]-257. LCP; APS; HSP

__________.
See also
The Gardener's Monthly And Horticultural Advertiser [pdcl.].
THOMAS, M., & SONS, Auctioneers, Philadelphia. For Sale: Springbrook .... [1857]. Incl. description of the gardens by Meehan.

1606. MEIGS, HENRY, 1782-1861.
An Address, On The Subject Of Agriculture And Horticulture Delivered In The Church Of The Messiah On Thursday, October 9th, 1845.
New York: James Van Norden & Co., Printers, 1845.
10 p. T.p. headed: American Institute. LCP

1607. __________.
Agricultural Address, Delivered In Castle Garden By Direction Of The American Institute, October 9th, 1846. Also, An Address To The Band Of Blind Musicians, October 19, 1846.
New York: Joseph H. Jennings, Printer, 1846.
11 p. T.p. headed: American Institute.--19th Annual Fair--1846. LCP

1608. MELISH, JOHN, 1771-1822.
Information And Advice To Emigrants To The United States: And From The Eastern To The Western States: Illustrated By A Map Of The United States, And A Chart Of The Atlantic Ocean.
Philadelphia: Printed For And Published By John Melish, 1819.
viii, 144, [6 (pub. adv.)] p.; fold. front. map; fold. map. LCP

1609. A Memorial On The Practicality Of Growing Vineyards In The State Of South-Carolina. Read By The Author To The Agricultural Society, At Their Extra Meeting, On Wednesday, The 7th Of November, 1798, And Published Under Their Patronage.
Charleston: Printed By W.P. Young And T.C. Cox, 1798.
21 p. HSP in LCP

MERCER COUNTY [PA.]. AGRICULTURAL SOCIETY.
See GOWEN, JAMES. Address Delivered Before .... 1853.

M'H--
See also under
MAC---
MC---

1610. M'HARRY, SAMUEL.
The Practical Distiller: Or An Introduction To Making Whiskey, Gin, Brandy, Spirits, &c. &c. Of Better Quality, And In Larger Quantities, Than Produced By The Present Mode Of Distilling, From The Produce Of The United States: Such As Rye, Corn, Buck-Wheat, Apples, Peaches, Potatoes, Pumpions And Turnips .... Together With Directions For Purifying, Clearing And Colouring Whiskey .... And Sundry Extracts Of Approved Receipts For Making Cider ....
Harrisburg, (Penn.): Published By John Wyeth, 1809.
184 p. HSP in LCP; LCP; HSP

1611. MICHIGAN. STATE AGRICULTURAL COLLEGE.
The Agricultural College Of The State Of Michigan.
Lansing, Michigan: Hosmer & Fitch, Book And Job Printers, 1857.
58 p. LCP (wrapper)

1612. MICHIGAN. STATE AGRICULTURAL SOCIETY.
Transactions.
Lansing.
LCP: 1849-54; 1856-58.

MIDDLESEX CO. [CONN.]. AGRICULTURAL SOCIETY.
See MATHER, THOMAS G. Address Delivered Before .... 1843.

MIDDLESEX CO. [MASS.]. SOCIETY OF HUSBANDMEN AND MANUFACTURERS.
See
BOUTWELL, GEORGE SEWALL, 1818-1905. Address Delivered At Concord ... Before .... 1850.
BRIGGS, CHARLES. A Discourse Delivered .... 1825.
CHENEY, JOHN M. Address Delivered Before .... 1831.
PHINNEY, ELIAS, 1780-1849. Address Delivered Before .... 1830.

1613. MILES, WILLIAM.
A Plain Treatise On Horse-Shoeing. With Illustrations.
Philadelphia: Henry Carey Baird, (Sucessor To E.L. Carey) [Stereotyped By L. Johnson & Co.; Printed By T.K. And P.G. Collins (t.p. verso)], 1856.
48, 16 (pub. adv.) p.; 8 pls. LCP (with pub. adv. p. bound: 1-8, 13-16, 9-12)

1614. MILLER, JOSEPH.
Der verborgene Arzt, oder Nutzliche Hausfreund; Ein neues System der Arzeney-Kunst: Hauptsachlich auf Krauter gegrundet. Enthaltend: Erstens: - Eine kurze Beschreibung des menschlichen Korpers .... Zweytens - Eine Abhandlung von den Entstehung der ... Krankheiten .... Aus den besten Schriften zusammengetragen, nach dem beruhmten Thompsonien System organisiert ....
New Berlin, Penn.: Gedruckt und zu haben bey Joseph Miller, 1830.
179, [1] p. HSP in LCP

MILLIS, JOHN, comp.
See MARKHAM, GERVASE, 1568?-1637. The Citizen And Countryman's Experienced Farrier .... 1764 etc.

1615. MILLS, JOHN, d. 1784.
The Modern System Of Farriery: Showing the most approved Methods Of Breeding, Rearing, and Fitting for Use, all kinds of Horses; with Directions for the proper Treatment of them in their several Disorders. To which is added, A successful Method of treating the Canine Species in that destructive Disease called the Distemper.
Boston: Printed and sold by William Spotswood, 1796.
[4], 274, [10] p.; last 2 p. pub. adv. HSP in LCP

1616. __________.
A Treatise On Cattle: Showing The Most Approved Methods Of Breeding, Rearing, and Fitting for Use, Asses, Mules, Horned Cattle, Sheep, Goats, and Swine. With Directions For The Proper Treatment Of Them In Their Several Disorders: To which is added, A Dissertation on their Contagious Diseases. Chiefly Collected from the best Authorities, and interspersed with Remarks.
Boston: Printed And Sold By William Spotswood, 1795.
iv, 215, [9] p. LCP

1617. MINER, T.B.
The American Bee Keeper's Manual; Being A Practical Treatise On The History And Domestic Economy Of The Honey-Bee, Embracing A Full Illustration Of The Whole Subject, With The Most Approved Methods Of Managing This Insect Through Every Branch Of Its Culture, The Result Of Many Years' Experience. Embellished By Thirty-Five Beautiful Engravings.
New York: Published By C.M. Saxton [C.W. Benedict, Stereotyper (t.p. verso)], 1849.
349, [1 (pub. adv.)] p.; front.; illus. LCP

1618. MINOR, FRANKLIN, 1812-1867.
An Address Delivered At The Third Annual Fair Of The Virginia State Agricultural Society. (Published by Order of the Society.)
Richmond: Printed By T. Bailie, 1855.
5 [i.e. 15] p. Pp. 14-15 misnumbered 4-5. LCP

MISSISSIPPI. GEOLOGICAL AND AGRICULTURAL SURVEYS.
See
HILGARD, EUGENE WOLDEMAR, 1833-1916. Report On The Geological & Agricultural Survey .... 1858 etc.
WAILES, BENJAMIN LEONARD COVINGTON, 1797-1862. Report On The Agriculture And Geology .... 1854.

1619. MITCHELL, ELISHA, 1793-1857.
Report On The Geology Of North-Carolina. Conducted under the direction of the Board of Agriculture. Part III. November, 1827.
[Raleigh: 1827].
60 p. Incl. (pp. 45-60): A Report On The Establishment Of Cotton And Woollen Manufactures ... by Charles Fisher. For Part II see OLMSTED, DENISON, 1791-1859. LCP

1620. [MITCHELL, JOHN, d. 1768].
The Present State Of Great Britain And North America, With Regard To Agriculture, Population, Trade, and Manufactures, impartially considered: Containing a particular Account of the dearth and scarcity of the necessaries of life in England; the want of staples commodities in the Colonies .... In Which The causes and consequences of these growing evils, and methods of preventing them, are suggested; The proper Regulations for the Colonies, and the taxes imposed upon them, are considered ....
London: Printed for T. Becket and P.A. de Hondt, 1767.
[8], xvi, 363, [1] p. HSP in LCP; LCP

__________, possible author.
See American Husbandry .... 1775.

MITCHELL, S. AUGUSTUS, possible author.
See Illinois In 1837 .... 1837.

1621. MITCHILL, SAMUEL LATHAM, 1764-1831.
Address Pronounced Before The New-York Horticultural Society, In The Literary And Philosophical Hall Of The Institution, On The Annual Celebration, August 29, 1826.
New York: Printed By E. Conrad, 1826.
32 p. APS

1622. MONROE CO. [N.Y.]. AGRICULTURAL SOCIETY.
Annual Report Of The Transactions Of The Monroe County Agricultural Society, Containing The Address At the Exhibition and Fair of the Society at Rochester, October 26, 1842, By Henry Colman, President of the Society; The Reports of the Committees, and of the Premiums awarded, and a list of the Officers And Members Of The Society, For 1842. Published by order of the Executive Committee.
Rochester: Strong & Dawson, Printers, 1842.
68 p. LCP (wrappers)

MONROE GARDEN AND NURSERIES, Greece, N.Y.
See ROWE, ASA. Annual Catalogue ... Cultivated And For Sale At .... 1841.

MONTGOMERY CO. [MD.]. AGRICULTURAL SOCIETY.
See
HALLOWELL, BENJAMIN, 1799-1877. Address Of ... At The Meeting Of .... 1852.
HOLCOMB, CHAUNCEY P. Address Delivered Before .... 1854.

MONTGOMERY CO. [N.Y.] AGRICULTURAL SOCIETY.
See PAIGE, ALONZO CHRISTOPHER, 1797-1868. Address Before .... 1860.

MONTGOMERY CO. [PA.]. AGRICULTURAL SOCIETY.
See
FISHER, SIDNEY GEORGE, 1809-1871. Address Delivered Before .... 1859.
TYSON, JOB ROBERTS, 1803-1858. Address Delivered Before .... 1856.
WALL, JAMES WALTERS, 1820-1872. Address Of ... Before .... 1860.

1623. [MOORE, CLEMENT CLARK, 1779-1863].
An Inquiry Into The Effects Of Our Foreign Carrying Trade Upon The Agriculture, Population, And Morals Of The Country. By Columella [pseud.].
New York: Printed By D. And G. Bruce, For E. Sargeant, 1806.
61 p. APS

1624. MOORE, THOMAS, 1760-1822.
An Essay On The Most Eligible Construction Of Ice-Houses. Also, A Description Of The Newly Invented Machine Called The Refrigerator.
Baltimore: Printed By Bonsal & Niles, 1803.
28 p. APS

1625. __________.
The Great Error Of American Agriculture Exposed: And Hints For Improvement Suggested.
Baltimore: Printed By Bonsal And Niles, For The Author, 1801.
72 p. On ploughing. APS; HSP

1626. [MOREAU DE SAINT MÉRY, MÉDÉRIC LOUIS ÉLIE, 1750-1819].
Essai Sur La Manière D'Améliorer L'Éducation des Chevaux En Amérique.
Philadelphia: Imprimé & vendu par Moreau De Saint-Méry, Imprimeur-Libraire, Novembre 1795.
iv, 40 p. LCP; APS

1627. [__________].
An Essay On The Manner Of Improving The Breed Of Horses In America.
Philadelphia: Printed & sold by Moreau De Saint-Méry, Printer and Bookseller, October 1795.
iv, 39 p. LCP

1628. MORFIT, CAMPBELL, 1820-1897.
Manures, Their Composition, Preparation, And Action Upon Soils; With The Quantities To Be Applied. Being A Field Companion For The Farmer. From The French Of Standard Authorities.
Philadelphia: Lindsay And Blakiston [Wm. S. Young, Printer (t.p. verso)], 1848.
Pp. [i]-vii, [13]-100. LCP

1629. MORRELL, LUKE A.
The American Shepherd: Being A History Of The Sheep, With Their Breeds, Management, And Diseases. Illustrated With Portraits Of Different Breeds, Sheep Barns, Sheds &c. With An Appendix, Embracing Upwards Of Twenty Letters From Eminent Wool-Growers And Sheep-Fatteners Of Different States, Detailing Their Respective Modes Of Management.
New York: Harper & Brothers, 1845.
Pp. [i]-xxii, [13]-437; illus. LCP

1630. __________.
The American Shepherd: Being A History Of The Sheep, With Their Breeds, Management, And Diseases .... With An Appendix ....
New York: Harper & Brothers, 1846.
Pp. [i]-xxii, [13]-437, [1], 1-6 (pub. adv.); illus. LCP

1631. MORRIS, JAMES PEMBERTON.
An Address, Delivered Before The Agricultural Society Of Bucks County [Pa.], On The Twenty-Eighth Of April, 1823. At the Request of the Society.
Doylestown: Printed By W.T. Rogers, 1823.
12 p. HSP

1632. __________.
An Address, Delivered Before The Agricultural Society of Bucks County, On the 28th of April, 1823. At The Request Of The Society.
Philadelphia: Clark & Raser, Printers, 1823.
12 p. HSP

1633. MORRIS, PASCHALL & CO., Nurserymen, West Chester, Pa.
Catalogue Of Fruit And Ornamental Trees, Evergreens, Flowering Shrubs & Plants, Roses, &c., Cultivated And For Sale By Morris, Paschall & Co., At Their Garden And Nursery, West Chester, Penn'a. [wrapper title]
West Chester: Register & Examiner Steam Press, 1852. [wrapper imprint]
16 p. APS

1634. MORSE, JEDIDIAH, 1761-1826.
A Description Of The Soil, Productions, Commercial, Agricultural and Local Advantages Of The Georgia Western Territory: Together With a summary and impartial View of the Claims of Georgia and of the United States to this territory .... Collected And Stated From Various Authentic Documents. Extracted, and published in this form, (by Permission) from Rev. Dr. Morse's American Gazetteer, A New Work. Illustrated with a New and correct Map.
Boston: Printed by Thomas & Andrews, 1797.
24 p.; fold. front. map. HSP in LCP

1635. MORSE, OLIVER ANDREW, 1815-1870.
Address, Delivered Before The Otsego County [N.Y.] Agricultural Society, At Cooperstown, October 6, 1842. Published By The Society.
Cooperstown: Printed By H. & E. Phinney, 1842.
16 p. LCP (wrapper)

1636. MOSELEY, EBENEZER, 1781-1854.
An Address To The Essex County [Mass.] Agricultural Society, At New Rowley, September 25, 1834; At Their Annual Cattle Show. Published by order of the Society.
Salem: Printed By Foote And Chisholm, 1835.
Pp. [1-2], [5]-23. LCP

MOUBRAY, B., pseud.
See LAWRENCE, JOHN, 1753-1839.

1637. MOULTON, JOSEPH WHITE, 1789-1875.
An Address Delivered At St. Paul's Church, Buffalo, On The Anniversary Celebration Of The Niagara And Erie Society For Promoting Agriculture and Domestic Manufactures, The 30th Day Of October, 1821.
Buffalo: Printed By D.M. Day, And H.A. Salisbury, 1821.
36 p. LCP

1638. MT. AIRY [PA.]. AGRICULTURAL INSTITUTE.
Catalogue Of The Mount Airy Agricultural Institute, Germantown, Pa., Designed For Instruction In Scientific And Practical Agriculture, Mathematics And The Natural Sciences.
Philadelphia: T.K. And P.G. Collins, 1849.
7, [1] p. HSP (wrappers)

__________.
See also HYATT, JAMES, d. 1904. Lime And Marl .... 1848.

MOUNT HOPE NURSERIES, Rochester, N.Y.
See ELLWANGER & BARRY, Nurserymen, Rochester, N.Y.

1639. MUNN, B.
A Practical Land Drainer. A Treatise on Draining Land. In Which The Most Approved Systems Of Drainage And The Scientific Principles On Which They Depend, Are Explained And Their Comparative Merits Discussed. With Full Directions For Cutting And Making Drains, And Remarks Upon The Various Materials Of Which They May Be Constructed. Numerously Illustrated.
New York: C.M. Saxton & Company, Agricultural Book Publishers [C.O. Jenkins, Printer & Stereotyper (t.p. verso)], 1856.
Pp. [1-2], vii-ix, [1], [7]-190; illus. LCP

1640. MUSE, JOSEPH E.
An Address Delivered At Their Tent, Before The Dorchester [Md.] Agricultural Society And A Large Assemblage Of The Citizens Of Dorchester, And Some Of The Adjoining Counties, At The Fourth Exhibition And Fair, Held In Cambridge, Nov. 2, 1837. Published By The Board.
Baltimore: Printed By John D. Toy, 1837.
29 p. LCP (wrappers)

1641. __________.
Address To The Agricultural Society, Of New-Castle County, At The Tenth Annual Meeting, Assembled At Wilmington, Del., on Sept. 18th, 1845, Delivered At The Request Of The Society. Published By Order Of The Society.
Wilmington: Evans & Vernon, Printers, 1845.
17 p.; errata slip tipped in at front. LCP

1642. MUSKINGUM [OHIO]. AGRICULTURAL AND MANUFACTURING SOCIETY. COMMITTEE.
Report To The Muskingum Agricultural and Manufacturing Society, On The Advantages Of An Experiment Farm.
Zanesville: Printed by Peters & Pelham, 1829.
[1] p. Signed p. 9: M.B. Cushing, Ch'm. With printed "Notice," signed by the president and 2 others, pasted on final blank leaf. LCP; HSP

1643. NASH, JOHN ADAMS, 1798-1877.
The Progressive Farmer: A Scientific Treatise On Agricultural Chemistry, The Geology Of Agriculture; On Plants, Animals, Manures, And Soils. Applied To Practical Agriculture.
New York: C.M. Saxton, Agricultural Book Publisher [E.O. Jenkins, Stereotyper And Printer (t.p. verso)], 1853.
254, 10 (pub. adv.) p. LCP

1644. NATIONAL CONVENTION OF FARMERS, GARDENERS AND SILK CULTURISTS, New York, Oct. 12-16, 1846.
Proceedings Of The National Convention Of Farmers, Gardeners And Silk Culturists, Held At Mechanics' Hall, In The City Of New York, On The 12th, 13th, And 16th Days Of October, 1846, In Connection With The Nineteenth Annual Fair Of The American Institute.
New York: Joseph H. Jennings, Printer, 1846.
50 p. LCP; APS

1645. NATIONAL CONVENTION OF FRUIT GROWERS, New York, Oct. 10, 1848.
Proceedings Of The National Convention Of Fruit Growers, Held In The City Of New-York, October 10, 1848, Under The Auspices Of The American Institute.
New York: Van Norden & Amerman, Printers, 1848.
52 p. LCP

1646. NATIONAL CONVENTION OF SILK GROWERS AND SILK MANUFACTURERS, New York, Oct. 13-14, 1843.
The Silk Question Settled. The Testimony Of One Hundred Fifty Witnesses. Report Of The Proceedings Of The National Convention Of Silk Growers And Manufacturers, Held In New York, Oct. 13th and 14th, 1843. Published Under The Direction Of The American Institute.
New York: Saxton And Miles, Publishers, 1844.
80 p. LCP; HSP

1647. __________.
The Silk Question Settled .... Report Of The Proceedings .... Second Edition, With Additions.
Boston: Printed By T.R. Marvin, 1844.
79, [1] p. APS (wrappers)

Necessaries; Best Product Of Land .... 1776.
See BORDLEY, JOHN BEALE, 1727-1804.

1648. NEFFLEN, JOHN.
A Method Of Increasing The Yield Of The Milch-Cow, By Selecting The Proper Animals For The Dairy; According To Guenon's Discovery. Tested And Verified By Many Years Of Observation And Of Experiments. Improved And Simplified, As Classified And Arranged, By John Nefflen. With Seventy-Eight Illustrations.
Philadelphia: C.B. Rogers; And For Sale By Lippincott, Grambo, And Co., 1853.
54, 36 (pub. adv.) p.; fold. front. HSP (wrappers)

1649. NEILL, PATRICK, 1776-1851.
The Fruit, Flower, And Kitchen Garden. Adapted To The United States, From The Fourth Edition, Revised And Improved By The Author.
Philadelphia: Henry Carey Baird [T.K. And P.G. Collins, Printers (t.p. verso)], 1851.
xii, 427, [1], 8+4 (pub. adv.) p. incl. front.; illus. LCP

1650. Neuer Erfahrner, Amerikanischer, Haus-und Stallarzt, mit den natürlichsten, und leichtesten Mitteln, wider alle Kranckheiten, und Schwachheiten, der Menschen und Viehe. Zum Nutzen der Deutschen Nation in den Vereinigten Staaten.
Frederick [Md.]: Gedruckt, verlegt und zu haben bey Matthias Bärtgis, 1794.
[15], 84, 36, [4] p. Sep. t.p. and pagination for: Zweyter Theil des neuen erfahrnen .... LCP

1651. The New American Pocket Farrier And Farmer's Guide In The Choice And Management Of Horses, Neat Cattle, Sheep And Swine: Including A Description Of Their Internal Structure Their Digestive System, The Diseases To Which They Are Liable, With Their Causes, Symptoms, And Most Approved Methods Of Cure: From The Writings Of Youatt, Lawrence, Hines, White, Clayter, And Others. To Which Is Added A Variety Of Agricultural And Miscellaneous Receipts.
Philadelphia: Published By Leary, Getz & Co. [Stereotyped by S. Douglas Wyeth (t.p. verso)], [1845].
284 p.; illus. LCP

1652. NEW CASTLE CO. [DEL.]. AGRICULTURAL SOCIETY.
Transactions.
Wilmington.
LCP: Vols. 8-10 (1843-45)
HSP: Vols. 8-9 (1843-44)

__________.
See also
EMERSON, GOUVERNEUR, 1796-1874. An Address Delivered Before .... 1855.
FISHER, SIDNEY GEORGE, 1809-1871. An Address Delivered Before .... 1860.
GIBBONS, WILLIAM, 1781-1845. Essay On The Culture Of Silk; Read Before .... 1836.
MUSE, JOSEPH E. Address To .... 1845.
SKINNER, JOHN STUART, 1788-1851. Address Delivered Before .... [1844].

1653. NEW CASTLE CO. [DEL.]. TEMPERANCE CONVENTION.
Report Of The Committee On "The Nature And Effects Of Still Slops As Food For Cattle And Hogs," To The New Castle County Temperance Convention. June 16, 1846.
[Wilmington, Del.: Printed By Washington Lowe (colop.)], [1846].
12 p. HSP

The New England Farrier; In Four Parts .... 1824.
See JEWETT, PAUL.

1654. The New Family Receipt Book, Containing Eight Hundred Truly Valuable Receipts In Various Branches Of Domestic Economy, Selected From The Works Of British And Foreign Writers Of Unquestionable Authority And Experience, And From The Attested Communications Of Scientific Friends. A New Edition, Corrected.
Philadelphia: Published By Collins & Croft; William Brown, Printer, 1818.
xxvii, [1], 402 p. Incl. chapters on Agriculture, Cattle, Dairy, Farriery, Gardening, etc. HSP in LCP

1655. The New Family Receipt-Book, Containing Eight Hundred Truly Valuable Receipts .... A New Edition, Corrected.
New Haven: Published By Howe & Spalding, And Samuel Wadsworth [A.H. Maltby & Co. Printers (t.p. verso)], 1819.
Pp. [i]-xxxi, [1], [37]-429, [1-7 (pub. adv.)]. LCP

1656. NEW HAMPSHIRE. STATE AGRICULTURAL SOCIETY.
Transactions.
Concord; Manchester; Dover.
APS: 1850-1860. All pub.

__________.
See also KING, WILLIAM STERLING, 1818-1882. Address Delivered Before .... 1853.

1657. NEW HAMPSHIRE. STATE BOARD OF AGRICULTURE.
The New-Hampshire Agricultural Repository. Published By The State Board Of Agriculture, Under The Patronage Of The Legislature Of The State of New-Hampshire.
Concord: Printed By Hill And Moore, 1822.
135 p. T.p. headed: No. I. All pub.? LCP

NEW HAVEN CO. [CONN.]. AGRICULTURAL SOCIETY.
See
BUEL, JESSE, 1778-1839. Address ... Delivered Before .... 1839.
__________. The Farmer's Companion .... With The Address ... Delivered Before The Agricultural And Horticultural Societies .... 1842.
COLMAN, HENRY, 1785-1849. Agricultural Addresses Delivered At .... 1840.

1658. [NEW HAVEN CO. (CONN.). HORTICULTURAL SOCIETY.]
Report Of The Managers Of The Horticultural Fair, Held September 25th, 1838. At The State-House, In New-Haven.
New Haven: Babcock & Galpin, Printers, 1838.
11, [1] p. Signed at end: Smith Dayton, Secretary. HSP

1659. __________.
Transactions.
New Haven.
LCP: Vol. I, no. 3 (1841)

__________.
See also
BACON, LEONARD, 1802-1881. An Address .... With Transactions Of The Society .... 1848.
BUEL, JESSE, 1778-1839. Address ... Delivered Before .... 1839.
__________. The Farmer's Companion .... With The Address ... Delivered Before The Agricultural And Horticultural Societies .... 1842.
COLMAN, HENRY, 1785-1849. Agricultural Addresses Delivered At .... 1840.
PARSONS, SAMUEL BOWNE, 1819-1906. An Address Delivered At The Annual Fair .... 1849.
SHEPARD, CHARLES UPHAM, 1804-1886. Address Delivered Before .... 1838.

1660. NEW JERSEY. STATE AGRICULTURAL SOCIETY.
Annual Report of the Agricultural Society, Of The State Of New Jersey, For The Year 1859. Ordered To Be Printed.
Trenton, N.J.: Printed At The "True American" Office, 1860.
100 p. LCP

1661. __________.
List Of Premiums And Regulations, For The Second Annual Exhibition, To Be Held In The City Of Newark, On the 9th, 10th, 11th, and 12th days of September, 1856.
Somerville, N.J.: J.H. Frazee, 1856.
24 p. HSP

1662. __________.
List Of Premiums And Regulations, For The Third Annual Exhibition, To Be Held In The City Of New Brunswick, Four Days. Commencing Tuesday, September 29, and Ending Friday, October 2, 1857.
[Somerville: J.H. Frazee, 1857].
24 p. HSP

__________.
See also WALL, JAMES WALTER, 1820-1872. Our Age, And Its Contributions To Agriculture: An Address, Delivered .... 1859.

NEW LONDON AND WINDHAM COS. [CONN.]. AGRICULTURAL AND MECHANIC INSTITUTE.
See also COLMAN, HENRY, 1785-1849. Agricultural Addresses Delivered At .... 1840.

NEW YORK agricultural societies.
See the following individual societies:
NEW YORK. SOCIETY FOR THE PROMOTION OF AGRICULTURE, ARTS AND MANUFACTURES [founded 1791].
NEW YORK. STATE AGRICULTURAL SOCIETY [founded 1832].

1663. NEW YORK CITY. AMERICAN INSTITUTE.
Charter Of The American Institute Of The City Of New York. Incorporated May 2, 1829. Acompanied With The By-Laws. Adopted May, 1830.
New York: Printed By Elliott And Hegeman, 1832.
8 p. HSP

1664. __________.
Third Annual Report Of the American Institute, on the subject of Agriculture. To The Legislature Of The State Of New-York. [cap. title]
[Albany: 1844].
343 p. Title headed: State Of New-York. No. 124. April 22, 1844. APS

1665. __________.
Annual Report Of .... To the N.Y. State Agricultural Society Feb. 1846.
Albany: Carroll And Cook, Printers To The Assembly, 1846.
239 p. Sep. State t.p. headed: State Of New-York. No. 200. In Assembly, March 31, 1846. LCP

1666. __________.
Fifth Annual Report Of .... Made to the Legislature April 20, 1847.
Albany: C. Van Benthuysen And Co., Public Printers, 1847.
xvi, [8], 558 p. Sep. State t.p. headed: State Of New-York. No. 151. In Assembly, April 20, 1847. APS

1667. __________.
Sixth Annual Report Of .... Made to the Legislature, March 25, 1848.
Albany: Charles Van Benthuysen, Public Printer, 1848.
815 p. Sep. State t.p. headed: State Of New-York. No. 216. In Assembly, March 25, 1848. APS

1668. __________.
Seventh Annual Report Of .... Made to the Legislature, March 29, 1849.
Albany: Weed, Parsons & Co., Public Printers, 1849.
420 p.; 2 fronts. Sep. State t.p. headed: State of New-York. No. 244. In Assembly, Mar. 29, 1849. APS

1669. __________.
Eighth Annual Report Of .... Made to the Legislature, February 26, 1850.
Albany: Weed, Parsons & Co., Public Printers, 1850.
447 p.; front, Sep. State t.p. headed: State of New-York. No. 199. In Assembly, Feb. 26, 1850. APS

1670. __________.
Journal.
New York.
LCP: Vols. 1-2 (Oct. 1835-Sept. 1837)
APS: Vols. 1-4 (Oct. 1835-Sept. 1839). All pub. of 1st series.

1671. __________.
List Of Premiums Awarded By The Managers of the Nineteenth Annual Fair Of The American Institute, October, 1846. [cap. title]
[New York: 1846].
24 p. HSP

1672. __________.
Report Of The Committee On Horticulture, In Conjunction With The Agricultural Board Of The American Institute, At Their Eighteenth Annual Fair, Held At Niblo's Garden, October, 1845.
New York: Published For The Managers Of The American Institute, By T. Bridgeman; Printed by D. Mitchell, 1845.
16 p. LCP

1673. __________.
Report Of The Third Annual Fair Of The ... Held At Masonic Hall, October, 1830; With A List of Premiums, And A Catalogue Of All The Articles Exhibited, By Whom Made And Sold, With The Duties Imposed On Similar Imported Commodities.
New York: Printed By J. Seymour, 1830.
32 p. HSP

1674. __________.
Transactions.
Albany.
APS: 1850-1864/5.

__________.
See also
BACHE, ALEXANDER DALLAS, 1806-1867. Anniversary Address ... During The Twenty-Eighth Annual Fair .... 1857.
BALDWIN, HENRY. Anniversary Address Before .... 1834.
BARNARD, DANIEL DEWEY, 1797-1861. Anniversary Address, Delivered Before .... 1843.
BURCHARD, NATHAN. Address ... Delivered At The Request Of .... 1847.
BURGES, TRISTAM, 1770-1853. Address Delivered ... At The Third Annual Fair .... 1830.
CHANDLER, ADONIRAM. Address, Delivered At .... 1846.
ELIOT, THOMAS DAWES, 1808-1870. Anniversary Address Delivered Before .... 1845.
EVERETT, EDWARD, 1794-1865. Address Delivered Before .... 1831.
KENNEDY, JOHN PENDLETON, 1795-1870. An Address Delivered Before .... 1833.
LYNCH, JAMES. Address, Delivered Before .... 1829.
MAPES, JAMES JAY, 1806-1866. Address Delivered ... At The 18th Annual Fair .... 1845.
MASON, CYRUS. The Oration On The Thirteenth Anniversary .... 1840.
MEIGS, HENRY, 1782-1861. An Address, On The Subject Of Agriculture .... 1845.
__________. Agricultural Address, Delivered .... 1846.
NATIONAL CONVENTION OF FARMERS, GARDENERS AND SILK CULTURISTS. Proceedings Of The ... Held ... In Connection With The Nineteenth Annual Fair .... 1846.
NATIONAL CONVENTION OF FRUIT GROWERS. Proceedings ... Under The Auspices Of .... 1848.
NATIONAL CONVENTION OF SILK GROWERS AND SILK MANUFACTURERS. The Silk Question Settled .... Published Under The Direction of .... 1844.
TALLMADGE, JAMES, 1778-1853. Address .... 1841 etc.
WAKEMAN, THADDEUS B. An Introductory Lecture, Delivered .... 1835.

New York Convention. Report On The Production And Manufacture Of Cotton. 1832.
See FRIENDS OF DOMESTIC INDUSTRY.

NEW YORK. CORRESPONDING ASSOCIATION FOR THE PROMOTION OF INTERNAL IMPROVEMENTS.
See [HAINES, CHARLES GLIDDEN, 1792-1825]. An Examination Into The Expediency Of Establishing A Board Of Agriculture .... Published By .... 1819.

NEW YORK. DOMESTIC HORTICULTURAL SOCIETY OF THE WESTERN PARTS OF THE STATE.
See
HOLLEY, MYRON, 1779-1841. An Initiatory Discourse, Delivered ... Before .... 1828.
STOUT, Z. BARTON. An Address, Delivered Before .... 1830.

1675. The New-York Farmer And Horticultural Repository .... Under The Patronage Of The New-York Horticultural Society.
New York.
LCP: Vols. 1-3 (Jan. 1828-Dec. 1830)

NEW YORK. FARMERS CLUB.
See BURCHARD, NATHAN. Address ... Accompanied By Remarks ... Made At The New York Farmers Club ... By Charles L. Fleischman .... 1847.

NEW YORK horticultural societies.
See the following individual societies:
NEW YORK. HORTICULTURAL SOCIETY [founded 1818].
NEW YORK. STATE HORTICULTURAL SOCIETY [proposed 1824].

1676. NEW YORK. HORTICULTURAL SOCIETY [founded 1818].
Proceedings Of The New-York Horticultural Society, At The Celebration Of Its Tenth Anniversary, August 26, 1828.
New York: Printed By E. Conrad, 1828.
24 p. + "Notice" slip regarding the Society's periodical. LCP

__________.
See also
CARTER, NATHANIEL HAZELTINE, 1787-1830. Address Read Before .... 1827.
FRANCIS, JOHN WAKEFIELD, 1789-1861. An Address Delivered Before .... 1830.
HALSEY, ABRAHAM, 1790-1857. An Address, Delivered Before .... 1830.
HOSACK, DAVID, 1769-1835. An Inaugural Discourse, Delivered Before ... 1824.
MITCHILL, SAMUEL LATHAM, 1764-1831. Address Pronounced Before .... 1826.
The New-York Farmer And Horticultural Repository .... Under The Patronage of .... [pdcl.].
SCHROEDER, JOHN FREDERICK, 1800-1857. The Intellectual And Moral Resources Of Horticulture. An Anniversary Discourse .... 1828 etc.

1677. NEW YORK. LEGISLATURE. ASSEMBLY. COMMISSIONERS APPOINTED TO MATURE AND REPORT A PLAN FOR AN AGRICULTURAL COLLEGE ....
Report Of the commissioners appointed to mature and report a plan for an Agricultural College and Experimental Farm. [cap. title]
[Albany: 1850].
18 p. Title headed: State of New-York. No. 30. In Assembly, Jan. 2, 1850. LCP

1678. NEW YORK. LEGISLATURE. ASSEMBLY. COMMITTEE ON TRADE AND MANUFACTURES.
Report of the Committee on trade and manufactures, on the memorial of sundry farmers of the counties of Madison and Otsego, relative to the alteration of the law regulating the inspection of Hops. [cap. title]
[Albany: 1846].
3 p. Title headed: State Of New-York. No. 92. In Assembly, February 11, 1846. LCP

NEW YORK. LYCEUM OF NATURAL HISTORY.
See VAN RENSSELAER, JEREMIAH. An Essay On Salt .... Delivered As A Lecture Before .... 1823.

NEW YORK. SOCIETY FOR THE PROMOTION OF AGRICULTURE, ARTS AND MANUFACTURES [founded 1791].
This society was re-incorporated in 1804 under the name Society for the Promotion of Useful Arts; in 1824 it was merged into the Albany Insitute.

1679. __________.
Agricultural Society, Of the State of New-York. Plan for collecting the grasses and other plants of the state of New-York into an Herbarium -- in a communication from Doctor David Hosack, addressed to Robert R. Livingston ....
[New York: 1795].
Brs. APS

1680. __________.
Transactions.
New York; Albany.
LCP: Vol. 1, part 2 (1794); vol. 1, Second Edition Revised (1801); vol. 2 (1807); vol. 4, part 2 (1819)
APS: Vol. 1, parts 1-2 (1792-94); vol. 1, 2nd edn.; 2; 3 (1814); 4, parts 1-2 (1816-1819). All pub.
HSP: Vol. 1, part 1; vol. 1, 2nd edn.; 2-4.
Later (1804-19) known as: Transactions of the Society for the Promotion of Useful Arts.

__________.
See also BROWNELL, THOMAS CHURCH, 1779-1865. Annual Address ... Delivered ... Before .... 1815.

1681. NEW YORK. STATE AGRICULTURAL COLLEGE.
Charter Ordinances, Regulations And Course Of Studies. 1859.
Albany: Charles Van Benthuysen, 1859.
16 p.; front. LCP (wrapper)

NEW YORK. STATE AGRICULTURAL CONVENTION, Albany, Feb. 1832.
See NEW YORK. STATE AGRICULTURAL SOCIETY. Proceedings Of The State Agricultural Convention .... With The Constitution Of A State Agricultural Society .... 1832.

1682. NEW YORK. STATE AGRICULTURAL SOCIETY.
Dedication Of The New York State Agricultural Rooms, Albany, February 12, 1857.
Albany: Van Benthuysen, Printer, 1857.
89, [2] p.; front. LCP

1683. __________.
Proceedings.
Albany.
LCP: Vols. 1-15 (1842-55); 17-19 (1857-59); 27 (1867)

1684. __________.
Proceedings Of The State Agricultural Convention, Held At the Capitol in the City of Albany, On the 14th, 15th and 16th February, 1832: With The Constitution Of A State Agricultural Society Agreed To And Adopted, By The Said Convention.
Albany: Printed by Webster and Skinners, 1832.
43 p. LCP

__________.
See also
ALLEN, LEWIS FALLEY, 1800-1890. Address Delivered Before .... 1849.
CHEEVER, SAMUEL. Address Delivered Before .... 1856.
The Cultivator .... [pdcl.]. Published by the Society.
DIX, JOHN ADAMS, 1798-1879. Address Delivered Before .... 1859.
DOUGLAS, STEPHEN ARNOLD, 1813-1861. Address Of ... At The Annual Fair .... 1852.
FITCH, ASA, 1809-1878. First Report On The ... Insects, Of The State of New York, Made To .... 1855.
__________. The Most Pernicious .... Delivered At The Annual Meeting .... 1860.
JENKINS, JOHN STILWELL, 1818-1852. The Life Of Silas Wright .... Containing ... His Address Read Before .... [1847].
McNAUGHTON, JAMES. Address Delivered Before .... 1837.
NEW YORK CITY. AMERICAN INSTITUTE. Annual Report Of ... To .... 1846.
NORTON, JOHN PITKIN, 1822-1852. Elements Of Scientific Agriculture .... Prize Essay Before .... 1850 etc.
SPENCER, JOHN CANFIELD, 1788-1855. Address Delivered At The Annual Exhibition .... 1848.

WRIGHT, JOSEPH ALBERT, 1810-1867. Address Of ... At The New York State Agricultural Fair .... 1855.

1685. NEW YORK. STATE BOARD OF AGRICULTURE.
Address Of The General Committee Of The Board Of Agriculture Of The State Of New-York, To The County Agricultural Societies, For 1820. With Accompanying Documents.
Albany: Printed For S. Southwick, At The Agricultural Press; By John O. Cole, 1820.
31, [1] p. LCP; APS

1686. __________.
Memoirs.
Albany.
LCP: Vols. 1-3 (1821-26)

__________.
See also
[HAINES, CHARLES GLIDDEN, 1792-1825.] An Examination Into The Expediency Of Establishing .... 1819.
The Plough Boy, And Journal Of The .... [pdcl.].
TIBBITS, GEORGE. A Memoir On The Expediency .... Read before .... 1827.

1687. NEW YORK. STATE HORTICULTURAL SOCIETY [proposed 1824].
Prospectus Of The New York State Horticultural Society. [cap. title]
[New York: 1824].
8 p. "The primary object of this Society will be, to establish a Garden ... for the express purpose of Horticultural and Botanical improvement ...." LCP

1688. NEWTOWN [PA.] AND PARTS ADJACENT IN DELAWARE CO. UNION ASSOCIATION FOR DETECTING HORSE THIEVES.
Constitution And Rules Of The Union Association of Newtown And Parts Adjacent In Delaware County, For Detecting Horse Thieves, And The Recovery Of Horses And Other Stolen Property. Adopted and subscribed January 31st, 1835.
Philadelphia: Printed For The Association; Wm. Sloanaker, 1835.
8 p. HSP

NIAGARA AND ERIE [N.Y.]. SOCIETY FOR PROMOTING AGRICULTURE AND DOMESTIC MANUFACTURES.
See MOULTON, JOSEPH WHITE, 1789-1875. An Address Delivered .... 1821.

1689. NICHOLS, ANDREW, 1785-1853.
An Address, To The Essex [Mass.] Agricultural Society, At Their First Cattle Show, At Topsfield, Oct. 5, 1820.
Salem: Printed By John D. Cushing, 1821.
24 p. LCP

1690. NICHOLSON, JOHN.
The Farmer's Assistant. The Subjects On Which This Work Treats Are Numerous .... [sic] Embracing Every Article, Relating To Agriculture, Arranged In Alphabetical Order.
Albany: Printed By H.C. Southwick, 1814.
334 [i.e. 338] p. P. 338 misnumbered 334. LCP

1691. __________.
The Farmer's Assistant: Being A Complete Treatise On Agriculture In General. Also, Upon The Various Subjects, Of Gardening, Fruit-Trees, Farriery, And The Dairy. Arranged In Alphabetical Order. To Which Is Added, An Appendix, And A Copious Index.
Albany: Published By Henry C. Southwick; New York: And T.C. Fay; H.C. Southwick, Printer, 1815.
Pp. [i]-x, [3]-327. LCP

1692. __________.
The Farmer's Assistant; Being A Digest Of All That Relates To Agriculture, And The Conducting Of Rural Affairs; Alphabetically Arranged, And Adapted For The United States. Second Edition, Corrected And Enlarged.
Philadelphia and Richmond, Virginia: Published By Benjamin Warner, And Sold At His Bookstores; Lancaster, (Pa.): William Dickson, Printer, 1820.
468 p. Prelims. erratically paginated. HSP in LCP; LCP (lacking 3rd and 4th leaves, i.e. Introduction)

1693. NILES, JOHN MILTON, 1787-1856.
Oration, Pronounced Before The Agricultural Society, Of Hartford County [Conn.], By The Hon. John M. Niles, Friday, October, 8, 1841. Published By A Vote Of The Society.
Hartford: Review Office Press, 1841.
16 p. LCP

1694. NOLTE, VINCENT & CO., Merchants, New Orleans.
[No title; text begins:] In following up our annual custom we beg leave to lay before you detailed statements of our exports of cotton and tobacco, since the beginning of October 1819 to the present date ....
[New Orleans: 1820].
[4] p. Dated at head: July 31st, 1820. APS (with MS. addition of figures "since the circular has been in print")

1695. NORFOLK [MASS.]. AGRICULTURAL SOCIETY.
The Second Annual Cattle Show & Fair Of The Norfolk Agricultural Society, Will Be Holden At Dedham, On Wednesday, September 25, 1850. The Trustees invite the Agriculturists, Mechanics, Manufacturers, Horticulturists, and Ladies of the County, to join their endeavors ....
Dedham, Mass.: H. Mann, Printer, 1850.
12 p. LCP

1696. __________.
Transactions.
Boston.
LCP: 1852; 1871.

__________.
See also
ELDRIDGE, JOHN S. Address ... Delivered Before .... 1858.
HILLARD, GEORGE STILLMAN, 1808-1879. Address Before .... [1860].
RUSSELL, GEORGE ROBERT, 1800-1866. An Address Before .... [1851].
WILDER, MARSHALL PINCKNEY, 1798-1886. Address Delivered Before .... 1849.

1697. NORTH AMERICAN LAND COMPANY, Philadelphia.
Observations On The North-American Land-Company, Lately Instituted In Philadelphia: Containing an Illustration of the Object of the Company's Plan, the Articles of Association, with a succinct Account of the States wherein their Lands lie: To Which Are Added, Remarks on American Lands in general, more particularly the Pine-Lands of the Southern and Western States, in Two Letters from Robert G. Harper, Esquire, Member of Congress, for South Carolina ....
London: Printed, By H.L. Galabin, For C. Barrell And H. Servante; Sold Also By J. Debrett; J. Johnson; And W. Richardson, 1796.
[4], 149, [2] p. LCP; HSP

NORTH CAROLINA. BOARD OF AGRICULTURE.
See
MITCHELL, ELISHA, 1793-1857. Report On The Geology .... Conducted under the direction of the .... 1827.
OLMSTED, DENISON, 1791-1859. Report On The Geology .... Conducted under the direction of the .... 1825.

1698. NORTH CAROLINA. GENERAL ASSEMBLY. SENATE. SELECT COMMITTEE ON COTTON AND WOOLLEN MANUFACTORIES.
Report On The Subject Of Cotton And Woollen Manufactories, And On The Growing Of Wool In North Carolina. 1828.
Raleigh: Printed By Lawrence And Lemay, Printers to the State; Philadelphia: Reprinted By William Brown, 1828.
16 p. Signed at end: Chs. Fisher, Chairman. HSP

NORTH CAROLINA. STATE AGRICULTURAL SOCIETY.
See RAYNER, KENNETH, 1808-1884. An Address Delivered Before .... 1854.

1699. NORTH-WESTERN FRUIT-GROWERS' ASSOCIATION.
Transactions.
Chicago.
LCP: Fourth session (Sept. 1855)

1700. NORTON, JOHN PITKIN, 1822-1852.
Address Delivered Before The Hampshire, Franklin, And Hampden Agricultural Societies, At Northampton, Mass. October, 1848.
Albany: C. Van Benthuysen, Printer, 1848.
23 p. LCP

1701. __________.
Elements Of Scientific Agriculture, Or The Connection Between Science And The Art Of Practical Farming. Prize Essay Of The New York Agricul. Society.
Albany: Erastus H. Pease & Company [J. Munsell, Printer And Stereotyper (t.p. verso)], 1850.
x, 208, 4 (pub. adv.) p. illus. LCP

1702. __________.
Elements Of Scientific Agriculture .... Third Edition.
Albany: Erastus H. Pease & Co.; Cincinnati: Jacob Ernst; New Haven: T. H. Pease [J. Munsell, Printer And Stereotyper (t.p. verso)], 1851.
x, 208, 6+2 (pub. adv.) p.; illus. LCP

1703. __________.
Elements Of Scientific Agriculture .... Adapted To The Use Of Schools.
New York: C.M. Saxton & Co., Agricultural Book Publishers, 1855.
x, 208, [2]+6+11+[1] (pub. adv.) p.; illus. T.p. headed: Tenth Thousand. APS

NORWICH CO. [CONN.]. AGRICULTURAL SOCIETY.
See COLMAN, HENRY, 1785-1849. Agricultural Addresses Delivered At .... 1840.

Notes On Farming. 1787.
See THOMSON, CHARLES, 1729-1824.

Notice On The Beet Sugar .... 1837.
See CHURCH, EDWARD.

Observations On The Sugar Beet .... 1840.
See RONALDSON, JAMES.

Observations On The Whale-Fishery. [1788].
See JEFFERSON, THOMAS, 1743-1826.

Off-Hand Sketches; A Companion For The Tourist .... 1854.
See BOWEN, ELI, b. 1824.

1704. OHIO. STATE BOARD OF AGRICULTURE.
Second Annual Report Of The Ohio State Board Of Agriculture, Made To The Forty-Sixth General Assembly, In Pursuance Of "An Act For The Encouragement Of Agriculture," Passed February 27, 1846. For the Year 1847.
Columbus: Chas. Scott's Steam Press, 1848.
101 p. LCP (wrapper)

1705. __________.
Third Annual Report Of The ... To The Forty-Seventh General Assembly. January, 1849.
Columbus: S. Medary, Printer, 1849.
222 p.; illus. LCP (wrapper)

1706. __________.
Fourth Annual Report Of The ... To The Forty-Eighth General Assembly Of The State Of Ohio, January, 1850.
Columbus: S. Medary, Printer, 1850.
304 p. LCP

1707. __________.
Fifth Annual Report Of The ... To The Forty-Ninth General Assembly, For The Year 1850.
[Columbus: 1851].
Pp. 157, [1-5], iii-xii, 173-966; illus. Title headed: Doc. No. 3. LCP

1708. __________.
Sixth Annual Report Of The ... To The Fiftieth General Assembly, For The Year 1851 ....
Columbus: Printed By S. Medary, 1852.
563 p.; front. map; 2 pls., 1 with unnumbered leaf of explanation; illus. LCP

1709. __________.
Seventh Annual Report Of The ... To The Fiftieth General Assembly, For The Year 1852 ....
[Columbus: 1853].
Pp. [1-8], 125-512, 521-528, 521-571; 2 fronts.; illus. Running head: Public Document No. 5. LCP

1710. __________.
Eighth Annual Report Of The ... To The Fifty-First General Assembly, For The Year 1853 ....
[Columbus: 1854].
Pp. 449-724; 2 fronts.; 22 pls. T.p. headed: Public Document No. 21. LCP

1711. __________.
Ninth Annual Report Of The ... To The Governor. For the Year 1854.
Columbus: S. Medary, Printer, 1855.
272 p.; front.; 5 pls.; illus. LCP

1712. __________.
Tenth Annual Report Of The ... To The Governor. For The Year 1855.
Chillicothe: Baker & Miller, Printers, 1856.
356 p.; 2 fronts.; 7 pls.; illus. LCP

1713. __________.
Eleventh Annual Report Of The ... To The Governor, For The Year 1856.
Columbus: Richard Nevins, State Printer, Statesman Steam Press, 1857.
383, [1] p.; 2 fronts. LCP

1714. __________.
Twelfth Annual Report Of The ... With An Abstract Of The Proceedings Of The County Agricultural Societies, To The General Assembly Of Ohio; For The Year 1857.
Columbus: Richard Nevins, State Printer, 1858.
823, [10] p.; 17 pls.; 6 maps. LCP

1715. __________.
Thirteenth Annual Report Of The ... With An Abstract Of The Proceedings Of The County Agricultural Societies: To The General Assembly Of Ohio, For The Year 1858.
Columbus: Richard Nevins, State Printer, 1859.
xlviii, 727, [1] p.; illus. LCP

1716. __________.
Fourteenth Annual Report Of The ... With An Abstract Of The Proceedings Of The County Agricultural Societies: To The General Assembly Of Ohio, For The Year 1859.
Columbus: Richard Nevins, State Printer, 1860.
xvi, 630 p. LCP

1717. OHIO. STATE POMOLOGICAL SOCIETY.
Transactions.
Columbus.
LCP: Sessions 7-8 (Sept. 18, 1856-Dec. 8-11, 1857)

1718. OLCOTT, HENRY STEEL, 1832-1907.
Outlines Of The First Course Of Yale Agricultural Lectures. With An Introduction By John A. Porter ....
New York: C.M. Saxton, Barker & Co. [Edward O. Jenkins, Printer & Stereotyper (t.p. verso)], 1860.
186, 6 (pub. adv.) p. LCP (wrappers)

1719. __________.
Sorgho And Imphee, The Chinese And African Sugar Canes. A Treatise Upon Their Origin, Varieties And Culture; Their Value As A Forage Crop; And The Manufacture Of Sugar, Syrup, Alcohol .... To Which Are Added, Copious Translations Of Valuable French Pamphlets. Fully Illustrated ....
New York: A.O. Moore, Agricultural Book Publisher, (Late C.M. Saxton & Co.) [Edward O. Jenkins, Printer and Stereotyper (t.p. verso)], 1857.
350, 10 (pub. adv.) p.; front.; 1 pl.; illus. Publisher's notice tipped in at front. LCP

1720. __________.
Sorgho And Imphee, The Chinese And African Sugar Canes ....
New York: A.O. Moore, Agricultural Book Publisher, (Late C.M. Saxton & Co.), 1858.
352, 8 (pub. adv.) p.; front.; 1 pl.; illus. T.p. headed: Sixth Edition. LCP

1721. __________.
Supplement To Sorgho And Imphee, The Chinese And African Sugar Canes. Containing The American Experiments Of 1857, And J.S. Lovering's Pamphlet. [wrapper title]
New York: A.O. Moore, Agricultural Book Publisher, 1858. [wrapper imprint]
38, 10 (pub. adv.) p. Pp. 6-38: A Detailed Account of Experiments and Observations upon the Sorghum Saccharatum, or Chinese Sugar Cane, made with the view of determining its value .... by Joseph S. Lovering. HSP (wrappers)

1722. OLMSTED, DENISON, 1791-1859.
Memoir Of Eli Whitney, Esq. First published in the American Journal of Science, for 1832.
New Haven: Durrie & Peck; Peck & Stafford, Printers, 1846.
80 p.; front.; 1 pl. LCP

1723. __________.
Report On The Geology Of North-Carolina, Conducted under the direction of the Board of Agriculture. Part II. November, 1825.
[Raleigh: 1827].
Pp. [1-4], [87]-155. Incl. (pp. 143-155): On Planting Cotton, by Robert Moore; dated at end: November 13, 1825. With general t.p.: Papers On Agricultural Subjects And Professor Olmsted's Report ....; Raleigh: Printed By J. Gales & Son, 1827. For Part III See MITCHELL, ELISHA, 1793-1857. LCP

ONEIDA CO. [N.Y.]. AGRICULTURAL SOCIETY.
See GOLD, THOMAS RUGGLES. Address Of ... Delivered Before .... 1820.

The Orchardist's Companion [pdcl.].
See Hoffy's Orchardist's Companion.

1724. ORLEANS TERRITORY. HOUSE OF REPRESENTATIVES.
Memorial Of The House Of Representatives Of The Territory Of Orleans, Praying That An Alteration May Be Made In The Law Of Last Session, Respecting The Titles To Lands, And For A Further Encouragement To The Culture Of Sugar. December 31, 1805. Read, And Referred To Mr. Anderson ....
Washington: Printed By William Duane And Son, 1806.
14 p. T.p. headed: In Senate Of The United States. LCP

OSWEGO CO. [N.Y.]. AGRICULTURAL SOCIETY.
See HOLMES, J.N. The Annual Address, Before .... 1855.

OTSEGO CO. [N.Y.]. AGRICULTURAL SOCIETY.
See MORSE, OLIVER ANDREW, 1815-1870. Address, Delivered Before .... 1842.

1725. OTTOLENGHE, JOSEPH.
Directions For Breeding Silk-Worms, Extracted from a Letter Of Joseph Ottolenghe, Esq.; Late Superintendent Of The Public Filature In Georgia.
Philadelphia: Printed by Joseph Crukshank, 1771.
8 p. HSP in LCP; LCP

Our Farm Of Four Acres And The Money We Made By It .... 1860.
See COULTON, MISS -----.

Outlines Of A Plan .... 1794.
See BORDLEY, JOHN BEALE, 1727-1804.

1726. OVERDURFF, EDWARD.
Edward Overdurff's Book Of Farriery.
Pittsburgh: Printed By Victor Scriba, 1840.
12 p. LCP

1727. OVERMAN, C.R.
Premium Essay On The Germination, Cultivation, Use And Value To The State Of Illinois Of The "Maclura" Or Osage Orange Hedge. Second Edition.
Springfield [Ill.]: Lanphier & Walker, Printers, 1855.
13 p. The premium was offered by the Illinois State Agricultural Society. LCP (wrappers)

OWEN, RICHARD 1810-1890, jt. auth.
See SAFFORD, JAMES MERRILL, 1822-1907, and __________.

1728. PAIGE, ALONZO CHRISTOPHER, 1797-1858.
Address Before The Montgomery Co. [N.Y.] Agricultural Society, On The Tenth Of October, 1860.
Canajoharie, N.Y.: Radii Print, 1860.
21 p. HSP (wrapper)

1729. [PAINE, ROBERT TREAT, 1731-1814].
The Art Of Making Common Salt. Particularly Adapted To The Use Of The American Colonies. With An Extract from Dr. Brownrigg's Treatise on the Art Of Making Bay-Salt. Detached from the Pennsylvania Magazine, for March 1776.
Philadelphia: Printed by R. Aitken, Printer and Bookseller, 1776.
7 p.; 1 pl. LCP

PALFREY, JOHN GORHAM, 1796-1881, Secretary of the Commonwealth of Massachusetts. Abstract ....
See MASSACHUSETTS. SECRETARY OF THE COMMONWEALTH. Transactions [pdcl.].

1730. PARDEE, RICHARD GAY, 1811-1869.
A Complete Manual For The Cultivation of the Strawberry; With A Description Of The Best Varieties. Also, Notices Of The Raspberry, Blackberry, Currant, Gooseberry, And Grape; With Directions For Their Cultivation, And The Selection Of The Best Varieties.
New York: C.M. Saxton, Agricultural Book Publisher [Edward O. Jenkins, Printer And Stereotyper (t.p. verso)], 1854.
144, 10 (pub. adv.) p.; illus. LCP

1731. PARKER, M.
The Arcana Of Arts And Sciences, Or, Farmers' & Mechanics' Manual; Containing A Great Variety Of Valuable Receipts And Useful Discoveries, In The Various Departments Of Human Knowledge; Many Of Which Were Never Before Published.
Washington, Pa.: Printed By J. Grayson, 1824.
348 p. LCP

1732. PARKER, NATHAN HOWE.
Iowa As It Is In 1855; A Gazetteer For Citizens, And A Hand-Book for Immigrants, Embracing A Full Description Of The State Of Iowa: Her Agricultural, Mineralogical, And Geological Character; Her Water Courses, Timber Lands, Soil And Climate .... Information For The Immigrant Respecting The Selection, Entry, And Cultivation Of Prairie Soil; A List Of Unentered Lands In The State, &c. With Numerious Illustrations.
Chicago, Ill.: Keen And Lee [Philadelphia: Stereotyped By J. Fagan; Printed By Smith & Peters (t.p. verso)], 1855.
264 p.; front.; 8 pls. LCP

1733. __________.
The Iowa Handbook, For 1856. With A New And Correct Map.
Boston: John P. Jewett And Company; Cleveland, Ohio: H.P.B. Jewett; New York: Sheldon, Blakeman And Company [Boston: Lithotyped By The American Stereotype Company; Andover: Printed By W.F. Draper (t.p. verso)], 1856.
188 p.; col. fold. front. map. LCP

1734. __________.
The Minnesota Handbook, For 1856-7. With A New And Accurate Map.
Boston: John P. Jewett And Company; Cleveland, Ohio: H.P.B. Jewett; New York: Sheldon, Blakeman And Company [Boston: Lithotyped By The American Stereotype Company; Andover: Printed By W.F. Draper (t.p. verso)], 1857.
148 [i.e. 160] p.; col. fold. front. map. P. 160 misnumbered 148. LCP

1735. PARKES, SAMUEL, 1759-1825.
A Letter To The Farmers And Graziers Of Great Britain, To Explain The Advantages Of Using Salt In The Various Branches Of Agriculture, And In Feeding All Kinds Of Farming Stock.
Philadelphia: Published By M. Carey And Son, 1819.
xi, [1], 88 p. HSP in LCP; LCP; APS; HSP

1736. PARKINSON, RICHARD, 1748-1815.
The Experienced Farmer, An Entire New Work, In Which The Whole System Of Agriculture, Husbandry, And Breeding Of Cattle, Is Explained And Copiously Enlarged Upon; And The Best Methods, With The Most Recent Improvements, Pointed Out. In Two Volumes. Vol. I [- II].
Philadelphia: Printed By Charles Cist, 1799.
2 vols. Vol. I: xx, 275 p. II: [2], 292 p. LCP

1737. __________.
The Experienced Farmer's Tour In America Exhibiting, In A Copious And Familiar View, The American System Of Agriculture And Breeding Of Cattle, With Its Recent Improvements. To Which Are Added, Sketches, Published By J.B. Boadley [i.e. Bordley], Esq. Philadelphia, An American Cultivator.
London: Printed For John Stockdale [Printed by T. Davison (colop.)], 1805.
[14], 735 p. Incl. (pp. 669-735) extracts from, and commentary on, Bordley's Sketches On Rotations Of Crops. Bordley is called Boadley throughout. LCP

1738. __________.
A Tour In America, In 1798, 1799, And 1800. Exhibiting Sketches Of Society And Manners, And A Particular Account Of The American System Of Agriculture, With Its Recent Improvements. Vol. I [- II].
London: Printed For J. Harding; And J. Murray [Printed by T. Davison (t.p. verso)], 1805.
2 vols. Vol. I: [8], 319 p. II: pp. [1-8], [319]-735. Incl. (pp. 669-735) extracts from, and observations on, Bordley's Sketches. LCP; APS (with [6] p. pub. adv. at end of Vol. I); HSP (with [6] p. pub. adv. at end of each vol.)

1739. PARSONS, SAMUEL BOWNE, 1819-1906.
An Address Delivered At The Annual Fair Of The New Haven County Horticultural Society, September 26, 1849. With Transactions Of The Society.
New Haven: Printed By B.L. Hamlen, Printer to Yale College, 1849.
49 p. LCP

1740. PARSONS & CO., Nurserymen, Flushing, N.Y.
Catalogue Of Fruit And Forest Trees, Ornamental Shrubs, Plants, &c. Cultivated And For Sale At The Commercial Garden And Nursery Of Parsons & Co. Flushing, Long Island, Near New-York.
New York: Press Of M. Day & Co.; James Egbert, Printer, 1843.
40 p. T.p. headed: 1843. LCP

PASCALIS, FELIX.
See
PASCALIS-OUVIÈRE, FELIX, 1750?-1833.

1741. PASCALIS-OUVIÈRE, FELIX, 1750?-1833.
Practical Instructions And Directions For Silk-worm Nurseries, And For The Culture Of The Mulberry Tree. Dedicated To The American Institute Of New-York. Vol. I [- II].
New York: J. Seymour, Printer, 1829.
2 vols. Vol. I: 112, [1] p. II: 105 p.; 1 pl. APS (2 vols. in 1); HSP (Vol. I only; wrappers)

1742. PATTERSON, JAMES WILLIS, 1823-1893.
An Address Delivered Before The Grafton County Agricultural Society, At Their Annual Fair. Held At Haverhill, New-Hampshire, October 1 & 2, 1857. Published By Request.
Hanover: Printed At The Dartmouth Press, 1857.
20 p. LCP (wrappers)

1743. PECK, WILLIAM DANDRIDGE, 1763-1822.
Natural History Of The Slug Worm. By Order Of The Massachusetts Agricultural Society [i.e. Society for Promoting Agriculture].
Boston: Printed By Young & Minns, Printers To The State, 1799.
14 p.; 1 pl. LCP; APS

1744. PEDDER, JAMES, 1775-1859, et al.
Copy of an Agreement entered into, for the introduction into this country, of such information as can be obtained in Europe, relative to the manufacture of Sugar from the Beet .... [Cap. title; text begins:] The following articles of agreement are this day entered into between James Ronaldson, John Vaughan and Jacob Snider, jr. ... and James Pedder ....
[Philadelphia: 1836].
[2] p. Pedder was to gather information on the culture of the sugar beet on behalf of the other three men. APS

1745. ________.
The Farmers' Land-Measurer, Or Pocket Companion, Showing, At One View, The Content Of Any Piece Of Land, From Dimensions Taken In Yards; With A Set Of Useful Agricultural Tables.
Philadelphia: Thomas, Cowperthwait & Co. [J. Fagan, Stereotyper; T.K. And P.G. Collins, Printers (t.p. verso)], 1842.
6, 144 p. LCP; HSP

1746. ________.
The Farmers' Land-Measurer, Or Pocket Companion ....
Philadelphia: Crissy & Markley; Thomas, Cowperthwait & Co.; New York: C.M. Saxton [Kite & Walton, Printers (t.p. verso)], 1850.
144 p. LCP; HSP

1747. ________.
The Farmers' Land-Measurer, Or Pocket Companion ....
Philadelphia: Crissey & Markley; Thomas, Cowperthwait & Co.; New York: C.M. Saxton, 1853.
144 p. HSP

1748. [________].
Frank; Or, Dialogues Between A Father And Son, On The Subjects Of Agriculture, Husbandry, And Rural Affairs. By The Author Of "The Yellow Shoestrings."
Philadelphia: Kimber & Sharpless [C. Sherman & Co. Printers (t.p. verso)], 1840.
248 p. LCP; APS; HSP

1749. ________.
Report Made To The Beet Sugar Society Of Philadelphia, On The Culture, In France, Of The Beet Root, And Manufacture Of Sugar Therefrom; With Miscellaneous Remarks On The Same, And On The Culture Of The Poppy, And Extraction Of Salad Oil From The Seed; For The Purpose Of Introducing, And Permanently Establishing These Valuable Branches Of Agriculture And Manufacture In The United States.
[Philadelphia:] Published By The Beet Sugar Society Of Philadelphia; C. Sherman & Co., Printers, 1836.
40 p. LCP (wrappers); APS

PENNSYLVANIA. AGRICULTURAL COLLEGE.
See PENNSYLVANIA. FARMERS' HIGH SCHOOL, Centre Co. Earliest (1855-1862) name of the above institution.

PENNSYLVANIA agricultural societies
See the following individual societies:
PENNSYLVANIA. AGRICULTURAL SOCIETY [founded 1822 by J.H. Powel, N. Biddle, and other disgruntled members of the Philadelphia Society for Promoting Agriculture (q.v.)].
PENNSYLVANIA. STATE AGRICULTURAL SOCIETY [founded 1851].

1750. PENNSYLVANIA. AGRICULTURAL SOCIETY [founded 1822].
Hints For American Husbandmen, With Communications To The Pennsylvania Agricultural Society. By Order Of The Directors.
Philadelphia: Printed by Clark & Raser, 1827.
[178] p.; fold. front.; 4 pls. Cover title: Hints For American Husbandmen, With Original Notices And Extracts, On Selecting And Breeding Neat Cattle And Sheep. HSP in LCP; LCP (lacking front. and 1 pl.); APS (lacking front. and 3 pls.); HSP (lacking 1 pl.)

1751. ________.
Memoirs. With Selections From The Most Approved Authors ....
Baltimore.
LCP; HSP: 1823 & 1824; 1824.
APS: 1824.
The 1st 255 p. of both nos. are identical; the 1824 issue then continues to p. 322.

________. Memoirs.
See also POWEL, JOHN HARE, 1786-1856. Reply to Col. Pickering's Attack .... 1825. Answers criticism of the Memoirs.

1752. ________.
Proceedings.
Philadelphia.
LCP: Vols. 1, 3 (1823, 1825)
APS; HSP: Vol. 1.

________.
See also [POWEL, JOHN HARE, 1786-1856]. Pennsylvania Agricultural Society. Strictures .... 1823.

1753. PENNSYLVANIA FARM AND LAND ASSOCIATION, Potter Co., Pa.
A Full Account Of The Farms Laid Out By The Pennsylvania Farm and Land Association, In Potter County, Pennsylvania, Between The Sunbury And New-York & Erie Railroad, As Also Of The Building Lots in the centre of the same For Founding The City Of Germania. Together with other particulars relating to the quality, fruitfulness and production of the soil ....
New York: To Be Had At The Office Of The Association; Henry Ludwig, Printer, 1855.
24 p. HSP

1754. Pennsylvania Farm Journal, Devoted to Agriculture, Horticulture And Rural Economy.
Lancaster; West Chester; Philadelphia.
LCP: Vols. 2-3; vol. 4, no. 1 (April 1852-Jan. 1854)
HSP: Vols. 1-3 (April 1852-Dec. 1853); vol. 4, nos. 1, 4, 6, 7, 9-12 (Jan., April, June, July, Sept.-Dec. 1854); vol. 5, nos. 1, 2, 6-9, 12 (Jan., Feb., June-Sept., Dec. 1855); vol. 6, nos. 1, 3, 6, 9-12 (Jan., March, June, Sept.-Dec. 1856); vol. 7, nos. 1, 3-7 (Jan., March-July 1857)
Also known as: Farm Journal and Progressive Farmer (1855-57) after merger with The Progressive Farmer (q.v.).

Pennsylvania Farmer and Gardener [pdcl.].
See Farmer and Gardener.

1755. Pennsylvania Farmer, And Scientific Journal. Devoted To Agriculture, Horticulture ....
Lancaster, Pa.
HSP: Vol. 1, no. 1 (Jan. 15, 1844). All pub.

1756. PENNSYLVANIA. FARMERS' HIGH SCHOOL.
An Address From The Board Of Trustees Of The Farmers' High School Of Pennsylvania. July, 1857. [wrapper title]
Carlisle, Pa.: Printed At The "Herald" Office, 1857. [wrapper imprint]
8 p. APS (wrappers); HSP (wrappers)

1757. __________.
An Address From The Board Of Trustees .... July, 1858. [wrapper title]
Carlisle, Pa.: Printed At The Herald Office, 1858. [wrapper imprint]
10 p. LCP (wrapper); HSP

1758. __________.
Annual Report Of The Board Of Trustees Of The ... For 1857. Three thousand copies ordered to be printed by the H.R., February 1, 1858.
[Harrisburg:] A. Boyd Hamilton, State Printer, [1858].
22 p. HSP

1759. __________.
Annual Report ... For 1857.
Harrisburg: A. Boyd Hamilton, Printer, 1858.
22 p. HSP

1760. __________.
Second Annual Report ... December 1, 1858.
Harrisburg: A. Boyd Hamilton, State Printer, 1859.
79 p. HSP

1761. __________.
Catalogue Of The Officers and Students .... For The Year 1859.
Philadelphia: King & Baird, Printers, 1859.
32 p. APS (wrapper); HSP (wrappers)

1762. __________.
Catalogue Of The Officers And Students, For the Second Annual Session ... Together With A Financial Report Of The Affairs Of The Institution.
Philadelphia: From Bryson's Caloric Power Printing Rooms, 1860.
32 p. HSP (wrappers)

1763. __________.
Memorial Of The Committee Of The Board Of Trustees Of The .... (1000 copies ordered to be printed by the House of Representatives, January 24, 1856.) [cap. title]
[Harrisburg: 1856].
5 p. Signed at end: H.N. McAllister, et al. LCP

1764. PENNSYLVANIA. GENERAL ASSEMBLY. HOUSE OF REPRESENTATIVES. COMMITTEE ON AGRICULTURE.
Bericht der Committee über Ackerbau, in dem Hause der Representanten, über den Beschluss, sie anweisend de Einfluss zu untersuchen, welchen die jetzt gegebene Aufmunterung zu den Manufaktur Interessen des Staats, auf die Interessen des Ackerbaues und auf den Kaufpreis und auf die Niederlassung von Unangebauten Ländereyen des Staats hat. Hr. Lardner, Vorsitzer. Gelesen, Februar 2, 1833.
Harrisburg: Gedruckt bey Joseph Miller, 1833.
27 p. LCP.

1765. __________.
Report Of The Committee On Agriculture, In The House Of Representatives, On The Resolution Instructing Them To Inquire Into The Influence Which The Encouragement Now Given To The Manufacturing Interests Of The State, Has Upon The Interests Of Agriculture And Upon The Purchase And Settlement Of The Unseated Lands Of The Commonwealth. Mr. Lardner, Chairman. Read, February 2, 1833.
Harrisburg: Printed By Henry Welsh, 1833.
24 p. LCP; APS; HSP

1766. __________.
Report Of The Committee on Agriculture, Read in the House of Representatives, March 15, 1823. Mr. N. Jones, Chairman. The Committee on Agriculture, Report: [cap. title; text begins:] That in their opinion but little practical information can be submitted ....
[Harrisburg: 1823].
3 p. APS; HSP

1767. PENNSYLVANIA. GENERAL ASSEMBLY. HOUSE OF REPRESENTATIVES. COMMITTEE ON EDUCATION.
Report Of The Committee On Education, On The Subject Of Manual Labor Academies, In Pursuance Of A Resolution Passed By The House Of Representatives, Dec. 14, 1832. By Mr. Matthias. Read In The House Of Representatives, February 21, 1833.
Harrisburg: Printed By Henry Welsh, 1833.
14 p. Emphasizes agricultural labor. LCP; APS

1768. PENNSYLVANIA. HORTICULTURAL SOCIETY.
By-Laws Of The Pennsylvania Horticultural Society. Instituted 1833.
Philadelphia: Stavely And M'Calla, Printers, 1844.
[2], 48 p. Wrapper title: By-Laws And Library Catalogue Of ....; incl. (pp. 31-48) catalogue. LCP; HSP (wrappers)

1769. ________.
Catalogue Of The Library Of The ....
Philadelphia: 1840.
24 p. LCP; APS

1770. ________.
Catalogue Of The Library Of The ....
Philadelphia: Stavely & McCalla, Printers, 1850.
48 p. LCP; HSP (wrappers)

1771. ________.
Constitution And By-Laws, Of The ....
Philadelphia: T.A. Conrad, Printer, 1831.
20 p. APS; HSP

1772. ________.
Constitution And By-Laws Of The ....
Philadelphia: Printed By A. Waldie, 1838.
19 p. LCP; HSP (wrappers)

1773. ________.
Constitution And By-Laws Of The ....
Philadelphia: William Stavely & Co., 1840.
24 p. LCP; APS (wrappers); HSP (wrappers)

1774. ________.
Constitution Of The ... At Philadelphia.
Philadelphia: Clark & Raser, Printers, 1827.
7 p. LCP; HSP

1775. ________.
The ... Established in Philadelphia, Offers The Following Premiums For Culinary Vegetables, Fruits, & Flowers, For 1839.
[Philadelphia: 1839].
Brs. LCP

1776. ________.
The ... Established In Philadelphia, Offers The Following Premiums For Culinary Vegetables, Fruits And Flowers, For 1839.
Philadelphia: Printed For The Society, 1839.
12 p. APS (wrapper); HSP

1777. ________.
Report Of The Committee Appointed By The ... For Visiting The Nurseries And Gardens In The Vicinity Of Philadelphia. July 13th, 1830.
Philadelphia: Wm. F. Geddes, Printer, 1831.
16 p. LCP; APS; HSP (wrappers)

1778. ________.
Report Of The Tenth Exhibition Of The .... To Which Is Added A List Of Premiums, Awarded In The Year Ending With September, 1837.
Philadelphia: Published By Order Of The Society, 1837.
16 p. APS; HSP (wrappers)

1779. ________.
Transactions.
Philadelphia.
LCP; APS; HSP: 1839.

PENNSYLVANIA poultry societies.
See the following individual societies:
PENNSYLVANIA. POULTRY SOCIETY [founded 1852?].
PENNSYLVANIA. STATE POULTRY SOCIETY [founded 1853].

1780. PENNSYLVANIA. POULTRY SOCIETY [founded 1852?].
Report Of The Annual Exhibition, Held At Philadelphia, November 24th, 25th and 26th, 1852.
Philadelphia: Printed By King & Baird, 1853.
16 p. HSP (wrappers)

1781. PENNSYLVANIA. POLYTECHNIC COLLEGE.
Second Annual Announcement Of The Polytechnic College, Of The State Of Pennsylvania. Session 1854-55. Engineering, Mining, Agriculture, And The Chemical And Mechanical Arts.
Philadelphia: Inquirer Book Press, 1854.
18 p.; front. LCP

1782. PENNSYLVANIA. SILK CONVENTION, Harrisburg, Feb. 22-23, 1839.
Proceedings Of The Pennsylvania Silk Convention, Held At Harrisburg On The Twenty-Second And Twenty-Third Days Of February, 1839. Printed By Order Of The Senate, March 26, 1839.
Harrisburg: Printed By E. Guyer, 1839.
36 p. LCP; HSP (wrappers)

PENNSYLVANIA. SILK SOCIETY.
See PENNSYLVANIA. SOCIETY FOR PROMOTING THE CULTURE OF THE MULBERRY, AND THE RAISING OF SILK WORMS.

1783. PENNSYLVANIA. SOCIETY FOR PROMOTING THE CULTURE OF THE MULBERRY, AND THE RAISING OF SILK WORMS.
Address Of The Pennsylvania Silk Society To The Public. [cap. title]
[Philadelphia: 1830].
4 p.; illus. Signed at end: Benjamin R. Morgan, President; dated: Jan. 1, 1830. APS

1784. ________.
At a Meeting of the Subscribers to the Plan for Promoting the Culture of the Mulberry Tree, and the raising of Silk Worms, held at the Indian Queen, April 2, 1828 .... [cap. title]
[Philadelphia: 1828].
4 p. Constitution and list of premiums. Signed at end: B.R. Morgan, President; M. Carey, Secretary. LCP; APS; HSP

1785. ________.
Directions For The Rearing Of Silk Worms, And The Culture Of The White Mulberry Tree. Published By The Pennsylvania Society Associated For The Promotion Of Those Objects.
Philadelphia: Clark & Raser, Printers, 1828.
25 p. "The Acting Committee of the Society ... have abstracted from the Manual prepared by order of the Secretary of the Treasury ... the following directions ...." (Preface). APS; HSP

1786. PENNSYLVANIA. SOCIETY FOR THE ENCOURAGEMENT OF AMERICAN MANUFACTURES.
Memorial. To the Senate and House of Representatives of the United States: The memorial of the Pennsylvania Society for the encouragement of American Manufactures, [cap. title; text begins:] Respectfully Sheweth, That your memorialists have read with the deepest regret, two remonstrances ... from agricultural societies in the state of Virginia .... [i.e. the United Agricultural Societies Of Virginia].
[Philadelphia: 1820].
14 p. Signed at end: Thomas Leiper, Vice President; James Mease, Secretary pro tempore. Dated: Philadelphia, April 3rd, 1820. Copy with organization's name in cap. title as: Pennsylvania Society for the encouragement .... LCP; APS

1787. __________.
Memorial. To the Senate and House ... The memorial of the Pennsylvania Society for the promotion of American Manufactures, [cap. title]
[Philadelphia: 1820].
14 p. Copy as above except with organization's name in cap. title as: Pennsylvania Society for the promotion .... LCP

PENNSYLVANIA. STATE AGRICULTURAL CONVENTION, Harrisburg, Jan. 21, 1851.
See PENNSYLVANIA. STATE AGRICULTURAL SOCIETY. Proceedings Of The Agricultural Convention .... 1851.

1788. [PENNSYLVANIA. STATE AGRICULTURAL SOCIETY (founded 1851)].
[Announcement of the Ninth Annual Exhibition].
[Philadelphia: 1859].
[28] p. Without t.p.; text headed: Preface. Chiefly adv. LCP

1789. __________.
List Of Judges Of The Pennsylvania State Agricultural Society For Its Fifth Annual Exhibition To be held at Harrisburg on the 25th, 26th, 27th and 28th of September, 1855. [wrapper title]
Philadelphia: August 15, 1855. [wrapper imprint]
8 p. APS (wrapper)

1790. __________.
List Of Premiums & Regulations For The Sixth Annual Exhibition, At Pittsburgh, September 30th, & October 1st, 2d & 3d, 1856. [wrapper title]
Pittsburgh: Printed By John T. Shryock, [1856]. [wrapper imprint]
19 p. APS (wrapper)

1791. __________.
List of Premiums & Regulations Of The Pennsylvania State Agricultural Society, At Its Seventh Annual Exhibition, To Be Held At Philadelphia, On September 29th, 30th, and October 1st and 2d, 1857. [wrapper title]
[Philadelphia: 1857].
34 p. LCP (wrappers)

1792. __________.
List Of Premiums And Rules And Regulations For The First Cattle Show And Agricultural & Horticultural Exhibition, To Be Held By The ... At Harrisburg, On Wednesday, Thursday And Friday, The 22d, 23d And 24th Of October, 1851.
[Lancaster?: 1851].
8 p. T.p. headed: Supplement To The Farm Journal. APS

1793. __________.
Officers, Charter, Constitution And Life Members Of The Pennsylvania State Agricultural Society, With The Regulations And Award of Premiums, At Its Annual Exhibitions At Pittsburg In 1858, And At Philadelphia In 1857.
Harrisburg: A. Boyd Hamilton, Printer, 1858.
84 p. LCP (wrapper)

1794. __________.
Proceedings Of The Agricultural Convention Held At Harrisburg, January 21, 1851.
[Harrisburg: 1851].
14 p. The Society was established at this convention. LCP; APS

1795. __________.
Proceedings Of The .... [cap. title]
[Philadelphia: 1852].
8 p. Text headed: Tuesday, Jan. 20, 1852. APS

1796. __________.
Reports Of Judges And Premiums Awarded At Second Annual Exhibition, Of The ... Held At Lancaster, On The 20th, 21st and 22d days of October, 1852.
[Lancaster:] Myers & Wiley, Prs.; Farm Journal Office, 1852.
32 p. APS (wrapper)

1797. __________.
Report Of Transactions.
Harrisburg.
LCP: Vols. 1-5 6-7 (1853-58); (1861-70); 10 (1874-75); 12 (1877)
Until 1858 known as: Annual Report Of Transactions.

__________.
See also,
BRINCKLE, WILLIAM DRAPER, 1798-1862. Remarks On Entomology .... 1852.
STEVENSON, ANDREW, 1784-1857. Address Delivered Before .... 1851.

1798. PENNSYLVANIA. STATE CONVENTION FOR THE PROMOTION OF AGRICULTURE AND MANUFACTURES, Harrisburg, June 27, 1827.
Journal Of A Convention of Delegates Of The State Of Pennsylvania, Held For The Promotion Of The State Agricultural And Manufacturing Interests, At the Capitol in Harrisburg, On Wednesday, The 27th of June, A.D. 1827.
[Harrisburg:] Atkinson & Alexander, Printers, 1827.
16 p. HSP

1799. PENNSYLVANIA. STATE POULTRY SOCIETY [founded 1853].
Constitution and By-Laws Of The State Poultry Society Of Pennsylvania, For the Promotion of the Improved Breeds Of Poultry In The United States.
Philadelphia: Printed by Bicking & Guilbert, 1853.
12 p. HSP

1800. PENNSYLVANIA. VINE COMPANY.
Pennsylvania Vine Company. [cap. title]
[Philadelphia: 1812].
10 p. Dated at beginning: Philadelphia, April 25th, 1812. Minutes of a stockholders' meeting with comments on the state of the vineyard. LCP; APS

1801. PERRY, GARDNER BRAMAN, 1783-1859.
An Address To The Essex County [Mass.] Agricultural Society, Delivered At Newbury, September 27, 1832, At Their Annual Cattle Show. Published By Order Of The Society.
Salem, Mass.: Foote & Chisholm - Gazette And Mercury Office, 1833.
Pp. [1-2], [5]-27. LCP (wrappers)

1802. The Pests Of The Farm; Being An Account Of The Various Depradating Animals, Birds, And Insects Which Annoy The American Farmer. With Directions For Their Destruction. With Illustrations On Wood.
New York: C.M. Saxton, Agricultural Book Publisher [S.W. Benedict, Stereotyper And Printer (t.p. verso)], 1852.
135, [1]+5 (pub. adv.) p.; illus. LCP

1803. PETERS, RICHARD, 1744-1828.
Address Delivered Before The Philadelphia Society for promoting Agriculture, At Its Annual Meeting, On The Twenty-Third Of January, 1823, By Richard Peters, Its President, At The Request Of The Society. Printed By Order Of The Society.
Philadelphia: Clark & Raser, Printers, 1823.
Pp. [1-4], [3]-25. LCP; APS; HSP

1804. __________.
Agricultural Enquiries On Plaister Of Paris. Also, Facts, Observations And Conjectures On That Subtance [sic], When Applied As Manure. Collected, chiefly from the practice of farmers in Pennsylvania, and published as much with a view to invite, as to give information.
Philadelphia: Printed By Charles Cist; And John Markland, 1797.
[2], 111, [3] p. HSP in LCP; LCP; APS

1805. __________.
Agricultural Inquiries On Plaister Of Paris .... With Some Additional Notes; And More Recent Facts And Information.
Philadelphia: Re-Printed By Jane Aitken, 1810.
Pp. [1-4], [3]-129, [1], i-ix, [1-10]; 1 pl. Wrapper title: Agricultural Inquiries On Plaister Of Paris. Taken From Vol. II Memoirs, Philadelphia Society For Promoting Agriculture. LCP (wrappers); APS (wrappers)

1806. __________.
A Discourse On Agriculture. Its Antiquity; And Importance, To Every Member Of The Community: The Obligations All Are Under To Encourage It; - Its Principles .... Delivered before The Philadelphia Society For Promoting Agriculture; at their request, on their annual meeting, the 9th of January, 1816.
Philadelphia: Published By Johnson And Warner, 1816.
40 p.; errata slip pasted p. 37. LCP; APS

__________.
See also
BRECK, SAMUEL, 1771-1862. Address ... On The Death .... 1828.
WASHINGTON, GEORGE, 1732-1799. Letters On Agriculture .... With Statistical Tables And Remarks, By Thomas Jefferson, Richard Peters .... 1847.

1807. PEYSSON, AUGUSTUS.
Public Health And Agriculture.
Philadelphia: J.H. Jones, Printer, 1851.
21, [2] p. Describes a plan "for the transmutation of feculent matter into inoderous [sic] and chemical manure" (p. 4). LCP (2 copies, one with fold. certificate, the Articles of Association of the Philadelphia Chemical Manure Co.; one with wrappers); HSP (wrappers)

1808. PHELPS, EBENEZER W.
Phelps' Bee-Keeper's Chart; Being A Brief Practical Treatise On The Instinct, Habits And Management Of The Honey-Bee, In All Its Various Branches. The Results Of Many Years' Practical Experience, To Render Bee-Keeping Less Difficult, And At The Same Time More Sure, Profitable And Pleasant Than It Has Formerly Been.
New York: A.O. Moore, Agricultural Book Publisher, (Late C.M. Saxton & Co.) [E.O. Jenkins, Printer And Stereotyper (t.p. verso)], 1858.
96 p.; illus. Wrapper headed: Moore's Rural Hand-Books. LCP (wrappers)

PHILADELPHIA. BEET SUGAR SOCIETY.
See PEDDER, JAMES, 1775-1859. Report Made To .... 1836.

PHILADELPHIA CHEMICAL MANURE MANUFACTURING CO.
See PEYSSON, AUGUSTUS. Public Health And Agriculture. 1851. Proposes establishment of the above company.

1809. PHILADELPHIA. COMMITTEE OF THE CITY AND LIBERTIES.
To The Public. [Text begins:] Whereas the Congress ... firmly agree and associate, to use their utmost Endeavours to improve the Breed of Sheep ... do most earnestly recommend ... neither to purchase ... any Ewe or Mutton Lamb ....
[Philadelphia: 1774].
Brs. Headed: In Committee, December 14, 1774. Resolved, that the Proceedings of this Committee on November 30th, concerning the Killing of Sheep, be republished .... Signed at end: Jonathan B. Smith, Secretary. LCP

PHILADELPHIA. EMIGRANTS' FRIEND SOCIETY.
See THOMASON, DENNY R. Hints To Emigrants .... 1848. Written for the Society.

The Philadelphia Florist And Horticultural Journal .... R. Robinson Scott, Editor [pdcl.].
See The Florist And Horticultural Journal .... H.C. Hanson, Editor. Later name of the above.

1810. PHILADELPHIA GUANO COMPANY.
Colombian And Other Phosphatic Guanos, Brought From The Guano Islands In The Caribbean Sea, Belonging To The Republic Of Venezuela, By The Philadelphia Guano Company. For Sale By Jas. Lee & Co., No. 49 Wall Street, New York, General Agents. Robert Bell, Liverpool. Richards & Miller, 92 South Front Street, Philadelphia ....
Philadelphia: McLaughlin Brothers' Book And Job Printing Establishment, 1856.
16 p. HSP

PHILADELPHIA. SOCIETY FOR PROMOTING AGRICULTURE [founded 1785].
See also PENNSYLVANIA. AGRICULTURAL SOCIETY [founded 1822 by disgruntled members of the above society].

1811. __________.
An Address, From The Philadelphia Society For Promoting Agriculture; With A Summary Of Its Laws; And Premiums Offered.
[Philadelphia:] 1785.
16 p. HSP in LCP (with tipped-in notice of the election to membership of Thomas McKean); LCP; APS

1812. __________.
Address To The Citizens Of Pennsylvania On The Importance of a more liberal Encouragement Of Agriculture. Accompanied With Inquiries On Agricultural Subjects Proposed By The ... With A View To Form An Expose Of The State Of Agriculture In Pennsylvania.
Philadelphia: Printed By Order Of The Society, 1818.
14 p. APS

1813. __________.
Catalogue Of The Annual Exhibition Of The ... Held At Powelton, West Philadelphia. From Sept. 11th To Sept. 14th, Inclusive.
Philadelphia: Printed By Henry B. Ashmead, 1855.
32, [2]+16+[4] (adv.) p.; fold front. HSP (wrappers)

1814. __________.
Catalogue Of The Annual Exhibition Of The ... Held At Powelton, West Philadelphia, On Tuesday, Wednesday, Thursday, and Friday, 25th, 26th, 27th, and 28th days of September, 1860. Published by S.E. Cohen, Under the sanction of the Society.
Philadelphia: C. Sherman & Son, Book And Job Printers, 1860.
34 p.; pp. 30-34 adv. HSP (wrappers)

1815. __________.
Charter And By-Laws Of The .... Instituted February 14, 1809.
Germantown, Pa.: Printed By P.R. Freas & Co., 1840.
11, [1] p. LCP; APS; HSP

1816. __________.
Charter And By-Laws Of The .... Instituted February 11, 1785. Incorporated February 14, 1809.
Philadelphia: Inquirer Printing Office, 1858.
24 p. HSP

1817. __________.
Laws Of The ... As revised and enacted by the said Society, January 10, 1786, With The Premiums Proposed, February 5, 1788. To Which Is Prefixed, A List Of The Members Of The Society.
[Philadelphia:] 1788.
24 p. APS

1818. __________.
Laws Of The ... As Revised and Enacted by the said Society, February 16th, 1789, with The Premiums Proposed, February 3d, 1789. To Which Is Prefixed, A List of the Members of the Society.
[Philadelphia:] 1789.
23 p. LCP; APS; HSP

1819. __________.
Laws Of The .... As Revised And Enacted At The Annual Meeting, Held On The 14th January, 1812. To Which Is Prefixed A List Of The Members Of The Society. Incorporated February 14, 1809.
Philadelphia: Printed By Lydia R. Bailey, 1812.
15 p. LCP

1820. [__________].
Lectures Upon Agriculture. [Text begins:] Lecture I. Section I. The natural history of Vegetables ....
[Philadelphia: 1788?].
Brs. LCP

1821. __________.
Memoirs .... Containing Communications On Various Subjects In Husbandry And Rural Affairs ....
Philadelphia.
LCP: Vols. 1-5 (1808; 1811; 1814; 1818; 1826); vol. 1, 2nd edn. (1815)
APS: Vols. 1-5; 6 (1839)
HSP: Vols. 1-6; vol. 1, 2nd edn.

1822. __________.
Memorial Of The ... To the Honourable Senate and House of Representatives of the Commonwealth of Pennsylvania ... [Text begins:] The Memorial respectfully sheweth, That your Memorial lists have been endeavouring for the last thirty-three years, to advance the interests of Agriculture ....
[Philadelphia: 1818].
8 p. APS

1823. __________.
Memorial Of The .... To the Honourable Senate and House ....
[Philadelphia: 1818].
12 p. Copy as above but with addition at end of: Petition To The Legislature. APS

1824. __________.
Minutes Of The ... From Its Institution in Feburary, 1785, to March, 1810.
Philadelphia: John C. Clark & Son, Printers, 1854.
[2], 124 p. LCP; APS (wrapper); HSP (wrappers)

1825. __________.
Philadelphia Society For Promoting Agriculture. [Cap. title; text begins:] The Society was incorporated in the year one thousand eight hundred and nine, for the term of ten years ....
[Philadelphia: 1819].
16 p. Incl. (pp. 1-2): An Act To incorporate; (pp. 3-9) Laws; (pp. 11-16) List Of Members. LCP; HSP

1826. __________.
[No title; text begins:] The Philadelphia Society For Promoting Agriculture, impressed with the belief, that the period has arrived when the encouragement of Agriculture has become the imperious duty of all who feel an interest in the prosperity of their country ....
[Philadelphia: 1818].
[2] p. Solicits contributions toward a pattern farm. Headed: Philadelphia, November 28, 1818. Signed at end: Richard Peters, President. Roberts Vaux, Secretary. HSP

1827. __________.
Proceedings Of The Annual Exhibition Of The ... Held At The Rising Sun Village, October 16-17, 1850.
Germantown, Pa: Printed at the Office of the Telegraph, 1850.
34 p. Incl. (pp. 13-34): Address Delivered Before The Philadelphia Society For Promoting Agriculture, At Their Annual Exhibition ... By Sidney G. Fisher, Esq. Of Philadelphia. LCP; HSP (wrappers)

1828. __________.
Proceedings Of The Annual Exhibition Of The ... Held At The Rising Sun Village, October 8-9, 1851.
Germantown, Pa.: Printed at the Office of the Telegraph, 1851.
27 p. HSP (wrappers)

1829. __________.
Proceedings Of The Annual Exhibition Of The ... Held At The Rising Sun Village, October 19-20, 1853.
Germantown, Pa.: Printed At The Office Of The Telegraph, 1853.
27 p. HSP (wrappers)

1830. __________.
Proceedings Of The ... For The Years 1854, 1855 and 1856, To Which Are Prefixed A Catalogue of the Exhibition of 1855, The Awards Of Premiums, And The Address Of Ex-Governor Bigler.
Philadelphia: Published By The Society; Henry B. Ashmead, Printer, 1857.
[4], 122 p.; fold. map. With second t.p.: Catalogue Of The Annual Exhibition ... Held At Powelton, West Philadelphia. From Sept. 11th To Sept. 14th, Inclusive. Philadelphia: Printed By Henry B. Ashmead, 1855. LCP (wrapper); HSP (wrappers)

1831. __________.
Regulations And List Of Premiums Of The Annual Exhibition Of The ... To Be Held At Powelton, West Philadelphia, On Tuesday, Wednesday, Thursday And Friday, 25th, 26th, 27th and 28th days of September, 1860.
Philadelphia: U.S. Steam-Power Book And Job Printing Office, 1860.
15, [1] p. HSP

1832. __________.
Report Of The Curators Of The ... On The Cultivation Of Flax. Published By Order Of The Society. [cap. title]
[Philadelphia: 1821].
4 p. Title headed: In Anticipation Of The National Recorder For March 3d, 1821. APS

1833. __________.
Reports And Letters To The Philadelphia Society For Promoting Agriculture, On The Proper Season For Cutting Timber For Fencing and other Agricultural Purposes.
Philadelphia: Printed By John C. Clark, 1850.
16 p. HSP (wrappers)

__________.
See also
ALMANACS. Philadelphia. 1817 etc. The Agricultural Almanac .... Patronized By ....
BIDDLE, CRAIG, 1823-1910. Address: Delivered Before .... 1860.
BIDDLE, NICHOLAS, 1786-1844. Address Delivered Before .... 1822.
BOOTH, JAMES CURTIS, 1810-1888. Practical Value ... (Read before ....) 1853.
BROOM, JAMES MADISON, 1778-1850. An Address Delivered Before .... 1821 [2 edns.].
CAREY, MATHEW, 1760-1839. Address Delivered Before .... 1824 etc.
DARLINGTON, WILLIAM, 1782-1863. Address Delivered Before .... 1844 (2 edns.)
EMERSON, GOUVERNEUR, 1796-1874. Address Delivered Before .... Together With The Reports .... 1849.
FISHER, SIDNEY GEORGE, 1809-1871. Address Delivered Before .... 1850.
[__________]. A Report On The Cultivation Of Native Grapes .... Read To .... 1856.
INGERSOLL, JOSEPH REED, 1786-1868. Address Delivered Before .... 1849.
LOGAN, GEORGE, 1753-1821. An Address On The Errors Of Husbandry .... Delivered Before .... 1818 .
__________. Fourteen Agricultural Experiments ... Addressed To .... 1797.
MEASE, JAMES, 1771-1846. Address ... Before .... 1817 [2 edns.].
__________. Address On The Subject ... Delivered At .... 1818.
__________. Introductory Lecture To A Course Of Lectures [delivered before ....] 1814.
PETERS, RICHARD, 1744-1828. Address Delivered Before .... 1823.
__________. A Discourse On Agriculture .... Delivered before .... 1816.
RAWLE, WILLIAM, 1759-1836. An Address Delivered Before .... 1819.
TILLGHMAN, WILLIAM, 1756-1827. An Address Delivered Before .... 1820.
TROOST GERARD, 1776-1850. Geological Survey Of The Environs Of Philadelphia: Performed By Order .... 1826.
VAUX, ROBERTS, 1786-1836. Address Delivered Before .... 1825.

1834. PHINNEY, ELIAS, 1780-1849.
Address Delivered Before The Middlesex [Mass.] Society Of Husbandmen And Manufacturers. At Their Annual Festival, October 7, 1830. Published at the Request of the Society.
Charlestown: Printed By William W. Wheildon [Press of the Bunker-Hill Aurora (t.p. verso)], 1830.
28 p. LCP

PHYSIOLOGUS, PHILOTHEOS, pseud.
See TRYON, THOMAS, 1634-1703.

1835. PICKERING, TIMOTHY, 1745-1829.
Address To The Essex [Mass.] Agricultural Society. May 5, 1818.
Salem: Printed By Thomas C. Cushing, 1818.
27 p. HSP

1836. __________.
A Discourse, Read Before The Essex Agricultural Society, In Massachusetts, February 21, 1820. Suggesting Some Improvements In The Agriculture Of The County.
Salem: John D. Cushing, Printer, 1820.
27 p. LCP; HSP

__________.
See also [POWEL, JOHN HARE, 1786-1856]. Reply To Col. Pickering's Attack .... 1825.

1837. PITKIN, TIMOTHY, 1765-1847.
A Statistical View Of The Commerce Of The United States Of America: Its Connection With Agriculture And Manufactures: And An Account Of Public Debt, Revenues, And Expenditures Of The United States. With A Brief Review Of The Trade, Agriculture, And Manufactures Of The Colonies, Previous To Their Independence. Accompanied With Tables ....
Hartford: Printed By Charles Hosmer, 1816.
xii, 407, [1], xix, [1] p. LCP; APS

1838. __________.
A Statistical View Of The Commerce Of The United States Of America: Including Also An Account Of Banks, Manufactures And Internal Improvements: Together With That Of The Revenues And Expenditures .... Accompanied With Numerous Tables.
New Haven: Durrie & Peck [Printed by Hezekiah Howe & Co. (t.p. verso)], 1835.
xvi, 600 p. LCP

1839. PITTSBURGH. HORTICULTURAL SOCIETY.
Constitution And By-Laws Of The Pittsburgh Horticultural Society.
Pittsburgh: Printed By Ephraim Lloyd, 1834.
12 p. HSP

1840. __________.
Constitution And By-Laws Of The .... 1848.
Pittsburgh: Printed By Geo. Parking & Co., 1848.
11 p. LCP; HSP

A Plan For encouraging Agriculture, and increasing the Value of Farms .... [1793].
See COXE, TENCH, 1755-1824.

1841. PLEASANTVILLE [PA.]. HORSE COMPANY.
Constitution And By-Laws Of The Pleasantville Horse Company.
Norristown, Pa.: Robert Iredell, Printer, 1857.
9 p. The Company was formed to recover stolen horses. HSP

1842. The Plough Boy, And Journal Of The [N.Y.] Board Of Agriculture.
Albany.
HSP in LCP: Vols. 1-4 (June 5, 1819-July 5, 1823). All pub.
LCP: Vols. 2-4 (June 3, 1820-July 1, 1823); lacking: vol. 2, nos. 3, 47; vol. 3, nos. 13-15; vol. 4, nos. 27, 36, 39-41, 47.

The Plough, The Loom And The Anvil.
See American Farmer's Magazine.

1843. PLUMER, WILLIAM, JR., 1789-1854.
An Address Delivered Before The Rockingham [N.H.] Agricultural Society, October 18th, 1821.
Exeter: Printed By John J. Williams, 1821.
24 p. LCP

1844. PLYMOUTH CO. [MASS.]. AGRICULTURAL SOCIETY.
Constitution Of The Agricultural Society In Plymouth County, Incorporated June 11, 1819. With A List Of The Officers And Members.
East Bridgewater: Printed by G. H. Brown, Music and Ornamental Printer, 1842.
16 p. LCP

1845. __________.
Transactions.
Plymouth; Boston.
LCP: 1847; 1860.

__________.
See also JACKSON, CHARLES THOMAS, 1805-1880. An Address Delivered Before .... 1850.

1846. The Pocket Farrier; Or, Approved Receipts; Collected From different authors with an intent to cure or assist any immediate accidents that may happen to a horse till further help can be had.
Philadelphia: Published By Jacob Johnson, 1807.
Pp. [1-2], [5]-27, [1-4]; fold. front. HSP in LCP; LCP (lacking front.)

Pocket Manual for Producing Silk .... [1840].
See CLARKE, THOMAS COTTRELL, 1801-1874.

1847. POELLNITZ, FREDERICK CARL HANS BRUNO, Baron von, 1734-1801.
Essay On Agriculture.
New York: Printed By Francis Childs And John Swaine, 1790.
24 p.; 4 fold. pls. On agricultural implements. HSP

1848. [PORTER, GEORGE RICHARDSON, 1792-1852].
A Treatise On The Origin, Progressive Improvement, And Present State Of The Silk Manufacture.
Philadelphia: Carey & Lea [Stereotyped By J. Howe (t.p. verso)], 1832.
276 p.; illus. With general t.p., headed: The Cabinet Cyclopaedia. Conducted By The Rev. Dionysius Lardner .... Assisted By Eminent Literary And Scientific Men. Useful Arts. LCP; APS; HSP (with 4 p. pub. adv. at back)

1849. POSTLETHWAYT, MALACHY, 1707?-1767.
Britain's Commercial Interest Explained and Improved; In a Series of Dissertations On Several Important Branches of her Trade and Police: Containing A Candid Enquiry into the secret Causes of the present Misfortunes of the Nation .... In Two Volumes [- Vol. II].
London: Printed for D. Browne; A. Millar; J. Whiston and B. White; and W. Sandby, 1757.
2 vols. Vol. I: xviii, [6], 548 p. II: vi, [6], 551 p. Incl. chapters on American agriculture and resources. LCP

1850. [__________].
The History Of America: Containing The Geographical, Political, and Commercial State of that Continent: In which is given A succinct Account of the Trade, Customs and Policy of the French Nation, and the foundation they have laid for obtaining the sole dominion in those Colonies. Translated from the French of Mons. Du Bosque, late Historographer [sic] to the French King. In Two Volumes. Vol. I [- II].
London: Printed by George Cave and William Jordon, 1770.
2 vols. Vol. I: [2], 387 p. II: [2], 384 p. Howes D-525 (s.v. Du Bosque) calls for 12 prelim. p. in Vol. I and 6 in Vol. II. This is an unattributed reprint with altered title of Britain's Commercial Interest .... (above). LCP

1851. The Poultry Breeder's Text Book. Comprising Full Information Respecting The Choicest Breeds Of Poultry, And The Mode Of Rearing Them. With Twenty-Five Illustrations.
Philadelphia: King & Baird, [ca. 1854].
Pp. [1-4], [iii]-iv, 3-76 [i.e. 66]; illus. Pp. 64-66 misnumbered 74-76. HSP (wrappers)

1852. POWEL, JOHN HARE, 1786-1856.
Mr. Powel's Stock Of Improved Durham Short Horned Cattle, Imported, Or Bred By Him, Will Be Offered For Sale On Wednesday The 16th Of June Next, At Powelton On The Schuylkill, Opposite To Philadelphia. May 25, 1830.
[Philadelphia: 1830].
16 p. Cap. title: Catalogue And Pedigrees Of Imported Durham Short Horned Cattle .... LCP

1853. [__________].
Pennsylvania Agricultural Society. Strictures Upon Arator's Attack. [cap. title]
[Philadelphia: 1823].
4 p. Signed at end: A Farmer; and dated: December 17, 1823. Refers to a letter, signed "Arator," in Poulson's American Daily Advertiser for Dec. 5, 1823 (p. 2, col. 5), which criticizes the Society's system of awarding premiums. LCP; HSP

1855. The Practical American Gardener; Exhibiting The Time For Every Kind Of Work In The Kitchen Garden, Fruit Garden, Orchard, Nursery, Shrubbery, Pleasure Ground .... For every Month in the Year. By An Old Gardener.
Baltimore: Published By Fielding Lucas, Jr.; J. Robinson, printer, 1819.
Pp. [i]-xii, [5]-425. LCP

1854. [__________].
Reply To Col. Pickering's Attack Upon A Pennsylvania Farmer.
Philadelphia: Clark & Raser, Printers, 1825.
24 p. Author's name at end. Refers to Timothy Pickering's criticism of the Pennsylvania Agricultural Society Memoirs, pub. in the Massachusetts Agricultural Repository and Journal (Vol. 8, no. 3 [Jan. 1825]). LCP (with last leaf misbound after 1st); APS; HSP

PRATT, HENRY.
See HILL, D. & C.A., Auctioneers, Philadelphia. Catalogue Of Splendid And Rare Green House And Hot House Plants, To Be Sold By Auction At ... The Seat Of .... [1838].

The Present State of Great Britain and North America, With Regard To Agriculture .... 1767.
See MITCHELL, JOHN, d. 1768.

PRIME, A.J., jt. ed.
See American Journal of Agriculture and Science [pdcl.].

PRINCE FAMILY, Nurserymen, Flushing, N.Y.
Five generations of Princes were involved in the nursery. Robert Prince founded a nursery in about 1730, which his son William the elder (ca. 1725-1802) later managed. The latter's sons William the younger (1766-1842) and Benjamin separately ran the paternal nursery and an adjacent one beginning in 1793, and merged them as the Linnaean Botanic Garden soon thereafter. Around 1835 William the younger passed the business into the hands of his sons, one of whom, William Robert (1795-1869), became the titular head. By 1842 Winter & Co. (q.v.), the firm of William the younger's brother-in-law, held financial control, but William Robert, and, in the late 1850s, his sons, operated the nursery until its sale after the Civil War. See Bailey and DAB. See also monographic works by William Prince the younger and William Robert Prince, below.

1856. __________.
Annual Catalogue Of Fruit And Ornamental Trees And Plants, Bulbous Flower Roots, Green-House Plants, &c. &c. Cultivated At The Linnaean Botanic Garden, William Prince, Proprietor .... To Which Is Added, A Short Treatise On Their Cultivation, &c. Twenty-Third Edition.
New York: Printed By T. And J. Swords, 1825.
148 p. LCP

1857. __________.
Annual Catalogue Of Fruit And Ornamental Trees And Plants, (All Of Which Are Hardy) Cultivated at The Linnaean Botanic Garden .... Twenty-Fifth Edition.
New York: Printed By T. And J. Swords, 1829.
Pp. [i]-xiv, [1-2], [5]-86. LCP

1858. __________.
Annual Catalogue Of Fruit And Ornamental Trees And Plants, Cultivated At The Linnaean Botanic Garden And Nurseries. William Prince & Sons, Proprietors, Flushing, Long-Island, Near New-York. Twenty-Ninth Edition.
Jamaica, L.I.: I.F. Jones & Co., Printers, 1835.
88 p. T.p. headed: 1835 & 1836.

1859. __________.
Catalogue Of Bulbous And Tuberous Rooted Flowers, Of The Most Choice And Splendid Varieties, Cultivated At The Linnaean Botanic Garden And Nurseries, Flushing, Near New-York. William R. Prince & Co. [cap. title]
[New York: 1842].
36 p. Title headed: 1842 & 1843. LCP

1860. __________.
A Catalogue Of Fruit & Forest Trees, Flowering Shrubs, and Plants, For Sale By William And Benjamin Prince, At Flushing-Landing, on Long-Island, near New-York.
New York: Printed by T. & J. Swords, February, 1808.
Brs. LCP

1861. __________.
Catalogue Of Fruit And Ornamental Trees And Plants, Bulbous Flower Roots, Green-House Plants, &c. &c. Cultivated At The Linnaean Botanic Garden, William Prince, Proprietor, Flushing, Long-Island, near New York. To Which Is Added, A Short Treatise On Their Cultivation, &c. Twenty-Second Edition.
New York: Printed By T. And J. Swords, 1823.
143 p. LCP

1862. __________.
Catalogue Of Roses In The Nurseries Of William R. Prince & Co. [cap. title]
[New York: 1846].
50 p. LCP

1863. __________.
Prince's Descriptive Catalogue Of Fruit And Ornamental Trees, Shrubbery And Plants, Cultivated And For Sale At The Linnaean Botanic Garden And Nurseries .... Thirty-Third Edition.
New York: Printed By William Osborn, 1844.
60 p. T.p. headed: Periodical Pamphlet. 1844. LCP

1864. __________.
To Be Sosd [sic], By William Prince, At Flushing-Landing, On Long-Island, near New-York, A large Collection of Fruit Trees, as follows.
New York: Printed by H. Gaine, [1771].
Brs. LCP

1865. __________.
Wm. Prince & Son's [sic] Abridged Catalogue Of Green House Plants, With Reduced Prices. [cap. title]
[New York: ca. 1835].
4 p. LCP

1866. PRINCE, WILLIAM, 1766-1842.
A Short Treatise On Horticulture: Embracing A Description Of A Great Variety Of Fruit and Ornamental Trees and Shrubs, Grape Vines, Bulbous Flowers, Green-House Trees and Plants, &c. Nearly All Of Which Are At Present Comprised In The Collection Of The Linnaean Botanic Garden, At Flushing, near New-York, With Directions for their Culture, Management, &c.
New York: Printed By T. And J. Swords, 1828.
ix, [1], 196 p. LCP

__________.
See also
PRINCE FAMILY, Nurserymen, Flushing, N.Y.
PRINCE, WILLIAM ROBERT, 1795-1869, and __________.

1867. PRINCE, WILLIAM ROBERT, 1795-1869.
Prince's Manual Of Roses, Comprising The Most Complete History Of The Rose, Including Every Class, And All The Most Admirable Varieties That Have Appeared In Europe And America; Together With Ample Information On Their Culture And Propagation.
New York: Published By The Author, And Clark & Austin, Saxton & Miles, Wiley & Putnam, And Stanford & Swords, 1846.
213 p. LCP

__________.
See also PRINCE FAMILY, Nurserymen, Flushing, N.Y.

1868. __________, and WILLIAM PRINCE, 1766-1842.
The Pomological Manual; Or, A Treatise On Fruits: Containing Descriptions Of A Great Number Of The Most Valuable Varieties For The Orchard And Garden. Part I [- Part II].
New York: Published By T. & J. Swords, G. & C. & H. Carvill, E. Bliss, Collins & Co., G. Thorburn & Sons; Philadelphia: Judah Dobson, et al. [J. Seymour, Printer (t.p. verso)], 1831.
2 vols. Part I: 200 p. Part II: pp. [i]-xvi, [9]-216 Part I imprint has 10 publishers; Part II incl. 6 additional firms. LCP

1869. __________.
The Pomological Manual; Or, A Treatise On Fruits .... Part I [- II]. Second Edition.
New York: Published By T. & J. Swords, G. & C. & H. Carvill, et al. [J. Seymour, Printer (t.p. verso)], 1832.
2 vols. Part I: 200 p. Part II: pp. [i]-xvi, [9]-216. Publishers' names in imprints as in 1st edn. (above). LCP (2 vols. in 1)

1870. __________.
A Treatise On The Vine; Embracing Its History From The Earliest Ages To The Present Day, With Descriptions Of Above Two Hundred Foreign, And Eighty American Varieties; Together With A Complete Dissertation On The Establishment, Culture, And Management Of Vineyards.
New York: Published By T. & J. Swords, G. & C. & H. Carvill, E. Bliss, Collins & Co., G. Thorburn; Philadelphia: Judah Dobson, et al. [J. Seymour, Printer (t.p. verso)], 1830.
355 p.; front. LCP

1871. PRINCE GEORGE'S CO. [MD.]. AGRICULTURAL SOCIETY.
Proceedings Of The Prince George's County Agricultural Society At Its Eleventh Annual Meeting. Also The Address By Col. George W. Hughes.
Upper Marlboro': Printed At The "Gazette" Office, 1851.
27 p. LCP

1872. The Progressive Farmer. Edited by A.M. Spangler.
Philadelphia.
HSP: Vol. 1, nos. 1-12 (Jan. 16-Dec. 18, 1855). All pub. Merged with Pennsylvania Farm Journal (q.v.)

Proposals For Establishing A Number Of Farms .... 1816.
See COXE, TENCH, 1755-1824.

PROUTY, D.O., Agricultural Warehouse and Seed Store, Philadelphia.
See ALMANACS. Philadelphia. 1842. Agricultural Almanac .... To Which Is Added, A General Catalogue .... Incl. (pp. 33-45) Catalogue Of Vegetable Seeds; (46-66) Catalogue Of Implements.

1873. PULLEIN, SAMUEL.
The Culture of Silk: Or, An Essay On Its Rational Practice and Improvement. In Four Parts. I. On the raising and planting of Mulberry Trees. II. On hatching and rearing the Silk-Worms. III. On obtaining their Silk and Breed. IV. On reeling their Silk-Pods. For the Use of the American Colonies.
London: Printed for A. Millar, 1758.
xv, [1], 299, [1 (pub. adv.)] p.; front.; 1 pl. LCP; HSP

1874. __________.
An Essay Towards A Method Of Preserving the Seeds of Plants in a State fit for Vegetation, during Long Voyages. For the Improvement of the British Colonies in America.
London: Sold by A. Millar; and T. Pote, 1759.
15 p. LCP

__________, jt. auth.
See BOISSIER DE SAUVAGES DE LA CROIX, PIERRE AUGUSTIN, 1706-1767, and __________.

Purport of a Letter on Sheep. [1789].
See BORDLEY, JOHN BEALE, 1727-1804.

1875. PUTNAM, HENRY, 1778-1827.
Touches On Agriculture, Including A Treatise On The Preservation Of The Apple Tree. Together With Family Recipes, Experiments On Insects, &c. Second Edition, Enlarged.
Salem: John D. Cushing, Printer, 1825.
64 p. LCP

1876. PUVIS, MARC ANTOINE, 1776-1851.
[Essay On The Use Of Lime As A Manure. Translated From The French, By E. Ruffin ....] [wrapper title]
New York: Scatcherd & Adams, Printers, 1836. [wrapper imprint]
25 p. APS (without wrapper; cap. title: On The Use Of Lime As A Manure). Wrapper copy taken from NUC Pre-56.

1877. Quarterly Journal Of Agriculture [of the United States Agricultural Society].
Boston; Washington.
LCP: Vol. 3 (1856); vol. 7, no. 3 (Oct. 1859)

1878. QUEENS CO. [N.Y.]. AGRICULTURAL SOCIETY.
Annual Report Of The Transactions Of The Queens County Agricultural Society, Containing The Address Delivered At The Exhibition Of The Society, At Hempstead, October 17th, 1843, By D.S. Dickinson: The Reports Of Committees, The Premiums Awarded, And A List Of The Officers And Members Of The Society For 1843.
Hempstead, L.I.: Printed By Charles Willets, 1844.
28 p. LCP

__________.
See also
FURMAN, GABRIEL, 1800-1845. Address Delivered Before .... 1845.
McCOUN, WILLIAM T. Address, Delivered Before .... 1843.

1879. Queries, On the present state of Husbandry and Agriculture in [blank] [cap. title]
[Philadelphia?: ca. 1790].
7, [1] p. List of 44 queries. Advertisement at back requests dried samples of plants to be sent to Mons. le Comte d'Augiviller [i.e. Charles Claude de Flahault de la Billardie, comte d'Angiviller], Directeur General des Batimens du Roy; and remarks to be addressed to Mons. l'abbé Ressier [i.e. François Rozier] of the Academy of Sciences [of Paris]. APS

1880. QUINCY, EDMUND, 1703-1788.
A Treatise Of Hemp-Husbandry; Being a Collection of approved Instructions, as to the Choice and Preparation of the Soils, most proper for the Growth of that useful and valuable material, and also as to the subsequent Management thereof, agreable [sic] to the Experience of several Countries wherein it has been produced, both in Europe and America. With some introductory Observations, upon the Necessity which the American British Colonies are under, generally to engage in the said Production ....
Boston: Printed by Green & Russell, by Order of the honorable House of Representatives, 1765.
32, [1] p.; 1 pl. HSP in LCP; LCP

1881. QUINCY, JOSIAH, 1772-1864.
An Address Delivered Before The Massachusetts Agricultural Society [i.e. Society for Promoting Agriculture], At The Brighton Cattle Show, October 12th, 1819. [cap. title]
[Boston: 1819].
15 p. LCP

1882. __________.
Essays On The Soiling Of Cattle, Illustrated From Experience; And An Address, Containing Suggestions Which May Be Useful To Farmers.
Boston: Printed By John Wilson And Son, 1859.
64 p. APS

1883. RAFINESQUE, CONSTANTINE SAMUEL, 1783-1840.
American Manual Of The Grape Vines And The Art Of Making Wine: Including An Account of 62 Species of Vines, with nearly 300 varieties. An account of the principal Wines, American and Foreign. Properties and uses of Wines and Grapes. Cultivation of Vines in America; and the Art to make good Wines. With 8 Figures. [wrapper title]
Philadelphia: Printed For The Author, 1830. [wrapper imprint]
64 p.; illus. Wrapper headed: Eight Figures, Twenty-Five Cents. LCP (without wrapper); APS and HSP (wrappers; with 1st leaf [illus.] pasted verso of wrapper)

1884. __________.
American Manual Of The Mulberry Trees. Their History, Cultivation, Properties, Diseases, Species and Varieties &c, with hints on the production of Silk from their barks &c.
Philadelphia: 1839.
96 p. On cover: Publications Of The Eleutherium Of Knowledge, No. 5. LCP; HSP

1885. RANDALL, HENRY STEPHENS, 1811-1876.
Sheep Husbandry; With An Account Of The Different Breeds, And General Directions In Regard To Summer And Winter Management, Breeding, And The Treatment Of Diseases. With Portraits and other Engravings.
New York: C.M. Saxton, Agricultural Book Publisher, 1854.
320 p.; illus. Bound at back: W. Youatt's Sheep: Their Breeds .... 1855 (q.v.). Spine title: The Shepherd's Own Book. HSP

1886. __________.
Sheep Husbandry; With An Account Of The Different Breeds ....
New York: A.O. Moore, Agricultural Book Publisher, 1859.
320 p.; tinted front.; illus. LCP

1887. __________.
Sheep Husbandry; With An Account Of The Different Breeds .... With His Letter To The Texas Almanac On Sheep Husbandry In Texas, And Geo. W. Kendall's On Sheep Raising In Texas.
New York: C.M. Saxton, Barker & Co.; San Francisco: H.H. Bancroft & Co., 1860.
338 p.; illus. LC calls for front. LCP

RANDOLPH, JOHN, 1727-1784. A Treatise On Gardening ....
See also GARDINER, JOHN, and DAVID HEPBURN. The American Gardener .... To Which Is Added A Treatise ... By A Citizen Of Virginia [i.e. Randolph] .... 1818.

1888. [RAWLE, FRANCIS, ca. 1662-1727].
Ways and Means For the Inhabitants of Delaware To become Rich: Wherein The several Growths and Products of these Countries are demonstrated to be sufficient Fund for a flourishing Trade. Humbly submitted to the Legislative Authority of these Colonies.
Philadelphia: Prnited [sic] and sold by S. Keimer, 1725.
65, [1], [1 (pub. adv.)] p. HSP in LCP; LCP

1889. RAWLE, WILLIAM, 1759-1836.
An Address Delivered Before The Philadelphia Society for Promoting Agriculture: At Its Anniversary Meeting, January 19, 1819.
Philadelphia: Published By Order Of The Society; William Fry, Printer, 1819.
35 p. LCP; APS; HSP

1890. RAYNER, KENNETH, 1808-1884.
An Address Delivered Before The North Carolina State Agricultural Society, By Hon. Kenneth Rayner, Of Hertford, At The Second Annual Fair Of The Society, Thursday, October 19th, 1854. Printed by order of the Society.
Raleigh: Steam Power Press Of The "Southern Weekly Post," 1854.
40 p. LCP

1891. REDD, GEORGE.
Eine Neue Entdeckung, auserst wichtig fur Pflanzer und Landbauer, betreffend die Fruchtbarmachung des Magern und Erschopften Landes, Auf eine leichte und wolfeile Weise; Nebst einigen Bemerkungen uber den Obst-und Garten-Bau. (Aus dem Englischen ubersetzt.)
Lancaster, Pennsylvania: Gedruckt fur den Verfasser, von W. Hamilton und Compagnie, 1810.
16 p. HSP in LCP

1892. REEMELIN, CHARLES, b. 1814.
The Vine-Dresser's Manual, An Illustrated Treatise On Vineyards And Wine-Making.
New York: C.M. Saxton & Company, 1856.
Pp. [1-4], [7]-103; 10+[2] p. pub. adv.; front.; illus. LCP

1893. __________.
The Vine-Dresser's Manual, An Illustrated Treatise On Vineyards And Wine-Making.
New York: A.O. Moore, Agricultural Book Publisher (Late C.M. Saxton & Co.), 1858.
103, [1], 12 (pub. adv.) p. incl. front.; illus. LCP

1894. REID, WILLIAM, Nurseryman, Elizabeth, N.J.
Catalogue For 1852 and 1853 Of Fruit And Ornamental Trees, Flowering Shrubs, &c. Cultivated And For Sale By William Reid At His Nurseries, Elizabethtown, N.J. Formerly Of Murray Hill, N.Y. [wrapper title]
New York: Folger & Turner, Printers, 1852. [wrapper imprint]
18 p. incl. wrapper. LCP (wrappers)

1895. __________.
Wholesale Price List For The Fall of 1854 & Spring of 1855, Of Fruit Trees &c., &c. Cultivated And For Sale By William Reid, Elizabethtown, N.J. [cap. title]
[New York:] [Folger & Turner, Steam Prs. (colop.)], [1854].
[4] p. LCP

1896. REINHOLD, C.G.
The Farmer's Promotion Book, A New And Scientific Manuring System, For The Cultivation And Increase Of All Kinds Of Grains, Grasses And Fodder And Pasture, Upon All Kinds Of Soil, Proved By Actual Experiments And Based On Evident Truths. Designed To Improve Agriculture In All Its Branches. Represented by upward of One Hundred and Fifty Engravings ....
Pittsburgh: Printed By W.S. Haven, 1856.
127, [1] p.; 3 fold. pls. LCP

1897. The Religious Farmer. Edited And Published By George Junkin, A.M.
Milton, Pa.
LCP: Vol. 1 (1828)

Remarks On The Letters From An American Farmer .... 1783.
See AYSCOUGH, SAMUEL, 1745-1804.

1898. Remarks On The Manufacturing Of Maple Sugar; With Directions For Its Further Improvement. Collected By A Society Of Gentlemen, In Philadelphia, And Published For The General Information And Benefit Of The Citizens Of The United States.
Philadelphia: Printed By James & Johnson, 1790.
24 p. HSP in LCP; HSP

1899. Remarks On The Manufacturing Of Maple Sugar: With Directions For Its Further Improvement ....
Philadelphia: Printed; New York: Re-Printed By William Morton, 1791.
23 p. HSP in LCP

1900. Remarks On The Manufacturing Of Maple Sugar: With Directions For Its Further Improvement ....
Philadelphia: Printed 1790; London: Reprinted By James Phillips, 1791.
16 p. HSP in LCP

RENSSELAER CO. [N.Y.]. AGRICULTURAL SOCIETY.
See GENET, EDMOND CHARLES, 1765-1834. Address .... Delivered Before .... 1821.

RENSSELAER INSTITUTE, Troy, N.Y.
Established in 1824 as Rensselaer School; in 1837 became known as Rensselaer Institute. Although the name Rensselaer Polytechnic Institute was used informally beginning around 1849, the change was not formally made until 1861.

1901. __________.
The Constitution And Laws Of Rensselaer School, In Troy, New-York; Adopted By The Board Of Trustees, March 11, 1825.
Troy: Printed By Tuttle And Richards, 1825.
23 p. APS; HSP

1902. __________.
Constitution And Laws Of Rensselaer School, Adopted By The Board Of Trustees, April 3, 1826. Together With A Catalogue Of Officers And Students.
Albany: Printed By Packard & Van Benthuysen, 1826.
28 p. LCP

1903. __________.
Rensselaer Institute: A Polytechnic School, For Instruction In The Application Of Mathematics, Physics, Chemistry, And Natural History, To Civil Engineering, The Arts, Manufactures, And Agriculture; Established By The Hon. Stephen Van Rensselaer, And Endowed By His Family And The Common Council Of Troy.
Troy, N.Y.: From The Press Of Prescott & Wilson, 1847.
14 p. LCP (wrapper)

1904. __________.
Rensselaer Institute: Established Especially To Instruct Young Men In The Application Of Experimental Chemistry, Mathematics, Philosophy, And Natural History, To Agriculture, Domestic Economy, The Arts ....
Troy, N.Y.: From The Press Of N. Tuttle, 1846.
8 p. HSP (wrappers)

RENSSELAER SCHOOL, Troy, N.Y.
See RENSSELAER INSTITUTE, Troy, N.Y. Later name of the above institution.

1905. REPPLOGEL, GEORGE.
A Book Comprising A New System For The Preservation Of Bees; Discovered Through A Series Of Practical Experiments
Greensburgh, Pa.: Printed For The Author, 1830.
24 p. HSP

Requisites For ... Hog-Teaching. [1802].
See LEIB, MICHAEL, 1761-1822.

1906. REYNOLDS, JOSEPH, d. 1872.
A Prize Essay On The Preparation And Application Of Manures.
Boston: J.H. Eastburn's Press, 1858.
21 p. T.p. headed: No. 4. New Series. LCP (wrappers)

RHODE ISLAND. GEOLOGICAL AND AGRICULTURAL SURVEY.
See JACKSON, CHARLES THOMAS, 1805-1880. Report .... 1840.

RHODE ISLAND. HORTICULTURAL SOCIETY.
See
RUSSELL, GEORGE ROBERT, 1800-1866. Address Before .... 1852.
WAYLAND, FRANCIS, 1796-1865. Address Before .... 1852.

1907. RHODE ISLAND. SOCIETY FOR THE ENCOURAGEMENT OF DOMESTIC INDUSTRY.
Report Of The Sub-Committee On Agriculture To The Standing Committee Of The Rhode Island Society for the Encouragement of Domestic Industry. January 14, 1857. On The Chinese Sugar Cane.
Providence: Knowles, Anthony & Co., City Printers, 1957.
45 p. LCP

1908. __________.
Transactions.
Providence.
APS: 1854-61; 1864-66; scattered issues to 1873.

__________.
See also
ALLEN, STEPHEN MERRIL, 1819-1894. Fibrilia, Or Flax Cotton: Three Addresses, Before .... 1860.
BURGES, TRISTAM, 1770-1853. Address To .... 1822.
ROBBINS, ASHER, 1757-1845. Address To .... 1822.
RUSSELL, GEORGE ROBERT, 1800-1866. Address Before .... 1852.
TILLINGHAST, JOSEPH LEONARD, 1790-1844. Address Delivered Before .... 1827.
WAYLAND, FRANCIS, 1796-1865. Address Before .... 1852.

1909. RICHARDSON, JOSIAH, comp.
The New-England Farrier, And Family Physician; Containing, Firstly, Paul Jewett's Farriery .... Secondly, A Collection Of Brutal Receipts .... Thirdly, A very valuable collection of Receipts for Human Diseases .... Fourthly, Receipts, By a very learned, skilful and pious English Author .... Fifthly, An Extraordinary English Author ... on neat Cattle, Sheep, Swine .... Sixthly, A large number of Valuable Receipts From Capt. Joseph Smith .... Seventhly, A very large number of Miscellaneous Receipts .... Last Of All, Doct. J. Williams' Family Physician ....
Exeter [N.H.]: Published By Josiah Richardson, 1828.
444, 24 p. LCP

1910. RIDGELY, HENRY.
Address Delivered By Henry Ridgely, M.D., Before The Agricultural Society of Kent County, Delaware, In Dover, on Thursday, the 6th of October, 1859.
Dover: Delaware State Reporter Book And Job Printing Office, 1859.
19 p. LCP

RIDGWAY FARM AND COAL COMPANY, Philadelphia.
Company publications appear under the above name as well as under RIDGWAY FARM AND AGRICULTURAL COMPANY; RIDGWAY FARM, AGRICULTURAL AND COAL COMPANY; and RIDGWAY FARM AND COAL COMPANY. Philadelphia city directories for 1856 list the name as RIDGEWAY [sic] FARM AND COAL CO.; the company is unlisted in 1857, and in 1858 appears as RIDGEWAY [sic] FARM CO.

1911. __________.
Eine Bauerey und Heymath In dem Erreich eines jeden Mannes. Ridgway Bauerey-und Kohlen-Companie. Sieben und zwanzig tausend Acker vom besten Land in Elk County, Pennsylvanien ....
Lancaster: Gedruckt von Johann Bar und Sohnen, 1856.
35 p. LCP (wrapper)

1912. __________.
A Farm And Home Within the Reach of Every Man. Ridgway Farm and Coal Company. Twenty-Seven Thousand Acres Of The Best Land In Elk County, Pennsylvania, For Sale, Giving A Farm Of 25 Acres For $200 - Payable $1 Per Week Or Monthly. - More Than 25 Acres To Be Had Together In The Same Proportion.
Lancaster, Pa.: Printed By John H. Pearsol, 1856.
30 p. HSP

1913. __________.
A Farm Within The Reach Of Every Man. Pennsylvania Land. Ridgway Farm And Agricultural Company.
Philadelphia: Joseph Severns, 1855.
22 p.; map. Wrapper title ends: ... Ridgway Farm, Agricultural and Coal Co.; wrapper imprint date lacks last digit. LCP (wrapper)

1914. __________.
Report Of The Ridgway Farm and Land Company, Together With The Geological Report Of Dr. Charles T. Jackson, Of Boston. State Geologist, Analytical Chemist, &c., &c.
Philadelphia: M'Laughlin Brothers' Steam-Power Book and Job Printing Establishment, 1856.
32 p.; map (front wrapper verso). LCP (wrappers); HSP (wrappers)

1915. __________.
Report Of The ... Together With The Geological Report ....
Philadelphia: Brown's Steam-Power Book And Job Printing Office, 1856.
29 p.; map (front wrapper verso). HSP (wrappers)

RILEY, THEODORE W., Agent, New York.
See ROBINSON, SOLON, 1803-1880. Guano .... Prepared And Published ... For .... 1852.

RITTENHOUSE COLLEGE, Philadelphia.
See [BETTON, SAMUEL]. Address Of The President .... [1837].

1916. RITTER, CHRISTIAN.
Christian Ritter's Europäisches Pferde-Pulver. Fur das Gelbe Wasser und andere Krankheiten der Pferde.
[Reading: ca. 1805].
Brs. LCP

1917. ROBBINS, ASHER, 1757-1845.
Address To The Rhode-Island Society For The Encouragement Of Domestick Industry, Delivered at Pawtuxet, October 16th, 1822.
Providence: Miller & Hutchens, Printers, 1822.
54 p. LCP (wrapper)

1918. ROBBINS, LOUIS S.
Patents Granted To Louis S. Robbins For An Improvement In Fertilizing Compounds, And Report Upon Its Merits.
Baltimore: Printed By Sherwood & Co., 1857.
15 p. LCP

1919. ROBERTS, EDWARD P.
A Manual, Containing Directions For Sowing, Transplanting And Raising Of The Mulberry Tree; Together With Proper Instructions For Propagating The Same .... As Also, Instructions For The Culture Of Silk: To Which Is Added, Calculations Shewing The Produce ....
Baltimore: Printed By Sands & Neilson, 1835.
72 p.; illus. APS; HSP (wrappers)

1920. __________.
A Manual, Containing Directions For Sowing, Transplanting And Raising The Mulberry Tree .... Third Edition With Improvements And Additions.
Baltimore: Printed By Samuel Sands, 1838.
100 p.; illus. APS (wrapper); HSP

1921. __________.
A Manual, Containing Directions For Sowing, Transplanting And Raising The Mulberry Tree .... Fourth Edition With Improvements And Additions.
Baltimore: Printed By Samuel Sands, 1839.
108 p.; illus. LCP (wrappers); HSP (wrappers)

1922. ROBERTS, JOB, 1756-1851.
The Pennsylvania Farmer; Being A Selection From The Most Approved Treatises On Husbandry, Interspersed With Observations And Experiments.
Philadelphia: Published And Sold By Jacob Johnson & Co.; Portsmouth: Sold Also By C. Peirce, And W. & D. Treadwell; Boston: W.P. & L. Blake; New Haven; Beers & How; Albany: Whiting, Backus & Whiting, And C.R. & G. Webster; Troy: O. Penniman & Co., et al; A. Bartram, Printer, 1804.
Pp. [i]-v, [1-3], [5]-224. HSP in LCP; LCP; APS; HSP

1923. ROBINSON, SOLON, 1803-1880.
Guano: A Treatise of Practical Information for Farmers; Containing Plain Directions How To Apply Peruvian Guano To The Various Crops And Soils Of America, With A Brief Synopsis Of Its History, Locality, Quality, Method Of Procuring, Prospect Of Continued Supply, And Price .... Prepared And Published By Solon Robinson, For Messrs. F. Barreda & Brother, Agents For The Peruvian Government At Baltimore; And Theodore W. Riley, Esq., Their Agent In New York.
New York: 1852.
96 p. LCP

ROCK CO. [WIS.]. AGRICULTURAL SOCIETY AND MECHANICS' INSTITUTE.
See GUERNSEY, ORRIN, and JOSIAH T. WILLARD, eds. and comps. History Of Rock County, And Transactions Of The ....

ROCKINGHAM [N.H.]. AGRICULTURAL SOCIETY.
See PLUMER, WILLIAM, JR., 1789-1854. Address Delivered Before .... 1821.

1924. RODGERS, MILES M.
Scientific Agriculture, Or The Elements Of Chemistry, Geology, Botany And Meteorology, Applied To Practical Agriculture. Illustrated By Numerous Engravings, & A Copious Glossary.
Rochester [N.Y.]: Published By Erastus Darrow [Benton & Fisher, Printers (t.p. verso)], 1848.
279 p.; front.; illus. LCP

1925. ROESSLE, THEOPHILUS.
How To Cultivate And Preserve Celery. Edited, With A Preface, By Henry S. Olcott.
Albany: Theophilus Roessle; New York: C.M. Saxton, Barker & Company; And By All Booksellers [Munsell & Rowland, Printers (t.p. verso)], 1860.
[6], 100 p.; col. front.; 2 col. pls.; 1 section. Title headed: Roessle's Gardener's Hand-Books. No. 1. LCP

1926. ROHLWES, JOHANN NICOLAUS.
Allgemeines Vieharzneibuch, oder Unterricht, wie der Landmann seine Pferde, sein Rindvieh, seine Schafe, Schweine, Ziegen und Hunde aufziehen, warten und füttern, und ihre Krankheiten erkennen und heilen soll; nebst einem Anhange. Eine von der Märklichen ökonom. Gesellschaft zu Potsdam gekrönte Preisschrift. Nebst einer Kupfertafel. Siebente verbesserte und vermehrte Auflage.
Berlin: In der Maurerschen Buchhandlung; Easton [Pa.]: bei Christian J. Hutter, 1816.
Pp. [1-2], [v]-xviii, 358; fold. pl. LCP

1927. __________.
Allgemeines Vieharzneibuch, oder Unterricht .... Erste amerikanische, mit nöthigen Abanderungen und Zusätzen versehene Auflage.
Shellsburg (Pennsylvania): Gedruckt und zu haben bey Friedrich Goeb, 1823.
xvi, 291 p. HSP in LCP; LCP (pp. 37-274 only)

1928. __________.
Vollständiges Gäuls-Doctor-Buch; oder gründlicher Unterricht wie der Bauer und jeder Pferde-Besitzer alle Krankheiten seiner Pferde erkennen und auf die leichteste Art heilen kann. Mit einem Anhange, welcher zeigt: wie man den Puls des Viehes beurtheilen, Aderlassen, Fontanell-Legen, Haarseilzeihen und Blutstillen kann. Nebst einer Abhandlung von den Krankheiten der Hunde.
Reading [Pa.]: gedruckt und herausgegeben von Heinrich B. Sage, 1817.
[2], ix, [3], 108 p.; illus. LCP

1929. [RONALDSON, JAMES].
Observations On The Sugar Beet, And Its Cultivation.
Philadelphia: William F. Geddes, Printer, 1840.
15 p. APS (with 1 p. adv. at back); HSP (wrappers)

__________.
See also PEDDER, JAMES, 1775-1859. Copy of an Agreement entered into .... between .... [1836].

1930. ROSE, JOHANNES.
Mineralisches Pferde-Pulver, [text begins:] Ist eine ganze neue durch die Chemie erfundene Arzney ....
[Lancaster, Pa.?: ca. 1800].
Brs. LCP

1931. ROSE, ROBERT H.
An Address Delivered Before The Agricultural Society Of Susquehanna County [Pa.], At its Organization, December 6, 1820.
Montrose [Pa.]: Printed at the Office of the Messenger, By Adam Waldie, 1820.
22 p. APS

1932. ROSS, GEORGE M. von.
Des Auswanderers Handbuch. Getreue Schilderung der Vereinigten Staaten von Nordamerika und zuverlässiger Rathgeber für dahin Auswandernde jeden Standes.
Elberfeld and Iserlohn: Julius Bädeker [Leipzig: Druck von Philipp Reclam jun. (colop.)], 1851.
xii, 509, [1], [2 (pub. adv.)] p. With general t.p.: Neueste Länderkunde mit besonderer Beziehung auf deutsche Auswanderung und Colonisation. Dritter Band: Die Vereinigten Staaten Von Nordamerika. HSP

1933. ROST, PIERRE ADOLPHE, 1797-1868.
Oration Delivered Before The Agricultural And Mechanics' Association Of Louisiana, On the 12th of May, 1845.
Philadelphia: Printed By J. Van Court, At The Office Of The Ladies' Garland, 1845.
16 p. LCP; APS

1934. ROWE, ASA, Nurseryman, Greece, N.Y.
Annual Catalogue Of Fruit And Ornamental Trees, Shrubs, Herbaceous Plants, Bulbous Roots; And Green-House Plants, Cultivated And For Sale At The Monroe Garden And Nurseries, By Asa Rowe, Greece, Monroe County, N.Y.
Rochester: Printed By Thomas H. Hyatt, At the Daily Advertiser Office, 1841.
60 p. T.p. headed: 1841 And 1842. LCP

1935. ROYAL SOCIETY OF ARTS, London.
Premiums By The Society, Established at London, For The Encouragement Of Arts, Manufactures, And Commerce.
London: Printed by Order of the Society, 1758.
31 p. Incl. premiums for producing olives, silk, wine, etc. in the colonies. LCP

1936. __________.
[Premiums By The Society ....]
[London: 1764].
HSP in LCP: [2], 100 p. only; lacking t.p. Incl. premiums for productions in the colonies.

1937. __________.
A Register Of The Premiums And Bounties Given By The Society Instituted At London For The Encouragement Of Arts, Manufactures, And Commerce, From The Original Institution In The Year MDCCLIV, To The Year MDCCLXXVI Inclusive.
London: Printed, By Order Of The Society, By James Phillips, 1778.
iv, 55, [1] p. LCP

1938. RUFFIN, EDMUND, 1794-1865.
An Address On The Opposite Results Of Exhausting And Fertilizing Systems Of Agriculture, Read Before The South-Carolina Institute, At Its Fourth Annual Fair, November 18th, 1852.
Charleston: Steam Power Press Of Walker And James, 1853.
52 p. LCP

1939. __________.
Address To The Virginia State Agricultural Society, On The Effects Of Domestic Slavery On The Manners, Habits And Welfare Of The Agricultural Population Of The Southern States; And The Slavery Of Class To Class In The Northern States. Read At The First Annual Meeting ... December 16, 1852. Published By Order Of The Virginia State Agricultural Society.
Richmond: P.D. Bernard, Printer, 1853.
20 p. LCP

1940. __________.
An Essay On Calcareous Manures.
Petersburg, (Va.): Published By J.W. Campbell [Philadelphia: Printed by L.R. Bailey (t.p. verso)], 1832.
242, [1 (pub. adv.)] p. LCP

1941. __________.
An Essay On Calcareous Manures. Second Edition.
Shellbanks, Va.: Published At The Office Of The Farmers' Register [Robert Ricketts, Printer (t.p. verso)], 1835.
116 p.; illus. LCP; APS; HSP

1942. __________.
An Essay On Calcareous Manures. Third Edition.
Petersburg [Va.]: Printed For The Author, 1842.
Pp. [i]-vii, [1], [13]-316; illus. LCP; APS (wrapper)

1943. __________.
Premium Essay On Agricultural Education. Submitted To The Executive Committee Of The Southern Central Agricultural Association. Second Edition.
Richmond, Va.: J.W. Randolph [Clemmitt, Fore & Campbell, Printers (t.p. verso)], 1853.
24 p. LCP (wrappers); HSP

1944. __________.
Report Of The Commencement And Progress Of The Agricultural Survey Of South-Carolina, For 1843.
Columbia: A.H. Pemberton, State Printer, 1843.
120, 55 [1] p.; fold. tbl. Sep.-paginated Appendix. LCP

__________, ed. and prop.
See also The Farmer's Register .... [pdcl.].

1945. RUPP, ISRAEL DANIEL, 1803-1878.
The Farmer's Complete Farrier, Comprising A Historical Description Of All The Varieties Of That Noble And Useful Animal, The Horse; Giving Instructions In All Things That Relate To Him ... With A Description Of All The Diseases To Which He Is Liable .... With A Prize Essay On Mules .... With An Appendix, Containing An Additional Number Of Original And Selected Receipts .... Also: Addenda, Describing One Hundred & Fifty Plants And Medicines .... Selected, compiled and translated from the best German and English works ....
Lancaster, Pa.: Published By Gilbert Hills [Harrisburg, Pa.: Fenn & Wallace, Printers (t.p. verso)], 1842.
416 p.; fold. front.; 5 pls. HSP

Rural Economy: Or Essays .... 1792.
Rural Oeconomy .... 1776.
See YOUNG, ARTHUR, 1741-1820.

1946. The Rural Magazine And Farmer's Monthly Museum ....
Hartford.
LCP: Vol. 1, no. 1 (Feb. 1819)

1947. The Rural Magazine, And Literary Evening Fire-Side.
Philadelphia.
LCP; HSP: Vol. 1 (1820). All pub.

1948. The Rural Visiter: A Literary And Miscellaneous Gazette.
Burlington, N.J.
HSP: Vol. 1, nos. 1-52 (July 30, 1810-July 22, 1811). All pub.

1949. RUSH, BENJAMIN, 1745-1813.
An Account Of The Sugar Maple-Tree, Of The United States, And of the methods of obtaining Sugar from it, together with Observations upon the Advantages both public and private of this Sugar. In a Letter to Thomas Jefferson, Esq. Secretary of State of the United States, and one of the Vice- Presidents of the American Philosophical Society. Read In The American Philosophical Society, On The 19, Of August, 1791, And Extracted From The Third Volume Of Their Transactions Now In Press.
Philadelphia: Printed By R. Aitken & Son, For A Society Of Gentlemen, 1792.
16 p. LCP; HSP

1950. __________.
An Account Of The Sugar Maple-Tree, Of The United States, And Of The Methods Of Obtaining Sugar ....
Philadelphia: Printed; London: Reprinted By J. Phillips, 1792.
[4], 24 p.; 2nd leaf pub. adv. APS

1951. RUSKIN, JOHN, 1819-1900.
Notes On The Construction Of Sheepfolds.
New York: John Wiley [S.W. Benedict, Print. (t.p. verso)], 1851.
38 p. LCP

1952. RUSSELL, CHARLES THEODORE, 1815-1896.
Agricultural Progress in Massachusetts for the Last Half Century. An Address Delivered Before The Agricultural Society, Of Westborough and Vicinity, September 25, 1850. Published By Request Of The Society.
Boston: Printed By Charles C.P. Moody, 1850.
22 p. LCP (wrappers)

1953. RUSSELL, GEORGE ROBERT, 1800-1866.
An Address Before The Norfolk [Mass.] Agricultural Society, At Dedham, September 24, 1851.
[Boston:] Published By The Society [J.M. Hewes & Co. (t.p. verso)], [1851].
30 p. LCP

1954. __________.
Address Before The Rhode-Island Society For The Encouragement Of Domestic Industry, And The Rhode-Island Horticultural Society, At Their Third Industrial Exhibition In Providence; Delivered September 17th, 1852.
Providence: Knowles, Anthony & Co., Printers, 1852.
27 p. LCP

1955. RUSSELL, JOHN B., Nurseryman, Boston.
Catalogue Of Kitchen Garden, Herb, Tree, Field And Flower Seeds, Bulbous Flower Roots, Agricultural Books, Fruit Trees, Shrubs, And Grape Vines; For Sale At The Seed Store Connected With The New England Farmer ... Boston .... With A List Of Agricultural Implements, For sale at the Warehouse in the same building; by Joseph R. Newell .... Second Edition.
Boston: Printed At The New England Farmer Office, 1828.
47, [1] p. LCP

1956. RUSSELL, JONATHAN, 1771-1832.
Address Delivered Before The Worcester [Mass.] Agricultural Society, September 27, 1821, Being Their Anniversary Cattle Show And Exhibition Of Manufactures. Published by order of the Trustees.
Worcester: Printed By Manning & Trumbull, Oct. 1821.
24, [2] p. LCP

1957. RUSSELL, ROBERT, d. 1871.
North America Its Agriculture And Climate Containing Observations On The Agriculture And Climate Of Canada, The United States, And The Island Of Cuba.
Edinburgh: Adam And Charles Black [Printed By R. And R. Clark (t.p. verso)], 1857.
[8], 390 p.; col. fold. front. map; 7 maps, 5 fold.; fold. tbl.; illus. LCP; APS (lacking ½ t.p.); HSP

RUTLAND CO. [VT.]. AGRICULTURAL SOCIETY.
See MARSH, GEORGE PERKINS, 1801-1882. Address Delivered Before .... 1848.

1958. SABINE, LORENZO, 1803-1877.
Report On The Principal Fisheries Of The American Seas: Prepared For The Treasury Department Of The United States By Lorenzo Sabine, Of Massachusetts; And Submitted By The Hon. Thomas Corwin, Secretary Of The Treasury, As A Part Of His Annual Report On The Finances, At The Second Session Of The Thirty-Second Congress.
Washington: Robert Armstrong, Printer, 1853.
317 p. LCP

1959. SACKET, AUGUSTUS.
For Sale, 400,000 Acres Of Land. In the counties of Rutherford, Mecklenburgh & Buncomb In the southwestern parts of the state of North Carolina - not inferior in any respect to the Alabama ....
[Sacket's Harbor, N.Y.?: 1819?].
Brs. LCP

1960. SAFFORD, JAMES MERRILL, 1822-1907, and RICHARD OWEN, 1810-1890.
Report Upon The Mineral And Agricultural Resources Of The Lands Owned By The Hopkins Mastodon Coal and Iron Mining and Manufacturing Company: The Result Of An Exploration ....
Nashville, Tenn.: Nashville Printing And Publishing Co., 1857.
32 p.; front.; col. fold. map. The lands lay chiefly in Hopkins Co., Ky., and also in Indiana and Illinois. LCP

ST. JOHN, J. HECTOR, [pseud.].
See CRÈVECOEUR, MICHEL GUILLAUME ST. JEAN DE, 1735-1814.

ST. JOHN, SAMUEL, 1813-1876.
See TRUMBULL CO. [OHIO]. AGRICULTURAL SOCIETY. Transactions Of ... With The Address, Delivered By .... 1850.

ST. JOHN'S COLLETON [CO., S.C.]. AGRICULTURAL SOCIETY.
See
SEABROOK, WHITEMARSH BENJAMIN, 1795-1855. A Memoir On ... Cotton ... Read Before ... And The State Agricultural Society Of South-Carolina .... 1844.
TOWNSEND, JOHN. An Address Delivered on Edisto Island, Before The .... 1828.

ST. LAWRENCE CO. [N.Y.]. AGRICULTURAL SOCIETY.
See FOOTE, HENRY G. Address Of .... 1859.

1961. ST. LOUIS. AGRICULTURAL & MECHANICAL ASSOCIATION.
Fifth Annual Fair Of The St. Louis Agricultural & Mechanical Association, To Commence Monday, 24th Day Of September, 1860, And Continue Six Days. Schedule Of Premiums Amounting To Over 20,000 Dollars. [wrapper title]
St. Louis, Mo.: George Knapp & Co., Printers And Binders, 1860. [wrapper imprint]
32 p. LCP (wrappers)

1962. ST. LOUIS [MO.]. VINE & FRUIT GROWERS' ASSOCIATION.
Explanatory Statement Of The Views And Proceedings Of The St. Louis Vine & Fruit Growers' Association, Together With The Charter Of The Company. [cap. title]
[St. Louis: 1859].
8 p. LCP

__________.
See also HAVEN, CHARLES H. Views On The Vine Growing Resources .... 1858. Incl. proposal for the formation of the above association.

ST. LUKE'S PARISH [BLUFFTON, S.C.]. AGRICULTURAL SOCIETY.
See WIGG, WILLIAM HAZZARD. An Address Delivered .... 1853.

1963. SALTONSTALL, LEVERETT, 1783-1845.
An Address Before The Essex [Mass.] Agricultural Society ... September 27, 1843. Published By Order Of The Society.
Salem: Printed At The Gazette Office, 1843.
24 p. LCP (wrapper)

1964. SANBORN, EDWIN DAVID, 1808-1885.
An Address Delivered Before The Connecticut River Valley Agricultural Society, At Their Annual Fair, At Orford, N.H., September 26, 1850.
Hanover: Printed At The Dartmouth Press, 1850.
16 p. LCP

SARATOGA CO. [N.Y.]. AGRICULTURAL SOCIETY.
See STIMSON, EARL. Practical Husbandry. An Address .... 1823.

SAUNDERS, WILLIAM, 1836-1914. An Essay On The Culture Of The Native And Exotic Grape.
See [SPANGLER, ANDREW M., pub.]. Both Sides Of The Grape Question .... 1860. Incl. Saunders's essay.

1965. SAYERS, EDWARD.
The American Flower Garden Companion. Adapted To The Northern States.
Boston: Joseph Breck And Company; New York: G.C. Thorburn [Press of Tuttle, Dennett & Chisholm (t.p. verso)], 1838.
179 p.; pp. 175-79 adv. LCP

1966. __________.
The American Fruit Garden Companion, Being A Practical Treatise On The Propagation And Culture Of Fruit, Adapted To The Northern And Middle States.
Boston: Weeks, Jordan And Company [Tuttle, Dennett & Chisholm, Printers (t.p. verso)], 1839.
176 p. LCP

1967. SCHENCK, PETER ADAM.
The Gardener's Text-Book: Containing Practical Directions Upon The Formation And Management Of The Kitchen Garden; And For The Culture And Domestic Use Of Its Vegetables Fruits, And Medicinal Herbs.
New York: C.M. Saxton, Barker & Co., 1860.
306 p.; illus. LCP

1968. SCHENECTADY CO. [N.Y.]. AGRICULTURAL SOCIETY.
Proceedings Of The Agricultural Exhibition, At Schenectady, October 10, 1822. With The President's Address. Published by the Society.
Schenectady: Published At The Cabinet Printing-House; Isaac Riggs, Printer, 1822.
19 p. Incl. (pp. 8-19): Address by the president, George W. Featherstonhaugh. APS

1969. SCHMIDT, DANIEL.
Das gemeinnützige Haus-Arzneybuch zur Erkennung und Heilung der meisten im menschlichen Leben vorkommenden Krankheiten; Enthaltend Eine Sammlung von Mitteln für fast alle Krankheitszufälle denen der menschliche Körper unterworfen ist; nebst einer Beschreibung von dem Gebrauch der vornehmsten Kräuter; auch viele kunstreiche Stücke. Aus den nützlichsten Schriften zusammen getragen.
Carlisle [Pa.]: Gedruckt für den Verfasser, 1826.
164, [4] p. LCP

1970. __________.
Das gemeinnützige Haus-Arzneybuch .... zweyte Auflage.
Lancaster, Pa.: Gedruckt und zu haben be [sic] H.W. Villee, 1829.
192 p. HSP in LCP; LCP

1971. SCHOHARIE VALLEY [N.Y.]. STOCK GROWERS' ASSOCIATION.
By-Laws Of The Schoharie Valley Stock Growers' Association, Incorporated By The Legislature Of The State Of New York, April, 1860. Together With A Copy Of The Act Of Incorporation.
Albany: Weed, Parsons And Company, Printers, 1860.
27 p. LCP

1972. SCHROEDER, JOHN FREDERICK, 1800-1857.
The Intellectual And Moral Resources Of Horticulture. An Anniversary Discourse, Pronounced Before The New-York Horticultural Society, At The Annual Celebration, August 26, 1828.
New York: Published At The Request Of The Society [E. Conrad, Printer (colop.)], 1828.
40 p. Incl. (pp. 31-40) Appendix, with: Books And Treatises ... By Resident Members ....; List Of Publick Gardens ... In And Near The City .... etc. LCP; APS

1973. __________.
The Intellectual And Moral Resources Of Horticulture. An Anniversary Discourse .... Third Edition.
New York: Printed By William Osborn, 1838.
40 p. HSP

SCHUYLKILL C.[PA.]. AGRICULTURAL SOCIETY.
See KEIM, GEORGE MAY. Address Before .... 1854.

SCOTT, R. ROBINSON, ed.
See The Florist And Horticultural Journal .... [pdcl.].

1974. SCOTT, THOMAS FIELDING, 1807-1867.
Address On Education, Delivered Before The Central Agricultural Society Of Georgia, At Macon, October 29th, 1851.
Macon, Ga.: Printed By Benjamin F. Griffin, 1852.
26 p. LCP

1975. SEABROOK, WHITEMARSH BENJAMIN, 1795-1855.
An Address, Delivered At The First Anniversary Meeting Of The United Agricultural Society Of South-Carolina, In The Hall Of The House Of Representatives, At Columbia, On Thursday, 6th Dec. 1827. Published At The Request Of The Society.
Charleston: Printed By A.E. Miller, 1828.
40 p. APS

1976. __________.
A Memoir On The Origin, Cultivation And Uses Of Cotton, From The Earliest Ages To The Present Time, With Especial Reference To The Sea-Island Cotton Plant, Including The Improvements In Its Cultivation, And The Preparation Of The Wool, &c. in Georgia and South-Carolina; Read Before The Agricultural Society Of St. John's Colleton, November 13th, 1843, And The State Agricultural Society Of South-Carolina, December 6th, 1843, And By both Societies ordered to be published.
Charleston: Printed By Miller & Brown, 1844.
[4], 62, [1] p. LCP; HSP (wrappers)

1977. SEDGWICK, THEODORE, 1780-1839.
An Address, Delivered Before The Berkshire [Mass.] Agricultural Society, October 7, 1830.
Pittsfield: Printed By Phinehas Allen And Son, November, 1830.
20 p. APS

1978. Select Essays: Containing: The Manner of raising and dressing Flax, and Hemp. Also, the whole Method of Bleaching or Whitening Linen-Cloth. Likewise, Observations on the Management of Cows and Sheep. The Manner of raising Radishes, Turnips, Cabbage, and other such Plants .... Collected from the Dictionary of Arts and Sciences, and from various Modern Authors.
Philadelphia: Printed, By Robert Bell, 1777.
[8], 159, [1 (pub. adv.)] p.; fold. pl. 4th leaf is also pub. adv. "Several of the following Essays are translated from ... Journal Oeconomique ...." (Preface). HSP in LCP

A Series Of Tables Of The Several Branches Of American Manufactures .... [1813].
See COXE, TENCH, 1755-1824. A Statement Of The Arts And Manufactures .... 1814. Incl. the above.

1979. SEWARD, WILLIAM HENRY, 1801-1872.
The Reaper: Argument Of William H. Seward, In the Circuit Court of the United States, October 24, 1854.
Auburn [N.Y.]: William L. Finn, 1854.
29 p. Cap. title: M'Cormick's Reaper Patent Case. Argument ... delivered at Albany, N.Y., Oct. 24th, 1854 ... on the trial of the case of Cyrus H. M'Cormick vs. William H. Seymour and Dayton S. Morgan, for infringement of M'Cormick's Patent for his Reaping Machine. LCP (wrappers)

1980. SEYD, ERNEST, 1833-1881.
California And Its Resources. A Work For The Merchant, The Capitalist And The Emigrant.
London: Trubner And Co. [J. And W. Rider, Printers (t.p. verso)], 1858.
[4], 168, [1 (adv.)] p.; tinted front.; 17 pls., 6 tinted, 1 fold; 2 fold. maps. LCP

1981. [SHARPLESS, JOHN T.].
An Essay On The Bombyx Mori, Or Silk Worm. (Extracted from the Franklin Journal, for July, August, and September, 1826.)
[Philadelphia: 1826].
[21] p. Cap. title continues: ... Read before the Maclurian Lyceum of Science of Philadelphia, June 11th, 1826. Author's name p. 2. LCP; HSP

SHEDD, J. HERBERT. Thorough Drainage. 1859.
See FOLLEN, CHARLES. Suggestions On Landscape Gardening. 1859. Incl. also Shedd's work.

1982. SHEPARD, CHARLES UPHAM, 1804-1886.
Address Delivered Before The New Haven Horticultural Society, September 25, 1838.
New Haven: Babcock & Galpin, Printers, 1838.
24 p. HSP (wrapper)

1983. SHIRREFF, PATRICK.
A Tour Through North America; Together With A Comprehensive View Of The Canadas And United States. As Adapted For Agricultural Emigration.
Edinburgh: Printed By Ballantyne And Company; Published By Oliver And Boyd; London: Simpkin, Marshall, & Co.; Glasgow: David Robertson; Dublin: William Curry, Jun. And Co., 1835.
[2], iv, v, [1], 473 p. HSP

A Short And Practical Essay, On Farming .... [1798].
See LAMBERT, JOHN.

1984. A Short And Practical Treatise On the culture of the Wine-Grapes in The United States of America, Adapted to those States situated to the southward of 41 degrees of North latitude. [cap. title]
Georgetown: Printed by Hanson & Briggs, [1794].
[2] p. APS

1985. SIBBALD, GEORGE.
Notes And Observations, On The Pine Lands Of Georgia, Shewing The Advantages They Possess, Particularly In The Culture Of Cotton. Addressed To Persons, Emigrating, And Those Disposed To Encourage Migration to this State. Together With A Plan Of Emigration, For Their Immediate Settlement. To which is added a Geographical sketch of the State ....
Augusta: Printed By William J. Bunce, 1801.
71, [7] p. 2nd issue, with 6-p. Postcript at end. HSP in LCP; APS

1986. Sicher und bewährt befundes Gäuls-Doktor Büchlein. Enthaltend die aller auserlesenste Mittel für Pferde, nebst einen Anhang für das Rindvieh, Schafe und der gleichen.
Ephrata [Pa.]: Zum 4tenmal Gedruckt bey Johannes Baumann, 1809.
70 p. LCP

The Silk Culture In The United States: Embracing Complete Accounts .... 1844.
See BARBOUR, I. RICHMOND, comp.

1987. The Silk Culturist, And Farmer's Manual. Published By The Executive Committee Of The Hartford County Silk Society.
Hartford; Wethersfield.
APS: Vols. 1-2 (April 1835-March 1837)
HSP: Vols. 1-4 (April 1835-March 1839)

1988. The Silkworm. S. Blydenburgh, editor and proprietor.
Albany.
LCP: Vols. 1-2 (May 1835-May 1837)
Last (?) issue (Aug. 1837) known as: Silk-Worm and Beet Sugar Manual.

1989. [SILLIMAN, BENJAMIN, 1779-1864.]
Manual On The Cultivation Of The Sugar Cane, And The Fabrication And Refinement Of Sugar. Prepared Under The Direction Of The Hon. Secretary Of The Treasury, In Compliance With A Resolution Of The House Of Representatives Of Jan. 25, 1830.
Washington: Printed By Francis Preston Blair, 1833.
122 p.; 4 fold. pls. Author's name p. 3. APS; HSP

__________.
See also MASSACHUSETTS. AGRICULTURAL SURVEY. Report .... Containing The Remarks .... 1840.

1990. SIMMONS, JOHN, comp.
The American Pocket Farrier; Comprehending a description of the Various Diseases Incident To Horses, And Prescriptions For Their Cure .... To Which Is Added An Appendix: Containing directions for the management of these animals, in Breeding, Weaning .... Carefully Selected and compiled from the most approved Authors ....
Philadelphia: Printed For, And Sold By The Compiler, 1825.
144 p. HSP in LCP

1991. SIMPSON, THOMAS JOSEPHSON.
Eine grosse Ersparniss für Bauern und Viehmäster. Die patentirte Keil-Maschine.
[Baltimore: ca. 1820.]
Brs. LCP

SINCLAIR, GEORGE, 1786-1834, contrib. auth.
See DAVY, HUMPHREY, 1778-1829. Elements Of Agricultural Chemistry .... 1815 etc. Incl. Appendix by Sinclair.

1992. SINCLAIR, JOHN, 1754-1835.
The Code Of Agriculture; Including Observations On Gardens, Orchards, Woods, And Plantations. First American Edition With Notes.
Hartford: Printed For Hudson And Co. And Cooke And Hale, 1818.
Pp. [i]-viii, [3]-424; front.; 6 pls. LCP

1993. A Sketch Of The Present State Of Commerce, Agriculture, And Manufactures, In The United States, With Some Observations On The Impolicy Of High Tariff Laws.
Philadelphia: E. Littell, 1827.
[2], 102 p. APS

1994. Sketch Of Western Virginia; For The Use Of British Settlers In That Country.
London: Edward Bull [Bradbury And Evans, Printers (t.p. verso)], 1837.
vi, [2], 117, [1], 6 p. Final sep.-paginated section: Some Hints Relative To The Yearly Produce Of The Lands In Virginia, Alluded To In This Work. HSP

Sketches On Rotations Of Crops .... 1792 etc.
See BORDLEY, JOHN BEALE, 1727-1804.

1995. SKINNER, JOHN STUART, 1788-1851.
Address Delivered Before The Agricultural Society Of Newcastle County, Delaware, At the Annual Meeting, Held September 11, 1844. Published For Distribution By Order Of The Society.
[N.p.: 1844].
28 p. LCP

1996. ________.
A Christmas Gift To The Young Agriculturists of the United States.
Washington: 1841.
16 p. LCP

________, ed.
See The American Farmer .... [pdcl.].

1997. SMALL, MATTHEW.
Every Man His Own Cattle Doctor: Or a Treatise on the Diseases of Horses, Cattle, Dogs, Sheep, And Swine; With Their Causes, Symptoms, And Cure. With Additional Receipts, From The Writings Of Youatt, White, Lawrence, Hines, Clater, Etc.
Philadelphia: John B. Perry; New York: Nafis & Cornish, [1844].
72 p.; illus. LCP

1998. SMEE, ALFRED, 1818-1877.
The Potatoe Plant, Its Uses And Properties: Together With The Cause Of The Present Malady. The Extension Of That Disease To Other Plants, The Question Of Famine Arising Therefrom, And The Best Means Of Averting That Calamity. Illustrated With Ten Lithographs.
New YorK: Wiley & Putnam [R. Craighead, Printer (t.p. verso)], 1847.
xxxvii, [1], 157 p.; front.; 9 pls.; illus. LCP

1999. SMITH, DANIEL, Nurseryman, Burlington, N.J.
Catalogue Of Fruit And Ornamental Trees And Plants, Cultivated By Daniel Smith, Burlington, N.J. To Which Is Added Observations, &c. On Their Treatment And Culture.
[Burlington:] Printed By D. Allinson, 1826.
31 p. HSP

2000. ________.
A Catalogue Of Fruit Trees, Being part of a large collection of the most rare and valuable kinds, which will be fit for transplanting in the month of November next, at the following prices ....
[Burlington: 1802].
Brs. Signed at end: Daniel Smith & Co., Burlington, Aug. 10th, 1802. HSP

SMITH, GIDEON B., ed.
See AMERICAN SILK SOCIETY, Baltimore. Journal .... [pdcl.].

2001. SMITH, JAMES GRAY, 1797-1875.
A Brief Historical, Statistical And Descriptive Review Of East Tennessee, United States Of America: Developing Its Immense Agricultural, Mining, And Manufacturing Advantages. With Remarks To Emigrants. Accompanied With A Map & Lithographed Sketch Of A Tennessee Farm, Mansion House, And Buildings.
London: Published By J. Leath; And Sold By All Booksellers [Peart And Co., Printers (colop.)], 1842.
xii, 71 p.; fold. pl.; fold. map. LCP; HSP (lacking map)

2002. SMITH, JEROME VAN CROWNINSHIELD, 1800-1879.
An Essay On The Practicability Of Cultivating The Honey Bee, In Maritime Towns And Cities, As A Source Of Domestic Economy And Profit.
Boston: Published By Perkins And Marvin; New York: J. Leavitt, 1831.
106 p. incl. front.; illus. LCP

2003. SMITH, JOSEPH A.
Productive Farming; Or A Familiar Digest Of The Recent Discoveries Of Liebig, Johnston, Davy, And Other Celebrated Writers On Vegetable Chemistry; Showing How The Results Of Tillage Might Be Greatly Augmented.
New York: D. Appleton & Co.; Philadelphia: Geo. S. Appleton [John F. Trow, Printer (t.p. verso)], 1843.
138 p.; pp. 1-2 + [13]-14 (pub. adv.). HSP (wrappers)

__________. Productive Farming ....
See also HEERMANCE, HENRY. The Farmer's Mine .... To Which Is Added, Productive Farming .... 1843.

2004. SMITH, OLIVER HAMPTON, 1794-1859.
Address Of The Hon. Oliver H. Smith, Delivered On The Fair Grounds, At Indianapolis, Before The Marion County [Ind.] Agricultural Society, At The Annual Fair, In The Year 1856.
Indianapolis: Elder & Harkness, Printers, 1856.
16 p. LCP (with pp. 13-16 misbound following p. 4)

2005. SMYTH, JOHN FERDINAND DALZIEL, 1745-1814.
A Tour In The United States Of America: Containing An Account of the Present Situation of that Country; The Population, Agriculture, Commerce .... With A Description of the Indian Nations, the general Face of the Country, Mountains, Forests, Rivers .... Likewise Improvements in Husbandry .... Vol. I [- II].
Dublin: Printed By G. Perrin, For Messrs. Price, Moncrieffe, Walker, Exshaw, Wilson, Burnet, Jenkin, et al., 1784.
2 vols. Vol. I: [4], v, [13], 263 p. II: [10], 288 p. APS

2006. __________.
A Tour In The United States Of America: Containing An Account of the Present Situation of that Country .... Vol. I [- II].
London: Printed for G. Robinson; J. Robson, and J. Sewell, 1784.
2 vols. Vol. I: [10], iv, [10], 400 p. II: [12], 455, [1] p. HSP in LCP; LCP

2007. __________.
Voyage Dans Les États-Unis De L'Amérique, Fait En 1784; Contenant une description de sa Situation présente, de sa Population, Agriculture, Commerce ... des Nations indiennes, et des principales Villes et Rivieres .... Traduit de l'Anglois par M. De B .... Tome Premier [- Second].
Paris: Chez Buisson, Imprimeur et Libraire, 1791.
2 vols. Vol. I: xii, 206 p. II: [4], 272 p. APS (2 vols. in 1)

SMITHSONIAN INSTITUTION, Washington, D.C.
See BLODGET, LORIN, 1823-1901. Climate Of The Summer .... (A Report To ....) 1853.

2008. SOCIETY FOR ENCOURAGING INDUSTRY AND EMPLOYING THE POOR, Boston.
[No title: text begins:] Whereas it is found by Experience that this Province is not adapted for raising Sheep ... but ... is very capable to Produce and Manufacture [Linen] ... a Number of Gentlemen have formed themselves into a voluntary Society ....
[Boston: 1754].
12 p. LCP

2009. SOCIETY FOR PROMOTING THE MANUFACTURE OF SUGAR FROM THE SUGAR MAPLE-TREE, AND FURTHERING THE INTEREST OF AGRICULTURE IN THE STATE OF PENNSYLVANIA.
Constitution Of The .... Established The 18th Of September, 1792.
Philadelphia: Printed By R. Aitken & Son, 1793.
8 p. APS

SOCIETY FOR THE ENCOURAGEMENT OF ARTS, MANUFACTURES, AND COMMERCE, London.
See ROYAL SOCIETY OF ARTS, London.

SOCIETY FOR THE PROMOTION OF USEFUL ARTS, Albany.
See NEW YORK. SOCIETY FOR THE PROMOTION OF AGRICULTURE, ARTS AND MANUFACTURES. Earliest (1791-1804) name of the above society, which was merged in 1824 into the Albany Institute.

2010. SOLES, ABRAM.
Original And Interesting Book On The Horse, Showing The Characteristics of this Noble Animal, Interspersed With Anecdotes, Etc., Together With A Large Number Of Valuable And Original Recipes, Collected During A Career Of Thirty Years, Among The Indians, And From Practical Experience, And Now Offered To The Public For The First Time.
Schnectady: Young & Graham, 1860.
104, [1 (adv.)] p. LCP (wrappers)

2011. SOLLY, EDWARD, 1819-1886.
Syllabus Of A Complete Course Of Lectures On Chemistry, Including Its Application To The Arts, Agriculture, And Mining. Prepared For The Use Of The Gentlemen Cadets At The Honorable East India Company's Military Seminary, Addiscombe. Revised, With Additions, By The Author Of "Chemical And Pharmaceutical Manipulations," Etc. Etc.
London: Longman, Brown, Green And Longmans; Philadelphia: Reprinted By Henry C. Baird, Successor To E.L. Carey [T.K. And P.G. Collins, Printers (t.p. verso)], 1850.
198, [14 pub. adv.)] p.; illus. LCP

SOMERSET CO. [PA.]. AGRICULTURAL SOCIETY.
See BLACK, JEREMIAH SULLIVAN, 1810-1883. Address, Delivered Before .... 1854.

2012. SORSBY, NICHOLAS T.
Horizontal Plowing And Hill-Side Ditching.
Mobile: S.H. Goetzel & Co., 1860.
5 [i.e. 45] p.; illus. P. 45 misnumbered 5. LCP

2013. SOUDER, DAVID.
The Rural Economist's Assistant, In The Management Of Bees: Principally taken from the German writings of the Reverend J.L. Christ, First Minister of Krohnberg ....
Lancaster, (Penn.): Printed By William Greear, 1807.
vi, 55 p.; illus. APS

SOUTH CAROLINA agricultural societies.
See the following individual societies:
SOUTH CAROLINA. AGRICULTURAL SOCIETY [founded 1818].
__________. SOCIETY FOR PROMOTING AND IMPROVING AGRICULTURE [founded 1785].
__________. STATE AGRICULTURAL SOCIETY [founded 1839].
__________. UNITED AGRICULTURAL SOCIETY [founded 1826].
The South Carolina Institute, founded 1847 (q.v.), was devoted to the encouragement of commerce and manufactures as well as agriculture.

SOUTH CAROLINA. AGRICULTURAL SOCIETY [founded 1818].
See
DAVIE, WILLIAM R. An Address Delivered Before .... 1819.
HORRY, ELIAS, 1743-1834. An Address .... Before ... 1828.
WHITAKER, DANIEL KIMBALL, 1801-1881. An Address Delivered ... Before .... 1833.

SOUTH CAROLINA. AGRICULTURAL SURVEYS.
See
RUFFIN, EDMUND, 1794-1865. Report On The Commencement And Progress Of The Agricultural Survey .... 1843.
TUOMEY, MICHAEL, 1808-1857. Report On The Geological And Agricultural Survey .... 1844.

SOUTH CAROLINA. INSTITUTE.
See
HAMMOND, JAMES HENRY, 1807-1864. An Address Delivered Before .... 1849.
LUMPKIN, JOSEPH HENRY, 1799-1867. An Address Delivered Before .... 1851.
RUFFIN, EDMUND, 1794-1865. An Address ... Read Before .... 1853.

SOUTH CAROLINA. SOCIETY FOR PROMOTING AND IMPROVING AGRICULTURE [founded 1785].
See A Memorial On The Practicability Of Growing Vineyards .... Read ... To .... 1798.

2014. SOUTH CAROLINA. STATE AGRICULTURAL SOCIETY [founded 1839].
The Proceedings Of The Agricultural Convention And Of The State Agricultural Society Of South Carolina, From 1839 to 1845--Inclusive. Together With Orations, Reports And Other Communications Read And Made At Different Times To Those Bodies. To Which Are Added A Memoir on the Subject of Slavery, By Chancellor Wm. Harper, And A Letter On Marl, By Ex-Governor James H. Hammond. Published Under The Direction Of The State Agricultural Society Of South Carolina.
Columbia: Summer & Carroll, Publishers, 1846.
430 p. HSP (wrapper)

__________.
See also SEABROOK, WHITEMARSH BENJAMIN, 1795-1855. A Memoir On ... Cotton ... Read Before The Agricultural Society Of St. John's Colleton ... And .... 1844.

SOUTH CAROLINA. UNITED AGRICULTURAL SOCIETY [founded 1826].
See SEABROOK, WHITEMARSH BENJAMIN, 1795-1855. An Address, Delivered At The First Anniversary Meeting Of .... 1828.

2015. The Southern Agriculturist.
Charleston.
LCP: Vol. 9, nos. 5-6 (May-June 1836)
APS: Vol. 7, no. 11 (Nov. 1834)

SOUTHERN CENTRAL AGRICULTURAL ASSOCIATION, Georgia.
See RUFFIN, EDMUND, 1794-1865. Premium Essay .... Submitted To .... 1853.

2016. [SPANGLER, ANDREW M., pub.].
Both Sides Of The Grape Question. Comprising I. "An Essay On The Culture Of The Native And Exotic Grape." By William Saunders .... II. "Physiography In Its Application To Grape Culture." By F.J. Cope .... III. "A Contribution To The Classification Of The Species And Varieties ... By J.M. McMinn ....
Philadelphia: J.B. Lippincott & Co.; A.M. Spangler; New York: C.M. Saxton, Barker & Co. [King & Baird, Printers (t.p. verso)], 1860.
96 p.; illus. Saunders's essay was the winner, and the others runners-up, in a content sponsored by Farmer and Gardener (see Note By The Publisher [i.e. Spangler]). LCP (wrapper)

2017. [__________].
The Year Book Of The Farm And Garden: A Reliable Guide To all Important Rural Operations, Embracing Concise Directions for the Improvement of the Soil by Draining ... Implements of Culture ... Rural Architecture ... Laying out and Cropping the Esculent Garden, Fruit Culture ... List of Fruits, Seeds, Plants ....
Philadelphia, Pa.: A.M. Spangler, 1860.
108, 48 (adv.) p.; illus. LCP (wrapper)

__________, ed.
See The Progressive Farmer [pdcl.].

2018. SPENCER, JOHN CANFIELD, 1788-1855.
Address Delivered At The Annual Exhibition Of The N.Y. State Agricultural Society, At Buffalo, September 7th, 1848.
Albany: Printed By C. Van Benthuysen, 1849.
24 p. LCP

2019. SPRINGFIELD, ROLLO.
The Horse And His Rider; Or, Sketches and Anecdotes of the Noble Quadruped, And Of Equestrian Nations.
New York: Wiley & Putnam [R. Craighead's Power Press (t.p. verso)], 1847.
iv, 203 p.; front.; 5 pls. LCP

2020. SPURRIER, JOHN.
The Practical Farmer: Being A New And Compendious System of Husbandry, Adapted To The Different Soils And Climates Of America. Containing The Mechanical, Chemical And Philosophical Elements Of Agriculture. With Many Other Useful And Interesting Subjects.
Wilmington: Printed By Brynberg And Andrews, 1793.
LCP: pp. [i]-viii, [1], [i]-x, [9]-360, [i]-xv; with x-page list of Subscribers' Names bound after prelims. (pp. ix-x bound as x-ix); HSP: 360, xv, [1], x p.; with Subscribers' Names bound at end. Evans 26198 calls for Subscribers' Names bound immediately after t.p.

2021. SQUAREY, CHARLES.
A Popular Treatise On Agricultural Chemistry: Intended For The Use Of The Practical Farmer.
Philadelphia: Lea & Blanchard, 1842.
156 p. LCP

2022. SQUIBB, ROBERT.
The Gardener's Calendar For The States Of North-Carolina, South-Carolina, And Georgia. With An Appendix, Containing a variety of practical and general information on husbandry and horticulture.
Charleston, S.C.: Published By P. Hoff And E. Gibbs; P. Hoff, Printer, 1827.
[8], 176 p.; front. LCP

2023. [STABLER, EDWARD, 1794-1883].
A Brief Narrative Of The Invention Of Reaping Machines; And An Examination Of The Claims For Priority of Invention: By A Maryland Farmer And Machinist. Also A Short Appendix, Consisting Of English Publications, As To The Operation And Success Of Reaping Machines In England, In The Years 1851, 1852 And 1853.
Baltimore: From The Press Of Mills & Cox, (Office Of The "American Farmer"), 1854.
32 p. LCP

2024. STEPHENS, HENRY, 1795-1874.
The Book Of The Farm: Detailing The Labors Of The Farmer, Steward, Plowman, Hedger, Cattle-Man, Shepherd, Field-Workers, And Dairymaid. With Numerous Illustrations. To Which Are Added, Explanatory Notes, Remarks, &c. By John S. Skinner, Editor Of The Farmers' Library. Volume II [i.e. I] [- II].
New York: Greeley & McElrath, 1847.
2 vols. Vol. I: 569 p.; 4 pls.; illus. II: 462, xvii p.; 24 pls. on 15 p.; illus. Pls. erratically numbered to xxxiii. Half t.pp.: The Farmers' Library, Volume II [- III]. LCP

2025. __________.
The Book Of The Farm: Detailing The Labors Of The Farmer, Steward .... To Which Are Added Explanatory Notes, Remarks, Etc. By John S. Skinner .... Vol. I [- II].
Auburn [N.Y.]: Alden, Beardsley & Co.; Rochester: Wanzer, Beardsley & Co., 1852.
2 vols. Vol. I: pp. [1-2], 5-569; front.; 10 pls.; illus. II: pp. [1-2], [5]-462; front.; 16 pls. on 8 p.; illus. Pls. erratically numbered to xxxiii. Without ½ t.pp. LCP (2 vols. in 1)

2026. __________.
The Book Of The Farm: Detailing The Labors Of The Farmer, Steward .... To Which Are Added Explanatory Notes, Remarks, Etc. By John S. Skinner .... Volume I [- II].
New York: C.M. Saxton, Barker & Co., 1860.
2 vols. Vol. I: pp. [1-2], [5]-569; 10 pls. II: pp. [1-2], [5]-462, [i]-xvii; front.; 16 pls. on 8 p. Pls. erratically numbered to xxxii. HSP

2027. STEUART, HENRY, 1759-1836.
The Planter's Guide; Or, A Practical Essay On The Best Method Of Giving Immediate Effect To Wood, By The Removal Of Large Trees And Underwood; Being An Attempt To Place The Art, And That Of General Arboriculture, On Fixed And Phytological Principles; Interspersed With Observations On General Planting, And The Improvement Of Real Landscape. Originally Intended For The Climate Of Scotland. First American, From The Second London And Edinburgh Edition.
New York: Published By G. Thorburn And Sons; Albany: And Sold By William Thorburn; Boston: J.B. Russell [George P. Scott And Co., Printers (t.p. verso)], 1832.
422 p.; front.; 3 pls. LCP

2028. STEVENSON, ANDREW, 1784-1857.
Address Delivered Before The Penn'a State Agricultural Society, At Harrisburg, By The Hon. Andrew Stevenson, (Of Virginia,) October 31, 1851. With An Appendix, Containing A List Of The Premiums Awarded.
Harrisburg: M'Kinley & Lescure, Printers, 1851.
32 p. + errata slip. LCP; APS

2029. __________.
Address Of Hon. Andrew Stevenson, Delivered Before The Agricultural Society Of Albemarle [Va.], At Its Annual Meeting, On the 29th. of October, 1847.
Charlottesville: James Alexander, Printer, 1847.
28 p. LCP

2030. STEWART, JOHN.
Stable Economy: A Treatise On The Management Of Horses, In Relation To Stabling, Grooming, Feeding, Watering, And Working. From The Third English Edition, With Notes And Additions, Adapting It To American Food And Climate, By A.B. Allen, Editor Of The American Agriculturist.
New York: D. Appleton & Co.; Philadelphia: George S. Appleton, 1846.
378, [6 (pub. adv.)] p.; illus. LCP

2031. STIMSON, EARL.
Practical Husbandry. An Address, Delivered At The Fifth Anniversary Meeting Of The Saratoga County [N.Y.] Agricultural Society, Held At Ballston Spa, On The Eighth Day Of October, 1823.
Ballston Spa: Printed For The Society, By J. Comstock, 1823.
14, [1] p.; errata slip pasted t.p. verso. LCP

2032. STOCKHARDT, JULIUS ADOLPH, 1809-1886.
Chemical Field Lectures For Agriculturists. Translated From The German. Edited, With Notes, By James E. Teschemacher.
Cambridge [Mass.]: John Bartlett [Metcalf And Company, Printers To The University (t.p. verso)], 1853.
[4 (pub. adv.)], xii, [2], 242 p. LCP

2033. STONE, EDWIN MARTIN, 1805-1883.
An Address, Delivered In Ipswich, Mass., Before The Essex Agricultural Society, At Its Twenty-Eighth Annual Exhibition, September 24th, 1815.
Salem: Printed At The Gazette Office, 1846.
44 p. Wrapper headed: Agriculture And Mental Culture. LCP (wrappers)

2034. [STORK, WILLIAM].
An Account of East-Florida, With A Journal, Kept By John Bartram of Philadelphia, Botanist to His Majesty For The Floridas; Upon A Journey from St. Augustine up the River St. John's.
London: Sold by W. Nicoll; and G. Woodfall, [1766].
[6], 90, viii, 70 p. Includes a chapter on the cultivation of cotton, rice, silk, sugar, etc. Sep. ½ t.p. for Bartram's Journal. HSP in LCP; LCP (with Introduction to Bartram's Journal misbound preceding Stork's Account)

2035. [__________].
A Description Of East-Florida, With A Journal, Kept By John Bartram .... The Third Edition, much enlarged and improved.
London: Sold by W. Nicoll; and T. Jefferies, 1769.
[4], viii, 40, [2], xii, 35, [1] p.; fold. front. map; 2 fold. maps. Sep. t.p. for Bartram's Journal. LCP; APS

2036. STORKE, ELLIOT GREY, 1811-1879?, ed.
The Farm And Gardens; Embracing I. The Farm. - Its Management And Products. II. The Kitchen-Garden. - What To Grow And How To Grow It. III. The Fruit-Garden. - How To Cultivate Choice Fruits. IV. The Flower-Garden. - How To Cultivate All Out-Door Flowers. From The Latest Authorities.
Auburn, N.Y.: The Auburn Publishing Company; E.G. Storke, Publishing Agent [New York: C.A. Alvord, Stereotyper And Printer (t.p. verso)], 1859.
261, [3 (pub. adv.) p.; front.; illus. LCP

2037. STOUT, Z. BARTON.
An Address, Delivered Before The Domestic Horticultural Society Of The Western Part Of New-York, At Its Annual Meeting, In Geneva, Sept. 28, 1830. (Published at the request of the Society.)
Canandaigua: Printed By Morse & Harvey, 1830.
15, [1] p. APS; HSP

STRAFFORD [N.H.]. AGRICULTURAL SOCIETY.
See
LYFORD, STEPHEN C. An Address Delivered Before .... 1825.
WALKER, LYMAN B. An Address Delivered Before .... 1825.

2038. STREET SWEEPING & FERTILIZING CO., Philadelphia.
The Street Sweeping & Fertilizing Co. Of Philadelphia, Incorporated By The State Of Pennsylvania. Its Objects, Facilities and Prospects of Operation in the City of Phildelphia, And Other Cities And Towns Of The United States. Charter Perpetual. Capital, $200,000. Shares, $25.00 Each.
Philadelphia: King & Baird, Printers, 1857.
24 p. Purposes incl. "the Manufacture and Sale of Chemical Manures, Chemical Compounds and Fertilizers ...." (p. 3) LCP (wrapper)

2039. STRONG, JEDIDIAH.
Treatise On The Growing Of The Silk Worm.
Philadelphia: S.C. Atkinson, Printer, 1837.
16 p. LCP; HSP

2040. [STRUTHERS, T.].
Important To Farmers, Miners, Iron Manufacturers And Lumbermen. 120,000 acres Of Good Arable Land, Embracing the principal part of the extensive and valuable Coal Fields and Iron Deposites of M'Kean, and North side of Elk county Pennsylvania, are now offered for Sale .... H.M. Watts and others, Proprietors.
[Philadelphia?: 1855].
16 p. Author's name at end. LCP; HSP

2041. STUART, ISAAC WILLIAM, 1809-1861.
Address Before The Hartford County [Conn.] Agricultural Society, October 3, 1845. Together With The Reports Of The Committees, And The Transactions Of The Society. Published by Order of the Society.
Hartford: Printed By E. Gleason, 1845.
83, [1] p. LCP

2042. SULLIVAN CO. [PA.]. FARM AND IMPROVEMENT COMPANY.
Constitution And By-Laws Of The Sullivan County Farm And Improvement Company.
Philadelphia: From The Printing Rooms Of James H. Bryson, 1855.
8 p. LCP

2043. __________.
The Sullivan County Farm, And Improvement Company. 50,000 Acres Of Land, In Farms Of 28 Acres, In Sullivan County, Pennsylvania. Office, No. 90 Walnut Street, Below 4th, South Side.
Philadelphia: From The Printing Rooms Of James H. Bryson, 1856.
16 p.; fold. front. map. LCP; HSP

2044. __________.
The Sullivan County Farm And Improvement Company. [Cap. title; text begins:] This Company having recently made a purchase of Fifty Thousand Acres! ....
[Philadelphia: 1856].
4 p. LCP

Summary Of The Principal Chinese Treatises Upon The Culture Of The Mulberry Tree .... 1838.
See JULIEN, STANISLAS, 17977-1873.

A Summary View Of The Courses Of Crops .... 1784.
See BORDLEY, JOHN BEALE, 1727-1804.

2045. SUMMER, THOMAS J.
Analysis Of The Cotton Plant And Seed, With Suggestions As To Manures, &c.
Columbia: Allen, M'Carter & Co.; Charleston: John Russell [Printed at the Office of the South Carolina (t.p. verso)], 1848.
17 p. LCP

SUSQUEHANNA CO. [PA.]. AGRICULTURAL SOCIETY. See
JONES, J. RICHTER. Address Delivered Before .... 1854.
ROSE, ROBERT H. An Address Delivered Before .... 1820.

2046. SWALLOW, GEORGE CLINTON, 1817-1899.
Grape Culture In Missouri. Ext. from Trans. Acad. Sci., St. Louis, Vol. I, No. 2.
St. Louis, Mo.: Geo. Knapp & Co., Printers, 1858.
15 p.; 1 pl. LCP (wrapper)

2047. SWAN, JAMES, 1754-1831.
An Address To The Senate And House Of Representatives Of The United States, On The Question For An Enquiry Into The State Of Agriculture, Manufactures And Commerce.
Boston: Printed By John Eliot, 1817.
24 p. LCP

Tabular Statements Of The Several Branches Of American Manufactures .... [1813].
See COXE, TENCH, 1755-1824. A Statement Of The Arts And Manufactures .... 1814. Incl. the above.

2048. TALLMADGE, JAMES, 1778-1853.
Address Before The American Institute, At The Close Of Its Fourteenth Annual Fair, On the 26th of October, 1841. Published By Order Of The Managers. Second Edition.
New York: Hopkins And Jennings, Printers, 1841.
15, [1] p. HSP (wraper)

2049. __________.
Address, Delivered At The Close of The Eighteenth Annual Fair Of The American Institute, Of The City Of New-York, At Niblo's Saloon, October 24, 1845. Published From The Notes Of The Reporter. Third Edition.
New York: James Van Norden & Co., Printers, 1845.
16 p. HSP (wrapper)

2050. __________.
Address, Delivered At The Close Of The Sixteenth Annual Fair of The American Institute, New-York, October, 1843.
New York: James Van Norden & Co., Printers, 1843.
20 p. LCP (wrappers)

2051. __________.
Address of Gen. James Tallmadge, Delivered At Castle Garden, In The City Of New York, At The Close Of The Twentieth Annual Fair Of The American Institute, October 23, 1847. Reported For The Institute By Augustus Maverick, Of The New York Tribune.
New York: R. Craighead, Printer, 1847.
16 p. HSP

2052. __________.
Address Of Gen. James Tallmadge, Delivered At Castle Garden, In The City Of New-York, At The Close Of The Twenty-First Annual Fair Of The American Institute, October 20, 1848. Reported By Augustus Maverick.
New York: Pudney & Russell, Printers, 1848.
13 p. HSP (wrapper)

2053. TAPLIN, WILLIAM, d. 1807.
A Compendium Of Practical And Experimental Farriery, Originally Suggested By Reason And Confirmed By Practice, Equally Adapted For The Convenience Of The Gentleman, The Farmer, The Groom, And The Smith. Interspersed With Such Remarks, And Elucidated With Such Cases, As Evidently Tend To Insure The Prevention, As Well As To Ascertain The Cure Of Disease.
Philadelphia: Printed For Robert Cambell [sic] & Co., 1797.
viii, 290, [6] p.; fold. front. LCP (front. impf.)

2054. __________.
A Compendium Of Practical And Experimental Farriery, Originally Suggested By Reason ....
Wilmington: Printed By Bonsal & Niles; Philadelphia: For Robert Campbell, Bookseller, 1797.
viii, 290, [6] p.; fold. front. LCP

2055. __________.
The Gentleman's Stable Directory: Or, Modern System of Farriery. Comprehending All the most valuable Prescriptions and approved Remedies ... interspersed with Occasional References to the dangerous and almost obsolete practice of Gibson, Bracken, Bartlet, Osmer, and others .... To Which Is Added, A Supplement, Containing Practical Observations Upon Thorn Wounds .... Two Volumes In One. Twelfth Edition.
Philadelphia: Printed By T. Dobson, 1794.
Pp. [i]-xvi, [1-4], [1]-295, [1], [i]-iv; with pp. 537-39 misnumbered 357, 238, 239. With sep. ½ t.p. and t.p. to 2nd vol. HSP in LCP (lacking 4-p. Tbl. of Contents after 1st xvi p.; with prelims to 2nd vol. misbound); LCP

2056. __________.
The Gentleman's Stable Directory: Or, Modern System of Farriery .... Two Volumes In One. A New Edition.
Philadelphia: Published By James Webster, 1812.
xvi, [4], 540 p. With pp. 537-39 misnumbered 357, 238, 239. With sep. ½ t.p. to 2nd vol. LCP

TAPPAN & McKILLOP, Commercial agents, New York.
See McKILLOP, JOHN & CO., Commercial agents, Baltimore.

2057. TATHAM, WILLIAM, 1752-1819.
An Historical And Practical Essay On The Culture And Commerce Of Tobacco.
London: Printed For Vernon And Hood; By T. Bensley, 1800.
xv, [1], 330 p.; 4 pls., 2 col. Chiefly concerns Virginia tobacco. LCP

2058. TAYLOR, GEORGE.
Annual Address Before The Agricultural Societies Of Blair and Cambria Counties [Pa.], Delivered At Their Public Exhibitions In October, 1859.
Philadelphia: Pile & M'Elroy, Printers, 1859.
21 p. HSP (wrappers)

2059. TAYLOR, JOHN, 1753-1824.
Arator; Being A Series Of Agricultural Essays, Practical & Political: In Sixty-One Numbers. Second Edition .... [sic] Revised and Enlarged.
Georgetown, [District of] Columbia: Printed And Published By J.M. Carter, 1814.
Pp. [1-4], [9]-279, [1]. APS

2060. __________.
Arator; Being A Series Of Agricultural Essays .... Third Edition, Revised And Enlarged.
Baltimore: Printed by J. Robinson, For John M. Carter, 1817.
220 p. HSP in LCP; LCP

2061. __________.
Arator, Being A Series Of Agricultural Essays ... In Sixty-Four Numbers. Fifth Edition, Revised And Enlarged.
Petersburg [Va.]: Printed by Whitworth & Yancey, For John M. Carter, 1818.
239, [1] p. LCP

TAYLOR, WILLIAM H.H., Colonel.
See HAMILTON CO. [OHIO]. AGRICULTURAL SOCIETY. First Year's
Proceedings .... With The Address Of ....

2062. Ten Minutes Advice To Every Gentleman Going to purchase a Horse Out Of A Dealer, Jockey, Or, Groom's Stables. In which are laid down established Rules for discovering the Perfections and Blemishes of that noble Animal. [WITH:] The Gentleman's Pocket-Farrier; Shewing How to Use your Horse On a Journey; And What Remedies are proper for Common Accidents that may befal him on the Road .... [by William Burdon].
Philadelphia: Printed and Sold by Joseph Crukshank, 1787.
2 sep. vols.: [6], 43 p. and 70 p. Austin 1867 and Bristol B6598 have the works bound together. HSP in LCP

Ten Minutes Advice ....
See also
BRACKEN, HENRY. Farriery Improved .... To Which Is Prefixed, Ten Minutes' Advice .... 1796 etc.
__________. Taplin Improved .... To Which Is Prefixed Ten Minutes Advice .... 1815.
Every Man His Own Farrier .... 1853. A reprint of Ten Minutes Advice ....

2063. TENNESSEE. STATE AGRICULTURAL BUREAU.
Second Biennial Report Of The Tennessee State Agricultural Bureau, To The Legislature of the State, Comprising The Transactions Of 1856 And 1857. Submitted, in pursuance of the Law, By E.G. Eastman, Secretary Of The Bureau.
Nashville: G.C. Torbett And Company, Public Printers, 1858.
588 p. LCP

2064. __________.
A Synopsis Of The Transactions Of The Tennessee State Agricultural Bureau For 1858 and 1859, Being The Third Biennial Report Of The Bureau, To The Legislature of the State.
Nashville, Tenn.: E.G. Eastman And Company, State Printers, 1860.
94 p. LCP

2065. __________.
A Synopsis Of The Transactions Of The ... For 1858 and 1859, Being The Third Biennial Report ....
Nashville, Tenn.: E.G. Eastman And Company, State Printers, 1860.
Pp. [1]-80, 719-732. Copy as above except for pagination. LCP

2066. TESCHEMACHER, JAMES ENGLEBERT, 1790-1853.
Essay On Guano; Describing Its Properties And The Best Methods Of Its Application In Agriculture And Horticulture; With The Value of Importations From Different Localities; Founded On Actual Analyses, And On Personal Experiments Upon Numerous Kinds Of Trees, Vegetables, Flowers, And Insects, In This Climate. By I.E. Teschemacher ....
Boston: A.D. Phelps; New York: Saxton & Huntington [Stereotyped At The Boston Type And Stereotype Foundry (t.p. verso)], 1845.
51 p. LCP (wrappers)

2067. TESSIER, ALEXANDRE HENRI, 1741-1837.
A Complete Treatise On Merinos And Other Sheep, With Plates. Recently Published At Paris, By Order Of The Government, Compiled By Mr. Tessier, Inspector Of The Rambouillet Establishment And Others, In France. Containing The Method Of Forming Good Flocks, Of Increasing Them, And of treating them properly both when healthy and when diseased .... Translated From The French, And Dedicated To The Agricultural Societies Of The United States.
New York: Printed At The Economical School Office [Printed By Joseph Desnoues (colop.)], 1811.
158, 17, [5] p.; 3 pls.; fold. tbl. HSP in LCP; LCP; APS

2068. THACHER, JAMES, 1754-1844.
The American Orchardist; Or A Practical Treatise On The Culture And Management Of Apple And Other Fruit Trees, With Observations On The Diseases To Which They Are Liable, And Their Remedies. To Which Is Added The Most Approved Method Of Manufacturing And Preserving Cider. Compiled From The Latest And Most Approved Authorities, And Adapted To The Use Of American Farmers.
Boston: Printed And Published By Joseph W. Ingraham, 1822.
226 p. LCP

2069. __________.
The American Orchardist; Or A Practical Treatise On The Culture And Management Of Apple And Other Fruit Trees .... Adapted to the use of American Farmers, And All Lovers And Cultivators Of Fine Fruit. Second Edition, Much Improved.
Plymouth, Mass.: Published By Ezra Collier [Allen Danforth, Printer (t.p. verso)], 1825.
136 [i.e. 236] p. P. 236 misnumbered 136. LCP

2070. THAER, ALBRECHT DANIEL, 1752-1828.
The Principles Of Agriculture. Translated By William Shaw, Esq. ... And Cuthbert W. Johnson, Esq. F.R.S. Fourth Edition.
New York: Sold By Bangs, Brother, & Co., 1852.
Pp. [1-4], [iii]-ix, [1], [11]-551; [5] pub. adv. p. LCP

2071. THOMAS, DAVID, 1776-1859.
Travels Through The Western Country In The Summer Of 1816. Including Notices Of The Natural History, Antiquities, Topography, Agriculture, Commerce And Manufactures. With A Map Of The Wabash Country, Now Settling.
Auburn, (N.Y.): Printed By David Rumsey, 1819.
[4], 320 p.; fold. map; errata slip pasted on blank end leaf. LCP; APS; HSP (lacking errata slip)

2072. THOMAS, JOHN JACOBS, 1810-1895.
The American Fruit Culturist, Containing Directions For The Propagation And Culture Of Fruit Trees In The Nursery, Orchard And Garden. With Descriptions Of The Principal American And Foreign Varieties Cultivated In The United States. Illustrated With Three Hundred Accurate Figures.
Auburn [N.Y.]: Derby, Miller & Company, 1851.
Pp. [i]-xiv, [13]-410, [8 (pub. adv.)]; illus. T.p. headed: Sixth Thousand. LCP

2073. __________.
The American Fruit Culturist, Containing Directions ....
Auburn [N.Y.]: Derby, Miller & Company, 1852.
421, [1], [8 (pub. adv.)] p.; illus. T.p. headed: Seventh Thousand. LCP

2074. __________.
The American Fruit Culturist, Containing Directions ....
New York: C.M. Saxton, Barker & Co., 1860.
242, [6 (pub. adv.)] p.; front.; illus. LCP

2075. __________.
Farm Implements, And The Principles Of Their Construction And Use; An Elementary And Familiar Treatise On Mechanics, And On Natural Philosophy Generally, As Applied To The Ordinary Practices Of Agriculture. With 200 Engraved Illustrations ....
New York: Harper & Brothers, Publishers, 1854.
Pp. [1-4], [7]-267; illus. LCP

2076. __________.
The Fruit Culturist, Adapted To The Climate Of The Northern States; Containing Directions For Raising Young Trees In The Nursery, And For The Management Of The Orchard And Fruit Garden. Illustrated With Engravings.
New York: Mark H. Newman, 1846.
220 p.; illus. LCP

2077. __________.
The Illustrated Annual Register Of Rural Affairs And Cultivator Almanac, For The Year 1858, Containing Practical Suggestions For the Farmer And Horticulturist, Embellished With One Hundred And Thirty Engravings, Including Houses, Farm Buildings ....
Albany, N.Y.: Luther Tucker & Son; New York: C.M. Saxton & Co., 1858.
Pp. [1]-8, [21]-124; illus. LCP

2078. THOMAS, M., & SONS, Auctioneers, Philadelphia.
Catalogue Of Splendid And Rare Green And Hot-House Plants, Including Special Collections Of Cactae, Orchidae, Camellias, Azaleas, Aquatics, Ferns, &c. Also Grape-Vines In Pots, Plants For Bedding Out, Etc. To Be Sold By Public Auction, At Springbrook, The Country Seat Of Caleb Cope, Situate On The Bristol Turnpike Near Tacony And Holmesburg, Commencing On Wednesday, the 20th of May, 1857 ....
Philadelphia: T.K. And P.G. Collins, Printers, 1857.
38 p. LCP (wrappers); HSP (wrappers)

2079. __________.
For Sale: Springbrook, With its Splendid Collection of Plants. [cap. title]
[Philadelphia: 1857].
8 p. Wrapper title: Springbrook And Its Conservatories. Incl. (pp. 5-8): Description of the Green-Houses and Conservatories at Springbrook. By Thomas Meehan, of Germantown, late Gardener thereat. The estate lay on the Delaware River near Holmesburg, Pa. HSP

2080. THOMASON, DENNY R.
Hints To Emigrants, Or To Those Who May Contemplate Emigrating To the United States of America.
Philadelphia: Isaac Ashmead, Printer, 1848.
124, [1] p. Written for the Philadelphia Emigrants' Friend Society. LCP

2081. THOMPSON, J.
Modern Practice Of Farriery; Or, Complete Horse Doctor. Explaining The Various Symptoms Of The Approaching Disorders and the Most Approved Cures For the several Diseases to which Horses are Subject. Likewise The Most Certain Rules For Chusing Horses ....
Philadelphia: Printed By Joseph Crukshank, 1807.
[4], 152 p. LCP

2082. [THOMSON, CHARLES, 1729-1824].
Notes On Farming.
New York: Printed in the Year 1787.
38 p. APS (with MS. note t.p.: said to be published by the order of Chs: Thomson Esqr.)

2083. THOMSON, ROBERT DUNDAS, 1810-1864.
Experimental Researches On The Food Of Animals, And The Fattening Of Cattle. With Remarks On The Food Of Man. Based Upon Experiments Undertaken By Order Of The British Government. From The Last London Edition.
New York: D. Appleton & Co.; Philadelphia: Geo. S. Appleton, 1846.
172 p. incl. front. tbl. LCP

THORPE, THOMAS BANGS, 1815-1878.
See LOUISIANA. AGRICULTURISTS' AND MECHANICS' CONVENTION. Proceedings ... Containing The Address Delivered By .... 1846.

2084. TIBBITS, GEORGE, 1763-1849.
Essay On The Expediency And Practicability Of Improving Or Creating Home Markets For The Sale Of Agricultural Productions And Raw Materials, By The Introduction Or Growth Of Artizans And Manufacturers. Being an Appendix and Illustration of a Memoir on the same subject, read before the Board of Agriculture of the State of New York, March 8, 1825.
Philadelphia: Printed By J.R.A. Skerrett, 1827.
56 p. Copy without footnote p. 19. LCP

2085. __________.
Essay On The Expediency And Practicability Of Improving Or Creating Home Markets ....
Philadelphia: Printed By J.R.A. Skerrett, 1827.
56 p. Copy with footnote p. 19. LCP; APS

2086. __________.
Essay On The Expediency And Practicability Of Improving Or Creating Home Markets ....
Philadelphia: Clark & Raser, Printers, 1829.
[4], 55 p. APS

2087. __________.
A Memoir On The Expediency And Practicability Of Improving Or Creating Home Markets .... Read before the Board of Agriculture of the State of New York, March 8, 1825. Third Edition. To Which Are Added, Four Appendices.
Albany: Printed; Philadelphia: Reprinted By J.R.A. Skerrett, 1827.
48 p. Another edn. of Essay, above. LCP

2088. TILGHMAN, WILLIAM, 1756-1827.
An Address Delivered Before The Philadelphia Society For Promoting Agriculture; At Its Anniversary Meeting, January 18, 1820.
Philadelphia: Published By Order Of The Society; William Fry, Printer, 1820.
36 p. LCP; APS; HSP

2089. TILLINGHAST, JOSEPH LEONARD, 1790-1844.
Address Delivered Before The Rhode-Island Society For The Encouragement of Domestic Industry, At Their Anniversary, October 9, 1826.
Providence: Printed By Smith & Parmenter, 1827.
26 p. APS

2090. TINELLI, LUIGI.
An Address Delivered By L. Tinelli, L.L.D., Member Of The American Institute, &c., Before An Assembly Of Silk Culturists, Held At Levy's Saloon, In New York, On the 2nd Of March, 1840.
New York: Printed By C. Vinten, 1840.
8, [1], p. LCP

2091. __________.
Hints On The Cultivation Of The Mulberry; With Some General Observations On The Production Of Silk. By Lewis Tinelli ....
New York: Printed By William J. Spence, 1837.
Pp. [1-6], [5]-54. APS

2092. To Persons, Inclining To Emigrate To America. [Cap. title; text begins:] The following communication was first published in America, in some of the most respectable newspapers of Pennsylvania .... it is now respectfully offered to the enterprising inhabitants of Europe ....
[Philadelphia?: 1831].
16 p. On the agricultural possibilities and advantages of Pennsylvania. LCP

2093. TOBACCO CONVENTION, Washington D.C., Dec. 15-16, 1840.
Proceedings Of The Tobacco Convention, Held in the City of Washington, on the 15th and 16th of the present month. December 22, 1840. Referred to the Select Committee on the subject of the Tobacco Trade. [cap. title]
[Washington: 1840].
5 p. Title headed: 26th Congress, 2d Session. Dec. No. 24. Ho. of Reps. LCP

__________.
See also JENIFER, DANIEL, 1791-1855. Remarks Of ... At The Opening of The Convention Of Tobacco Planters .... 1840.

2094. TODD, CHARLES STEWART, 1791-1871.
Annual Address Before The Kentucky State Agricultural Society, Delivered At The Capitol In Frankfort, January, 14, 1839, On The Dignity Of The Profession Of Agriculture, And The Propriety Of Legislation For Its Improvement.
Frankfort, Ky.: Printed By F.D. Pettit, At The Franklin Farmer Office, 1839.
Pp. [1-4], [3]-15. LCP

2095. TODD, SERENO EDWARDS, 1820-1898.
The Young Farmer's Manual: Detailing The Manipulations Of The Farm In A Plain And Intelligent Manner. With Practical Directions For Laying Out A Farm, And Erecting Buildings, Fences, And Farm Gates. Embracing Also The Young Farmer's Workshop ....
New York: C.M. Saxton, Barker & Company [Edward O. Jenkins, Printer and Stereotyper (t.p. verso)], 1860.
Pp. [i]-vi, [13]-459; front.; illus. LCP

2096. TOPHAM, JOHN.
Chemistry Made Easy, For The Use Of Agriculturists. From The Second Edition.
Philadelphia: G.B. Zeiber & Co.; New York: Burgess, Stringer & Co.; Boston: Reading & Co.; Baltimore: Shurtz & Tayler; Stereotyped By L. Johnson, 1845.
85 p. LCP

2097. [TOWAR, ALEXANDER].
Der Amerikanische Pferdearzt, enthaltend eine Anweisung zur Erziehung und Haltung der Pferde, so wie auch eine Beschreibung aller Krankheiten .... Nebst einem Anhang, worin alle vorkommenden Krankheiten des Rindviehs, der Schaafe und der Schweine beschrieben .... Aus den besten Amerikanischen und Englischen Schriftstellern zusammen getragen und übersetzt.
Philadelphia: Herausgegeben von Alex. Towar; für Hogan und Thompson; Georg W. Mentz und Sohn; Lancaster, Pa.: A.F. Osterloh, und Andere [Stereotyped By J. Howe (t.p. verso)], 1832.
240 p. LCP (lacking pp. 99-102)

2098. TOWNLEY, EDWARD.
A Practical Treatise On Humanity To Honey Bees; Or Practical Directions For The Management Of Honey Bees, Upon An Improved And Humane Plan, By Which The Lives Of Bees May Be Preserved, And Abundance Of Honey Of A Superior Quality Obtained.
New York: Printed By William S. Dorr, 1843.
162 p. LCP

2099. TOWNSEND, JOHN.
An Address Delivered on Edisto Island, Before The Agricultural Society Of St. John's, Colleton [S.C.], At Its Anniversary Meeting, On Wednesday, The 9th Of July, 1828. Published by Order of the Society.
Charleston: Printed By A.E. Miller, 1828.
51 p. HSP

Traité Général Du Commerce D'Amérique .... Par M. C** .... 1783.
See CHAMBON, -----.

2100. TRASK, JOHN BOARDMAN, 1824-1879.
Report On The Geology Of Northern And Southern California, Embracing The Mineral And Agricultural Resources Of Those Sections; With Statistics Of The Northern, Southern And Middle Mines.
[Sacramento:] James Allen, State Printer, [1856].
64 p. T.p. headed: Document No. 14. In Assembly.) (Session Of 1856. LCP

2101. __________.
Report On The Geology Of The Coast Mountains, And Part Of The Sierra Nevada: Embracing Their Industrial Resources In Agriculture And Mining.
[Sacramento:] B.B. Redding, State Printer, [1854].
95 p. T.p. headed: Document No. 9. In the Assembly.) (Session 1854. LCP; APS

A Treatise On Agriculture ... [1819] etc.
See ARMSTRONG, JOHN, 1758-1843.

A Treatise On The Origin, Progressive Improvement ... Silk Manufacture. 1832.
See PORTER, GEORGE RICHARDSON, 1792-1852.

2102. TRIMBLE, CAREY ALLEN, 1813-1887.
Speech Of Hon. Carey A. Trimble, Of Ohio. Delivered in the U.S. House of Representatives, May 5, 1860. [cap. title]
[Washington:] [Published by the Republican Congressional Committee (colop.)], [1860].
8 p. Title headed: Mutual Interest of the Farmer and Manufacturer. LCP

2103. TROOST, GERARD, 1776-1850.
Geological Survey Of The Environs Of Philadelphia: Performed By Order Of The Philadelphia Society for promoting Agriculture.
Philadelphia: Published By H.S. Tanner; Clark & Raser, Printers, 1826.
39, [1] p.; col. fold. front. map. LCP; APS; HSP (lacking map)

2104. TRUMBULL CO. [OHIO]. AGRICULTURAL SOCIETY.
Transactions Of The Trumbull County Agricultural Society, At The Fourth Annual Meeting, Held At Warren, On the 25th, 26th and 27th of September, 1849: With The Address, Delivered By Samuel St. John .... Published by Order of the Society.
Warren: Printed At The "Trumbull Whig" Job Office, 1850.
62 p. LCP (wrapper)

2105. [TRYON, THOMAS, 1634-1703].
The Country-Man's Companion: Or, A New Method Of Ordering Horses & Sheep So as to preserve them both From Diseases and Causalities [sic] .... And particularly to preserve Sheep from that Monstrous, Mortifying distemper, The Rot. By Philotheos Physiologus [pseud.] ....
London: Printed and Sold by Andrew Sowle, [1684].
[8], 173, [3 (pub. adv.)] p. Incl. (pp. 100-141): The Planters Speech To his Neighbours & Country-men in Pennsylvania, East and West-Jersey, &c. .... HSP in LCP

2106. TUBBS, E.M.
The New Hampshire Kitchen, Fruit, And Floral Gardener. With Numerous Illustrations.
Peterboro': Published By K.C. Scott, 1852.
103 p.; illus. LCP

2107. TUFTS, GEORGE AARON.
Address, Delivered Before The Worcester [Mass.] Agricultural Society, October 12, 1825: Being Their Seventh Anniversary Cattle Show and Exhibition of Manufactures. Published By Order Of The Trustees.
Worcester: Charles Griffin, Printer, 1827.
16 p. LCP

2108. TULLOH, ANDREW.
To Farmers. [Text begins:] Large Tracts of Land in Loyalsock township, Lycoming County [Pa.], are now parcelled off in Lots for Settlement .... settlers, will meet with generous encouragement by applying to Mr. Andrew Tulloh .... Williamsport, September 23, 1800.
[Northumberland?: 1800].
Brs. LCP (with extensive MS. notations by Tulloh to Benjamin Rush)

2109. TUOMEY, MICHAEL, 1805-1857.
Report On The Geological And Agricultural Survey Of The State Of South Carolina. 1844.
Columbia, S.C.: Printed By A.S. Johnston, 1844.
63 p.; illus. LCP

2110. TURNER, JOSEPH ADDISON, 1826-1868.
The Cotton Planter's Manual: Being A Compilation Of Facts From The Best Authorities On The Culture Of Cotton; Its Natural History, Chemical Analysis, Trade, And Consumption; And Embracing A History Of Cotton And The Cotton Gin.
New York: C.M. Saxton And Company, Agricultural Book Publishers [Edward O. Jenkins, Printer and Stereotyper (t.p. verso)], 1857.
320, 11 (pub. adv.) p. LCP

2111. TYSON, JOB ROBERTS, 1803-1858.
Address Delivered Before The Montgomery County [Pa.] Agricultural Society.
Philadelphia: Printed By C. Sherman & Son, 1856.
31 p. LCP (wrappers); APS; HSP (wrapper)

2112. TYSON, PHILIP THOMAS, 1799-1877.
First Report Of Philip T. Tyson, State Agricultural Chemist, To The House Of Delegates Of Maryland, January, 1860.
Annapolis: Printed By Elihu S. Riley, 1860. [wrapper imprint]
Pp. [i]-xi, [1], [5]-45, [1], [1]-20; col. fold. map. T.p. headed: (Document C.) By The House Of Delegates, January 10, 1860. 5,000 copies ordered to be printed. LCP (wrapper)

2113. __________.
First Report Of ... State Agricultural Chemist, To The House Of Delegates Of Maryland, January, 1860.
Annapolis: B.H. Richardson, 1860 [wrapper imprint]
Pp. [i]-xii, [1], [5]-145, [1], [1]-20; col. fold. front. map. T.p. headed: (Document C.) By The Senate, February 14, 1860. 3,000 additional copies ordered to be printed. HSP (wrapper)

UNION AGRICULTURAL SOCIETY OF VIRGINIA AND NORTH CAROLINA, Petersburg, Va.
See GHOLSON, THOMAS SAUNDERS, 1809-1868. Valedictory Address ... Delivered Before .... 1854.

2114. UNION MANUFACTURING COMPANY, Camden, N.J.
Proposition To Build A Steam Flouring Mill, And Barrel Manufactory, At Machine Shop Wharf, Gloucester City, N.J., Nearly Opposite Philadelphia. By a Company to organize under the General Manufacturing Law of New Jersey ....
Philadelphia: 1854.
6 p. Wrapper title: Prospectus For The Building Of A Flouring Mill .... LCP (wrapper)

2115. UNITED STATES. AGRICULTURAL SOCIETY.
First National Field Trial Of Reapers, Mowers, &c., By The United States Agricultural Society. Insd. 1852. At Syracuse, N.Y., July 13, 1857.
Washington, D.C.: 1857.
10 p. LCP (wrappers)

2116. __________.
Fourth National Exhibition By The ... To Be Held In The City Of Philadelphia, Oct. 7th, 8th, 9th, 10th, and 11th, 1856. Twelve Thousand Dollars Offered In Premiums.
Washington: 1856.
36 p. LCP

2117. __________.
Fourth National Exhibition By The ... To Be Held In The City Of Philadelphia, October 7th, 8th, 9th, 10th and 11th, 1856. Fourteen Thousand Dollars Offered In Premiums.
Washington: 1856.
44 p. HSP

2118. __________.
Fifth National Exhibition By The ... To Be Held In The City Of Louisville, Aug. 31st, and Sept. 1st, 2d, 3d, 4th, and 5th, 1857. Twelve Thousand Dollars Offered In Premiums.
Louisville: C. Settle, 1857.
42 p. LCP

2119. __________.
Seventh National Exhibition By The ... To Be Held In The City Of Chicago, September 12th, 13th, 14th, 15th, 16th and 17th, 1859. Twenty Thousand Dollars Offered In Premiums.
Chicago: Press & Tribune Job Printing Office, 1859.
70 p. LCP

__________. Journal.
See Quarterly Journal Of Agriculture.

2120. UNITED STATES. CITIZENS.
Memorial To Congress Of Sundry Citizens Of The United States, Praying Relief From The Oppressive Operation Of Oliver Evans' Patent; Vesting in him an exclusive right to certain Machinery, now in general use, in the Manufacture of Flour; with evidence to shew that the said Evans was neither the Original Inventor nor first Applier ....
Baltimore: Printed By J. Robinson, 1813.
50 p. Incl. particulars of Evans's machinery. LCP

UNITED STATES. DEPARTMENT OF AGRICULTURE. Report Of The Secretary ....
See UNITED STATES. PATENT OFFICE. Report Of The Commissioner Of Patents .... 1852 etc. Before the Dept. of Agriculture was established in 1862, yearly reports on agriculture were issued as part of the Patent Office report.

2121. UNITED STATES. PATENT OFFICE.
Report Of The Commissioner Of Patents For The Year 1851. Part II. Agriculture. Contents. I. Agriculture And Agricultural Education ....
Washington: Robert Armstrong, Printer, 1852.
xi, [3], 676 p.; 9 pls. T.p. headed: 32d Congress 1st Session. (Ho. Of Reps.) Ex. Doc. No. 102. APS

2122. __________.
Report Of The ... For The Year 1852. Part II. Agriculture ....
Washington: Robert Armstrong, Printer, 1853.
vi, [2], 448 p.; 1 pl. T.p. headed: 32d Congress, 2d Session. (Ho. Of Reps.) Ex. Doc. No. 65. APS

2123. __________.
Report Of The ... For The Year 1853. Agriculture.
Washington: Beverley Tucker, Senate Printer, 1854.
vi, [2], 448 p.; 4 col. pls. T.p. headed: 33d Congress, 1st Session. (Senate.) Ex. Doc. No. 27. APS

2124. _________.
Report Of The ... For The Year 1854. Agriculture.
Washington: Beverley Tucker, Printer, 1855.
xl, 520 p.; 8 pls., 2 col. T.p. headed: 33d Congress, 2nd Session. Senate. Ex. Doc. No. 42. APS

2125. _________.
Report Of The ... For The Year 1855 Agriculture.
Washington: Cornelius Wendell, Printer, 1856.
lxiii, [1], 488 p.; 12 pls. on 7 sheets, 3 col. T.p. headed: 34th Congress, 1st Session. House Of Representatives. Ex. Doc. No. 12. APS

2126. _________.
Report Of The ... For The Year 1856. Agriculture.
Washington: A.O.P. Nicholson, Printer, 1857.
xvi, 536 p.; 50 pls. on 27 sheets, 4 col. T.p. headed: 34th Congress, 3d Session. Senate. Ex. Doc. No. 53. APS

2127. _________.
Report Of The ... For The Year 1857. Agriculture.
Washington: James B. Steedman, Printer, 1858.
viii, 551, [1] p.; 8 pls., 2 col. T.p. headed: 35th Congress, 1st Session. House Of Representatives. Ex. Doc. No. 32. APS

2128. _________.
Report Of The ... For The Year 1859. Agriculture.
Washington: George W. Bowman, Printer, 1860.
viii, 590 p.; 2 pls. T.p. headed: 36th Congress, 1st Session. House Of Representatives. Ex. Doc. APS

2129. UNITED STATES. SECRETARY OF STATE.
Report Of The Secretary Of State, On The Subject Of The Cod and Whale Fisheries, Made Conformably To an Order of the House of Representatives of the United States, Referring to him the Representation of the General Court of the Commonwealth of Massachusetts on those Subjects; February 1st, 1791. Published By Order Of The Senate ....
Philadelphia: Printed by John Fenno, 1791.
34 p. Secretary at the time was Thomas Jefferson (q.v.). APS

2130. _________.
Report Of The Secretary Of State, On The Subject Of The Cod And Whale Fisheries ....
Philadelphia: Printed By Francis Childs And John Swaine, 1792.
45 p.; fold. tbl. APS

2131. The United States Practical Receipt Book: Or, Complete Book Of Reference, For The Manufacturer, Tradesman, Agriculturist Or Housekeeper; Containing Many Thousand Valuable Receipts, In All The Useful And Domestic Arts, By A Practical Chemist.
Philadelphia: Lindsay & Blakiston [J. Fagan, Stereotyper; C. Sherman, Printer (t.p. verso)], 1844.
359 p. LCP

2132. URE, ANDREW, 1778-1857.
A Dictionary Of Arts, Manufactures, And Mines; Containing A Clear Exposition Of Their Principles And Practice. Illustrated With Nearly Sixteen Hundred Engravings On Wood. Reprinted From The Fourth English Edition. Corrected And Greatly Enlarged. In Two Volumes. Vol. I [- II].
New York: D. Appleton And Company, 1853.
2 vols. Vol. I: xiv, 1118 p.; illus. II: [2], 998 p.; illus. Incl. articles on husbandry. LCP

2133. _________.
A Dictionary Of Arts, Manufactures, And Mines .... Reprinted Entire From The Last Corrected And Greatly Enlarged English Edition. In Two Volumes. Vol. I [- II].
New York: D. Appleton & Company, 1854.
2 vols. Vol. I: xiv, 1118 p.; illus. II: [2], 998 p.; illus. LCP

2134. _________.
A Dictionary Of Chemistry, On The Basis Of Mr. Nicholson's; In Which The Principles Of The Science Are Investigated Anew, And Its Applications To The Phenomena Of Nature, Medicine, Mineralogy, Agriculture .... First American Edition; With Some Additions, Notes, And Corrections, By Robert Hare ... Assisted By Franklin Bache .... In Two Volumes. Vol. I [- II].
Philadelphia: Published By Robert Desilver [Paper, Manufactured By John And Thomas Gilpin (t.p. verso)], 1824.
2 vols. Vol. I: xix, [363] p. II: [372] p.; 19 pls., 3 fold. LCP

2135. _________.
Recent Improvements In Arts, Manufactures, And Mines: Being A Supplement To His Dictionary. Illustrated With One Hundred And Ninety Engravings.
New York: D. Appleton & Company; Philadelphia: George S. Appleton, 1845.
Pp. [1-2], [5]-304; illus. LCP

2136. VALLEY AGRICULTURAL SOCIETY, Virginia.
The Valley Agricultural Society Will Hold Its Fifth Annual Cattle Show And Fair, At Winchester, On Tuesday, Wednesday, Thursday, and Friday, October 16th, 17th, 18th, and 19th, 1860. [cap title]
[Winchester: 1860].
24 p. Title p. 2: List Of Premiums. LCP

_________.
See also BOWIE, WALTER WILLIAM WEEMS, 1814-1891. Address Delivered ... Before .... 1853.

2137. VANIERE, JACQUES, 1664-1739.
The Bees, From The Latin Of J. Vaniere: Being The Fourteenth Book Of His Praedium Rusticum. [Translated] By Arthur Murphy, Esq.
Middletown, (Conn.): Printed For I. Riley, New York [Richard Alsop, Printer (t.p. verso)], 1808.
[2], 59 p. Bound with: VERGILIUS MARO, PUBLIUS, 70-19 B.C. The Georgics .... (q.v.). General ½ t.p.: The Georgics, And The Bees. LCP; HSP

2138. [VAN RANST, C.W.].
An Authentic History Of The Celebrated Horse American Eclipse, Containing An Account Of His Pedigree And Performances, With Some General Information On The Subject Of The Turf, And The Means Adopted At Various Times To Improve The Breed Of The Horses ....
New York: Printed By E. Conrad, 1823.
37 p.; front. Author's name p. 5. P. 4 misnumbered 6. LCP; HSP (wrappers)

2139. VAN RENSSELAER, JEREMIAH.
An Essay On Salt, Containing Notices Of Its Origin, Formation, Geological Position And Principal Localities, Embracing A Particular Description Of The American Salines; With A View Of Its Uses In The Arts, Manufactures And Agriculture. Delivered as a Lecture before the New-York Lyceum of Natural History.
New York: Published By O. Wilder And J.M. Campbell; C.S. Van Winkle, Printer, 1823.
80 p. LCP; APS

VAN RENSSELAER, STEPHEN, 1765-1839.
See
[EATON, AMOS, 1776-1842]. A Geological And Agricultural Survey Of Rensselaer County .... Taken Under The Direction Of .... 1822.
[__________]. A Geological And Agricultural Survey Of The District .... Taken Under The Direction Of .... 1824.

2140. VARLO, CHARLES, 1725-1795?
A New System Of Husbandry. From many Years Experience, With Tables Shewing The Expense and Profit of each Crop. That a Farm of a 150 [sic] Acres will clear 402£.4s. sterl. a Year. How to stock Farms .... Of Trench-Ploughing .... Cheap Food For Cattle .... new discovered Grains and Grasses suitable for the Land and Climate of America .... Volume I [- II].
Philadelphia: Printed for the Author, 1785.
2 vols. Vol. I: pp. [1-16], [i]-iv, [17]-364; fold. front. tbl. II: pp. [1-8], [5]-368. HSP in LCP (with 2 extra leaves of N.Y. area subscribers Vol. I, and lacking last leaf of Philadelphia subscribers -- i.e. [18] prelim. p.); LCP; APS

VAUGHAN, BENJAMIN, 1751-1835, ed. The Rural Socrates.
See HIRZEL, HANS KASPAR, 1725-1803.

2141. VAUX, ROBERTS, 1786-1836.
Address Delivered Before The Philadelphia Society for promoting Agriculture, At Its Annual Meeting On The Eighteenth Of January, 1825. Published By Order Of The Society.
Philadelphia: Printed at the Port Folio Office [R. Wright, Printer (colop.)], 1825.
28 p. LCP; APS; HSP

2142. VERGILIUS MARO, PUBLIUS, 70-19 B.C.
The Georgics Of Virgil: Translated By William Sotheby, Esq. F.R.S. And A.S.S.
Middletown, (Conn.): Printed for I. Riley, New York [Richard Alsop, Printer (t.p. verso)], 1808.
[10], 120 p.; front. Bound with: VANIÈRE, JACQUES. The Bees .... (q.v.). General ½ t.p.: The Georgics, And The Bees. LCP; HSP

2143. VERNON, WILLIAM HENRY.
A Methodical Treatise On The Cultivation of the Mulberry Tree, On The Raising Of Silk Worms, And On Winding The Silk From The Cocoons. United to an accurate description of the Winding Mill. With Plates. Abridged from the French of M. De la Brousse: with Notes and an Appendix.
Boston: Published By Hilliard, Gray, & Co.; Providence: Eastman & Bridgham, Printers, 1828.
Pp. [i]-vi, [9]-174; 4 fold. pls. Probably refers to ----- De la Brousse's Mélanges d'agriculture sur les mûrieres et d'education des vers à soie .... Nismes 1789. APS

2144. VILLEE, H.W.
Der Bauer als Landmesser, oder die Practische Feldmesskunst; enthaltend: Eine gründliche Unterweisung, Ländereyen, Höhen, Mauerwerke, Baumstämme und alle übrige körperliche Dinge zu messen und zu berechnen; das Gewicht des lebendigen Viehes auszufinden und Fässer zu eichen .... Und viele andere Lehren. Mit mehreren nützlichen Tabellen und 6 Kupfertafeln.
Reading: Gedruckt fur den Verfasser, bey Johanne Ritter, 1824.
xi, [1], 154, [2] p.; 6 pls. on 2 fold. sheets. HSP in LCP; LCP; HSP

VIRGINIA agricultural societies.
See the following individual societies:
VIRGINIA. SOCIETY FOR PROMOTING AGRICULTURE [founded 1811].
__________. STATE AGRICULTURAL SOCIETY [founded 1845].
__________. UNITED AGRICULTURAL SOCIETIES [founded 1818].

VIRGINIA COMPANY, London.
See BONOEIL, JOHN. His Maiesties Graciovs Letter ... to the Councell and Company of Virginia heere .... 1622.

2145. VIRGINIA. SOCIETY FOR PROMOTING AGRICULTURE [founded 1811].
Memoirs Of The "Society Of Virginia For Promoting Agriculture:" Containing Communications On Various Subjects In Husbandry And Rural Affairs.
Richmond: Printed By Shepherd & Pollard, 1818.
Pp. [i]-xiii, [1], [9]-132. APS

2146. VIRGINIA. STATE AGRICULTURAL SOCIETY [founded 1845].
Report Of The President Of The Virginia State Agricultural Society, Made To The Farmers' Assembly, At The First Annual Meeting, held in the City of Richmond, October 28, 1856.
Richmond, Va: J.W. Randolph [Clemitt, Printer (t.p. verso)], 1856.
24 p. LCP

__________.
See also
MINOR, FRANKLIN, 1812-1867. Address Delivered At The Third Annual Fair .... 1855.
RUFFIN, EDMUND, 1794-1865. Address To .... At The First Annual Meeting .... 1853.

2147. VIRGINIA. UNITED AGRICULTURAL SOCIETIES.
An Address To The United Agricultural Societies, And A Petition To Congress, From The Delegation Of The United Agricultural Societies Of Virginia.
Petersburg: Intelligencer Press, December 1820.
16 p. Signed at end: Richard Feild [sic], Prest. of the delegation pro tem. HSP in LCP; APS

2148. __________.
A Memorial To Congress, And An Address To The Public, From The Delegation Of The ....
Petersburg, Va.: Intelligencer Press, 1820.
19 p. APS

Virginia's Discovery of Silke-Wormes .... 1650.
See WILLIAMS, EDWARD.

W., E.
See WILLIAMS, EDWARD.

2149. WAILES, BENJAMIN LEONARD COVINGTON, 1797-1862.
Report On The Agriculture And Geology Of Mississsippi, Embracing A Sketch Of The Social And Natural History Of The State. Published By Order Of The Legislature.
[Jackson:] E. Barksdale, State Printer, 1854.
Pp. [i]-xx, [17]-371; 13 pls., 4 col., 9 tinted; 4 col. sections. APS

2150. __________.
Report On The Agriculture And Geology Of Mississippi. Embracing ....
[Philadelphia:] Lippincott, Grambo, And Co., For E. Barksdale, State Printer, 1854.
Pp. [i]-xx, [17]-371; 13 pls., 4 col., 9 tinted; fold. map; 4 col. sections. LCP

2151. WAKEMAN, THADDEUS B.
An Introductory Lecture, Delivered Before The American Institute Of The City Of New-York, At Clinton Hall, On The Eighth Of January, 1835.
New York: Printed At The Office Of The Mechanics' Magazine, 1835.
15 p. APS

2152. WALDEN, J.H.
Soil Culture; Containing A Comprehensive View Of Agriculture, Horticulture, Pomology, And Agricultural Literature. Illustrated By Numerous Engravings.
New York: Published By Robert Sears [Savage & McCrea, Stereotypers; C.A. Alvord, Printers (t.p. verso)], 1858.
444, [2 (pub. adv. for A.O. Moore)] p.; front.; illus. LCP

WALKER, AMASA, 1799-1875, Secretary of the Commonwealth of Massachusetts.
See MASSACHUSETTS. SECRETARY OF THE COMMONWEALTH. A Synopsis Of The Several Communications On ... Potato Rot ... Published By .... 1852.

2153. WALKER, LYMAN B.
An Address Delivered Before The Strafford [N.H.] Agricultural Society, At Their Annual Fair And Cattle Show, Holden On The 19th And 20th Of October, 1825, At Sandwich. Published In Pursuance To The Vote Of Sd. Society.
Sandbornton: Strafford Gazette Press, D.V. Moulton, 1825.
27 p. LCP

2154. WALL, JAMES WALTER, 1820-1872.
An Address, Delivered Before The Agricultural Fair Held at Flemington, Hunterdon Co., New Jersey, September 27th, 1860. Published By Request Hunterdon County Agricultural Society.
Philadelphia: King & Baird, Printers, 1860.
31 p. HSP (wrapper)

2155. __________.
Address Of The Hon. James W. Wall Of New Jersey, Before The Montgomery County [Pa.] Agricultural Society, October 4, 1860.
Philadelphia: J.B. Chandler, Printer, 1860.
23 p. LCP

2156. __________.
Our Age, And Its Contributions to Agriculture: An Address, Delivered September 16th, 1859, By James W. Wall., Esq. At the New Jersey Agricultural State Fair, Held At Elizabethport. Published By Request Of The State Agricultural Society.
Philadelphia: King & Baird, Printers, 1859.
36 p. LCP; HSP

2157. __________.
Speech Of Col. Jas. W. Wall, Before The Agricultural Fair, Held [by the Hunterdon Co. Agricultural Society] At Flemington, Hunterdon County, New Jersey, September 27, 1860.
Burlington [N.J.]: Franklin Ferguson, Printer, 1860.
23 p. LCP

2158. WARD, MALTHUS AUGUSTUS.
An Address Pronounced Before The Massachusetts Horticultural Society, In Commemoration Of Its Third Annual Festival, September 21, 1831.
Boston: Printed By J.T. & E. Buckingham, 1831.
56 p. Incl. (pp. 35-56) accounts of the anniversary festival and meeting. LCP (wrapper)

2159. WARDER, JOHN ASTON, 1812-1833.
Hedges And Evergreens. A Complete Manual For The Cultivation, Pruning, and Management Of All Plants Suitable For American Hedging; Especially The Maclura, Or Osage Orange .... To Which Is Added, A Treatise On Evergreens; Their Different Varieties -- Their Propagation, Transplanting, And Culture In The United States.
New York: A.O. Moore, Agricultural Book Publisher, 1858.
Pp. [1-4], [3]-296, [1], [1-4] + 1-12 (pub. adv.); front.; 11 pls.; illus. LCP; APS

2160. WARING, GEORGE EDWIN, 1833-1898.
The Elements Of Agriculture: A Book for Young Farmers, With Questions Prepared For The Use Of Schools.
New York: D. Appleton And Company, 1854.
Pp. [1-2 (pub. adv.)], [1-6], [5]-288, [1-8 (pub. adv.)]; illus. LCP

2161. WARING, WILLIAM G.
The Fruit Grower's Hand Book. A Concise Manual Of Directions For The Selection And Culture Of The Best Hardy Fruits In The Garden Or Orchard.
Boalsburg, Centre County, Pa.: [Harrisburg: Lutz & Scheffer, Prs. (t.p. verso)], 1851.
134, [1 (adv.)] p.; illus. HSP

WASHINGTON AND ORANGE [IND.]. DISTRICT AGRICULTURAL SOCIETY.
See
ENGLISH, WILLIAM HAYDEN, 1822-1896. Address Of ... Delivered At .... 1858.
WRIGHT, JOSEPH ALBERT, 1810-1867. An Address Delivered .... 1854.

2162. WASHINGTON [D.C.]. HORTICULTURAL SOCIETY.
Constitution Of The Washington Horticultural Society; With The List Of Members. Instituted April 15, 1857.
Washington: William H. Moore, Printer, 1857.
21 p. LCP

2163. WASHINGTON, GEORGE, 1732-1799.
Fac Similes Of Letters From His Excellency George Washington, President Of The United States Of America, To Sir John Sinclair, Bart, M.P. On Agricultural And Other Interesting Topics, Engraved From The Original Letters ....
Philadelphia: Published By J.T. Bowen; T.K. & P.G. Collins, Printers, 1839.
Pp. [1]-18, 17-57. Copy with the pages of facsimile letters numbered in series with text pages. See also below. HSP in LCP

2164. __________.
Fac Similes Of Letters From His Excellency ... To Sir John Sinclair ....
Philadelphia: Published By J.T. Bowen; T.K. & P.G. Collins, Printers, 1839.
LCP: pp. [1]-6, 9-18, [1]-35, [1], [53]-57; lacking pp. 7-8 (i.e. Table of Contents). Copy with pages of facsimile letters numbered in a sep. series.

2165. __________.
Letters From His Excellency George Washington, President Of The United States Of America, To Sir John Sinclair, Bart. M.P. On Agricultural, And Other Interesting Topics. Engraved From The Original Letters ....
London: The Letter-Press Printed By W. Bulmer And Co.; The Letters Engraved By S.J. Neele; And The Work Sold, By G. And W. Nicol; Cadell And Davies; Robinsons; Sewell; Edinburgh: W. Creech; Dublin: And J. Archer, 1800.
Pp. [1]-16, [1-2], 17-57 incl. 18 leaves of eng. facsimile letters. Cover label: Washington's Letters. Second Impressions. LCP

2166. __________.
Letters From His Excellency General Washington, To Arthur Young, Esq. F.R.S. Containing An Account Of His Husbandry, With A Map Of His Farm; His Opinions On Various Questions In Agriculture; And Many Particulars Of The Rural Economy Of The United States.
London: Printed By B. McMillan; Sold By W.J. And J. Richardson; And J. Hatchard, 1801.
vi, 172, [1 (pub. adv.)] p.; map. LCP; APS

2167. __________.
Letters From His Excellency George Washington To Arthur Young, Esq. F.R.S. And Sir John Sinclair, Bart. M.P. Containing An Account Of His Husbandry, With His Opinions On Various Questions in Agriculture ....
Alexandria [Va.]: Printed By Cottom And Stewart, And Sold At Their Book Stores In Alexandria And Fredericksburg, 1803.
128 p. HSP in LCP; LCP; APS

2168. __________.
Letters On Agriculture From His Excellency, George Washington President Of The United States To Arthur Young, Esq. F.R.S. And Sir John Sinclair, Bart., M.P., With Statistical Tables And Remarks, By Thomas Jefferson, Richard Peters, And Other Gentlemen, On The Economy And Management Of Farms In The United States. Edited By Franklin Knight.
Washington: Published By The Editor; Philadelphia: William S. Martien; New York: Baker & Scribner, And William S. Martien [William S. Martien, Printer (t.p. verso)], 1847.
198 p. incl., between pp. 136 and 173, 36 p. of facsimile letters numbered 17-51; front.; 4 pls.; map. Half t.p.: Farmers' Souvenir .... LCP

2169. __________.
Selections From The Correspondence Of George Washington And James Anderson, L.L.D. F.R.S. F.A.S.E. &c. &c. Author of several Performances, and Editor of "Recreations In Agriculture, Natural History, Arts, And Miscellaneous Literature."
Charlestown: Printed And Sold By Samuel Etheridge, 1800.
79, [1] p. LCP

2170. WATSON, ALEXANDER.
The American Home Garden. Being Principles And Rules For The Culture Of Vegetables, Fruits, Flowers, And Shrubbery. To Which Are Added Brief Notes On Farm Crops, With A Table Of Their Average Product And Chemical Constituents. Illustrated.
New York: Harper & Brothers, Publishers, 1859.
531, [1], [8 (pub. adv.)] p.; illus. LCP

2171. WATSON, ELKANAH, 1758-1842.
History Of The Rise, Progress, and existing Condition Of The Western Canals In The State Of New York .... Together With The Rise, Progress, and existing State Of Modern Agricultural Societies, On the Berkshire System, From 1807, to the establishment of the Board of Agriculture in the State of New-York, January 10, 1820.
Albany: Published by D. Steele; Packard & Van Benthuysen, Printers, 1820.
Pp. [1-2], [1]-104, [1-4], [111]-210, [1-2]; front.; 4 pls., 2 fold.; 2 maps, 1 fold. Sep. t.p. for 2nd part. LCP (lacking 1 pl.); HSP

__________.
See also GOLD, THOMAS, 1759-1827. Address Of .... Together With The Report of the Committee on Agriculture, And The Address Of .... [1816].

2172. WATTERSTON, GEORGE, 1783-1854.
An Address Delivered Before The Columbian Horticultural Society, At The First Annual Exhibition, June 6th, 1834.
Washington: Printed By Wm. Davis Jr., 1834.
50 p.; errata slip tipped in after t.p. LCP

2173. __________.
A Memoir On The History, Culture, Manufactures, Uses, &c. Of The Tobacco Plant.
Washington: Printed By Jonathan Elliott, 1817.
12 p. LCP

2174. WAYLAND, FRANCIS, 1796-1865.
Address Before The Rhode-Island Society, For The Encouragement Of Domestic Industry, And The Rhode-Island Horticultural Society, At The Second Industrial Exhibition In Providence; Delivered September 12th, 1851.
Providence: Printed By Knowles, Anthony & Co., 1852.
27 p. LCP; HSP (wrapper)

WAYNE CO. [IND.]. AGRICULTURAL SOCIETY.
See WRIGHT, JOSEPH ALBERT, 1810-1867. Address ... Delivered Before .... 1851.

Ways and Means For The Inhabitants of Delaware To become Rich .... 1725.
See RAWLE, FRANCIS, 1660-1727.

WEBSTER, DANIEL, 1782-1852.
See MASSACHUSETTS. AGRICULTURAL SURVEY OF THE STATE. Report ... Containing The Remarks .... 1840.

2175. WEBSTER, NOAH, 1758-1843.
An Address, Delivered Before The Hampshire, Franklin And Hampden [Mass.] Agricultural Society, At Their Annual Meeting In Northampton, Oct. 14, 1818. By Noah Webster, Esq. Vice President Of The Society.
Northampton: Printed By Thomas W. Shepard & Co., 1818.
28 p. LCP

WEBSTER, PELETIAH, 1725-1795, possible auth.
See An Essay On The Culture Of Silk .... 1790.

2176. WEEKS, JOHN MOSELY, 1788-1858.
A Manual: Or An Easy Method Of Managing Bees, In The Most Profitable Manner To Their Owner, With Infallible Rules To Prevent Their Destruction By The Moth.
Middlebury [Vt.]: Knapp And Jewett, Printers, 1836.
73 p. LCP; APS

2177. __________.
A Manual: Or An Easy Method Of Managing Bees .... Second Edition.
Middlebury: Elam R. Jewett, Printer, 1837.
72 p. APS

2178. __________.
A Manual: On An Easy Method Of Managing Bees .... New Edition. Revised and Enlarged.
Boston: Weeks, Jordan & Co. [Printed By William A. Hall & Co. (t.p. verso)], 1840.
iv, [2], 128, [2 (pub. adv.)] p. LCP

2179. WELLS, DAVID AMES, 1828-1898.
The Year-Book of Agriculture; Or, The Annual Of Agricultural Progress And Discovery, For 1855 and 1856. Exhibiting The Most Important Discoveries And Improvements In Agricultural Mechanics, Agricultural Chemistry .... With A Comprehensive Review, By The Editor, Of The Progress Of American And Foreign Agriculture For The Year 1855 ....
Philadelphia: Childs & Peterson [Stereotyped By L. Johnson & Co.; Printed By Deacon & Peterson (t.p. verso)], 1856.
399 p.; col. front.; 4 pls., 3 col.; illus. LCP

2180. [WELLS, SAMUEL ROBERTS, 1820-1875].
The Farm: A Pocket Manual Of Practical Agriculture; Or, How To Cultivate All The Field Crops: Embracing A Thorough Exposition Of The Nature And Action Of Soils And Manures ... Etc., Etc.; With A Most Valuable Essay on Farm Management. By The Author Of "How To Do Business," "How To Behave," Etc.
New York: Fowler And Wells, Publishers [Davies And Roberts, Stereotypers (t.p. verso)], 1858.
156; 4+4+4 (pub. adv.) p.; illus. T.p. headed: Rural Hand-Books. -No. 2. LCP

WESTBOROUGH AND VICINITY [MASS.]. AGRICULTURAL SOCIETY.
See RUSSELL, CHARLES THEODORE, 1815-1896. Agricultural Progress .... An Address Delivered Before .... 1850.

2181. Western Journal and Civilian, Devoted to Agriculture, Manufactures, Mechanic Arts ....
St. Louis.
LCP: Vol. 1, no. 5 (May 1848); frag.
HSP: Vols. 2-9, no. 6 (Jan. 1849--March 1853)
Also (1848-1851) known as: The Western Journal, Of Agriculture, Manufactures ....

2182. WESTERN VIRGINIA FARM AND AGRICULTURAL COMPANY, Pocohontas Co., [W.] Va.
The Road to Competency and Independence. A Farm Of 20 Acres Of Good Land For $50. Terms Easy, And Within The Reach Of All. [cap. title]
[Philadelphia?: 1855?].
LCP: 12 p. only; text incomplete.

2183. WESTON, EZRA, 1771-1842.
An Address Delivered Before The Massachusetts Horticultural Society, At Their Eighth Anniversary, September 17, 1836.
Boston: Printed By Tuttle, Weeks & Dennett, 1836.
39 p. LCP (wrappers)

2184. WEYMOUTH FARM AND AGRICULTURAL COMPANY, Camden and Atlantic Counties, N.J.
Brief Of Title Of Lands Belonging To The Weymouth Farm And Agricultural Company.
Philadelphia: 1856.
16 p. HSP (wrappers; with business card of Henry Tunison, Jr. Real Estate Agent pasted on front wrapper)

__________.
See also LANDIS, CHARLES KLINE, 1835-1900. Practical Hints ... Written For The Weymouth Stockholders. 1855.

2185. WHITAKER, DANIEL KIMBALL, 1801-1881.
An Address Delivered in Charleston, Before The Agricultural Society of So. Caro. At Its Anniversary Meeting, August 20th, 1833. Published By Order Of The Society.
Charleston: Printed By A.E. Miller, 1833.
35 p. T.p. headed: The Claims Of Agriculture To Be Regarded As A Distinct Science. APS

2186. WHITE, JAMES, d. 1825.
A Complete System Of Farriery, And Veterinary Medicine. Containing A Compendium Of The Veterinary Art, Or An Accurate Description Of The Diseases Of Horses, And Their Mode Of Treatment; The Anatomy And Physiology Of The Foot, And The Principles And Practice Of Shoeing .... First American Edition, Newly Arranged From The 10th London Edition.
Pittsburgh: Published By R. Patterson And Lambdin; Butler & Lambdin, Printers, 1818.
Pp. [1-4], [3]-311; front.; 17 pls. HSP

2187. WHITE, S.
The Secret Art Of Taming And Training Colts, & Wild & Vicious Horses. Together With A Concise And Complete Farrier. (Eleventh Edition).
St. Clairsville, Ohio: Printed At The Chronicle Office, 1857.
96 p. LCP

2188. WHITMARSH, SAMUEL.
Eight Years Experience And Observations In The Culture Of The Mulberry Tree, And In The Care Of The Silk Worm. With Remarks Adapted To The American System Of Producing Raw Silk For Exportation.
Northampton [Mass.]: Published By J.H. Butler; For The Author, 1839.
156, [1], [1 (adv.)] p.; fold. front. LCP; APS

WHITNEY, ELI, 1765-1825.
See OLMSTED, DENISON, 1791-1859. Memoir Of .... 1846.

2189. WIESTLING, JOHN S.
Der vollständige Bienen-Wärter, oder Nützliche Anweisungen zur Bienen-Zucht. Zusammen getragen aus den Schriften von Christ, Columella, Wheit und M. Harasti.
Harrisburg, (Pennsylvania): gedruckt bey John S. Wiestling, 1819.
50 p.; illus. LCP

2190. WIGG, WILLIAM HAZZARD.
An Address Delivered On The Anniversary Of The Agricultural Society Of St. Luke's Parish, At Bluffton, S.C., May 13, 1846. Second Edition: Revised And Corrected From The "Southern Agriculturist."
Charleston: Steam Power Press Of Walker & James, 1853.
34 p. LCP

2191. WIGGINS, FRANCIS S., d. 1840.
The American Farmer's Instructor, Or Practical Agriculturist; Comprehending The Cultivation Of Plants, The Husbandry Of The Domestic Animals, And The Economy Of The Farm; Together With A Variety Of Information Which Will Be Found Important To The Farmer.
Philadelphia: Orrin Rogers; E.G. Dorsey, Printer, 1840.
Pp. [i]-iv, [vii]-x, [13]-504, 1-4 (pub. adv.); illus. LCP; HSP

2192. WILDER, MARSHALL PINCKNEY, 1798-1886.
Address Delivered At The Fourth Session Of The American Pomological Society, Held In Rochester, N.Y., September 24, 25, And 26, 1856.
Boston: Press Of The Franklin Printing House, 1856.
17 p. LCP (wrapper)

2193. __________.
Address Delivered Before The Norfolk [Mass.] Agricultural Society, On The Occasion Of Its First Annual Exhibition, At Dedham, Sept. 26, 1849.
[Boston:] Published By The Society [Coollidge & Wiley, Printers (t.p. verso)], 1849.
36 p. LCP

WILLARD, JOSIAH F., jt. ed.
See GUERNSEY, ORRIN, and __________.

2194. [WILLIAMS, EDWARD].
Virginia: More especially the South part thereof, Richly and truly valued: viz. The fertile Carolana .... The second Edition, with Addition of The Discovery Of Silkworms, with their benefit. And Implanting of Mulberry Trees. Also The Dressing of Vines .... By E.W. Gent.
London: Printed by T.H. for John Stephenson, 1650.
[12], 47, [9] p. HSP in LCP; LCP

2195. [__________].
Virginia's Discovery of Silke-VVormes, with their benefit. And The Implanting of Mulberry Trees. Also The dressing and keeping of Vines, for the rich Trade of making Wines there. Together with The making of the Saw-mill, very usefull in Virginia, for cutting of Timber and Clapboard, to build withall, and its conversion to other as profitable Uses.
London: Printed by T.H. for John Stephenson, 1650.
[6], 75, [3] p.; illus. HSP in LCP; LCP

2196. WILLIAMS, EDWIN, 1797-1854.
The Wheat Trade Of The United States And Europe.
New York: Published At The Office Of The National Magazine, [1846].
24 p. LCP

2197. WILLIAMS, THOMAS B.
Farmer's Guide In The Management Of Domestic Animals, And The Treatment Of Their Diseases A Treatise on Horses, Mules, Neat Cattle, Sheep, Swine, Poultry, Bees, etc. Embellished With Engravings.
New York: Ensign, Bridgman, & Fanning, [1849].
Pp. [1-2], [9]-100; front.; illus. LCP

2198. WILLIAMSON, PETER, 1730-1799.
The Travels Of Peter Williamson, Among the different Nations and Tribes of Savage Indians in America .... Also, A Description of all the different Serpents ... Together With A Plate, Description, Use, Advantage, manner of making and handling the curious Machine for reaping of Oats, Barley, &c. .... In Three Parts. Written by Himself.
Edinburgh: Printed for the Author, and sold by him, and by R. Fleming, and by other Booksellers ..., 1768.
viii, 184 p. Sep. t.pp. for each part. With chapters on Indian and pioneer husbandry. APS (lacking all 3 pls. called for by Sabin 104492)

2199. WILSON, WILLIAM.
Economy Of The Kitchen-Garden, The Orchard, And The Vinery, With Plain Practical Directions, For Their Management.
New York: Published By Anderson, Davis And Co., 1828.
206 p.; 1 pl. LCP

2200. WINTER, GEORG SIMON.
George Simon Winters wohlerfahrener Pferde-Arzt, welcher gründlich lehrt wie man die Complexion der Pferde, ihre Natur und Alter erkennen, alle innerlichen und äusserlichen Krankheiten heilen, so wie auch wie man dieselben vor bevorstehenden Uebeln verwahren könne, nebst einigen, höchst seltenen, für junge Fohlen vielfach bewährt gefundenen Arzenei-Mitteln. Durchgesehen, verbessert, mit neuen Zusätzen und erleichterten Kupfern versehen von Valentin Trichter.
Philadelphia: Herausbegeben von Edmund Y. Schelly, 1840.
LCP: 839 p.; 43 fold. pls. only (of 44; last pl. frag. only)

2201. __________.
George Simon Winters wohlerfahrner Pferde-Arzt .... Durchgesehen ... von Valentin Trichter.
Northampton County [Pa.]: Herausgegeben von Kleckner, Nolf, und Williams, 1841.
LCP: 839 p.; 39 pls. Henderson p. 194 gives pls. as in 1840 edn., above.

2202. WINTER & CO., Nurserymen, Flushing, N.Y.
Descriptive Catalogue, Of Fruit And Ornamental Trees, Shrubs, Vines, Plants, Etc., Cultivated And For Sale, At The Ancient And Celebrated Linnaean Botanic Garden And Nursery, (Late Prince's,) Flushing, Long Island, near New York, With Directions For Their Culture, Etc.
Flushing: Printed At The University Press, By C.R. Lincoln, [1845].
[4], 92 p. T.p. headed: 1844-1845. LCP (wrappers; with MS. note on front wrapper: Orders received by Henry A. Dreer, Seedsman ...)

__________.
See also PRINCE FAMILY, Nurserymen, Flushing, N.Y. Winter & Co. took over the mortgage of the Prince nursery around 1842.

2203. WISCONSIN. STATE AGRICULTURAL SOCIETY.
Annual Report Of The Wisconsin State Agricultural Society, For The Year 1855.
Madison: Calkins & Proudfit, Printers, 1856.
6 p. LCP

2204. __________.
Annual Report Of The .... For The Year 1856.
Madison: Atwood & Rublee, State Printers, 1857.
9 p. APS

2205. __________.
Transactions.
Madison.
LCP: Vols. 1-3 (1851-53); 4 (1854-57); 5 (1858-59); 7 (1861-68); 8 (1869)

2206. Wohl-eingerichtetes Vieh-Arzney-Buch, worin enhalten die Wartung und Pflege, sowol als die Krankheiten und Heilungsmittel I. Der Pferde, II. Des Rindviehes, III. Der Schaafe, IV. Der Schweine; und V. Der Gänse und Hüner ....
Philadelphia: Gedruckt und zu finden bey Heinrich Miller, 1771.
[10], 184 p. HSP in LCP; LCP

2207. Der wohlerfahrne Baum-Gärtner, oder gründliche Anweisung zur Behandlung der Obst-Bäume. Nebst einer Anleitung zur Bienenzucht.
Lebanon [Pa.]: Gedruckt bey Jacob Schnee, [1799].
[2], 16 p. HSP in LCP

2208. WOLBERT, C.J., Auctioneer, Philadelphia.
Bucks County Farms At Public Sale. Will be Sold at Public Sale, On Thursday The 11th of November next 1830, at 7 o'clock in the Evening at the Merchants Coffee House, in the City of Philadelphia.
[Philadelphia:] J.P. Justice, Printer, September 17th, 1830.
Brs. Describes 12 farms. HSP

2209. WOODMAN, DAVID, JR.
Guide To Texas Emigrants.
Boston: Printed By M. Hawes, For The Publishers, 1835.
Pp. [i]-vi, [13]-192; fold. front. map; 1 pl. LCP

WOODSIDE INSTITUTE, near Philadelphia.
See KEAGY, JOHN MILLER, 1792-1837. A Prospectus Of Woodside Institute, For Practical Education .... 1828.

WORCESTER [MASS.]. AGRICULTURAL SOCIETY.
See
BIGELOW, LEWIS, 1785-1838. Address, Delivered Before .... 1821.
DENNY, NATHANIEL P., Address Delivered Before .... [1822].
FISKE, OLIVER, 1762-1836. An Address Delivered Before .... 1831.
GOODWIN, ISAAC, 1786-1832. Address Delivered Before .... 1824.
RUSSELL, JONATHAN, 1771-1832. Address Delivered Before .... 1821.
LINCOLN, LEVI, 1782-1868. Address, Delivered Before .... 1819.
TUFTS, GEORGE A. Address, Delivered Before .... 1827.

2210. WRIGHT, JOSEPH ALBERT, 1810-1867.
An Address Delivered By Gov. Joseph A. Wright, On The 6th Day Of October, 1853, At Livonia, Washington County, Indiana, To The District Agricultural Society, Composed Of Washington And Orange.
Indianapolis: Austin H. Brown & Co., Printers, 1854.
20 p.; illus. LCP

2211. ________.
Address Of Governor Joseph A. Wright, Delivered Before The Wayne County [Ind.] Agricultural Fair, Held At Richmond, On Tuesday, Wednesday, And Thursday, October Seventh, Eighth, And Ninth, 1851.
Indianapolis: Printed By Austin H. Brown, 1851.
7 p. LCP

2212. ________.
Address Of Governor Wright To The Citizens of Cannelton, Indiana, on Saturday, May 10th, 1851. [cap. title]
[Indianapolis: 1851].
8 p. On agricultural resources. LCP

2213. ________.
Address Of His Excellency, Jos. A. Wright, Governor Of The State Of Indiana, Pronounced At The New York Agricultural State Fair, At Elmira, October, 1855.
Indianapolis: Elder & Harkness, Printers, 1855.
16 p. LCP

WRIGHT, ROBERT EMMET, 1810-1886.
See LEHIGH CO. [PA.]. AGRICULTURAL SOCIETY. Report Of The Managers .... 1854. Incl. address by Wright.

WRIGHT, SILAS, JR., 1795-1847. Address Read Before The New York State Agricultural Society .... [1847].
See JENKINS, JOHN STILLWELL, 1818-1852. The Life Of .... Containing ... His Address .... [1847].

2214. YALE COLLEGE. SCHOOL OF SCIENCE.
Appeal In Behalf Of The Yale Scientific School. With An Appendix.
New Haven: 1856.
31 p. Incl. (pp. 23-31): Plan Of An Agricultural School, by John A. Porter. LCP (wrapper)

2215. ________.
Proposed Plan For A Complete Organization Of The School Of Science Connected With Yale College. Printed For Private Distribution.
New Haven: Printed By Ezekiel Hayes, 1856.
32 p. Proposed depts. incl.: Geology, Mineralogy, And General Zoology; Entomology; Agricultural And Organic Chemistry; Zootechny; Practical Agriculture; Botany. LCP (wrapper)

The Year Book Of The Farm And Garden .... 1860.
See SPANGLER, ANDREW M.

YORK CO. [PA.]. FARMERS & DISTILLERS.
See A Hasty Defence .... [1833].

2216. YOUATT, WILLIAM, 1776-1847.
The Horse. A New Edition, With Numerous Illustrations. Together With A General History Of The Horse; A Dissertation On The American Trotting Horse, How Trained And Jockeyed, An Account Of His Remarkable Performances; And An Essay On The Ass And The Mule, By J.S. Skinner ....
Philadelphia: Lea And Blanchard, 1846.
448 p. incl. front.; illus. LCP

2217. ________.
The Pig: A Treatise On The Breeds, Management, Feeding, And Medical Treatment Of Swine; With Directions For Salting Pork, And Curing Bacon And Hams. Illustrated With Engravings Drawn From Life By William Harvey, Esq.
Philadelphia: Lea & Blanchard [Griggs & Co., Printers (t.p. verso)], 1847.
175 p. incl. front.; illus. LCP

2218. ________.
Sheep: Their Breeds, Management, And Diseases. Published Under The Superintendance Of The Society For The Diffusion Of Useful Knowledge. With Illustrative Engravings. To Which Are Added, Remarks On The Breeds And Management Of Sheep In The United States, And On The Culture Of Fine Wool In Silesia.
New York: C.M. Saxton, Agricultural Book Publisher [Stereotyped By Redfield & Savage (t.p. verso)], 1855.
159, [1] p.; illus. Bound with: H.S. Randall's Sheep Husbandry .... 1854 (q.v.). Spine title: The Shepherd's Own Book. HSP

________.
See also CLATER, FRANCIS, 1756-1823. Every Man His Own Cattle Doctor .... Edited, Revised, And Almost Rewritten By .... 1844.

2219. [YOUNG, ARTHUR, 1741-1820].
Rural Oeconomy: Or Essays On The Practical Parts of Husbandry. Designed to explain several of the most important Methods of conducting Farms of various kinds; including many Useful Hints to Gentlemen Farmers, relative to the oeconomical Management of their Business. Containing, among other Enquiries, Of that Proportioned Farm, which is of all others the most profitable .... To Which Is Added, The Rural Socrates .... The Second Edition.
London: Printed; Philadelphia: Re-Printed And Sold By James Humphreys, Junr., 1776.
245, [1 (pub. adv.)] p. Incl. (pp. 155-245): The Rural Socrates: Or, A Description of the Oeconomical and moral Conduct Of A Country Philosopher .... By Mr. [H.K.] Hirzel .... LCP; APS

2220. [________].
Rural Economy, Or Essays On The Practical Parts Of Husbandry .... To Which Is Added, The Rural Socrates .... The Third Edition.
Burlington [N.J.]: Printed By Isaac Neale, 1792.
299, [1] p. Sep. t.p. for 2nd part. HSP in LCP; LCP

________, possible author.
See American Husbandry .... 1775.

2221. ZÜND, JOHANN JOSEPH.
Handbuch der Pferde-und Vieharzney-Kunde, in besonderer Beziehung auf innerliche Krankheiten, Heilmittlellehre, Wundarzney, Geburtshulfe u.s.w. für den Landmann und Pferde-Besitzer. Erste Auflage.
Philadelphia: gedruckt für den Verfasser von Henrich Hory, 1832.
376, [8] p. LCP

Addresses
Abbot, A. Address
Allen, I. Address
Allen, L. Address
Allen, S. Address
Bache, A. Address
Bacon, L. Address
Baldwin, H. Address
Ball, L. Address
Barnard, D. Address
Bates, I. Address
Betton, S. Address
Biddle, C. Address
Biddle, J. Address
Biddle, N. Address
Bigelow, L. Address
Black, J. Address
Boutwell, G. Address
Bowen, J. Address
Bowie, W. Address
Breck, S. Address
Briggs, C. Discourse
Broom, J. Address
Brownell, T. Address
Buel, J. Address
Burchard, N. Address
Burges, T. Address
Carey, M. Address
Carter, N. Address
Cass, L. Address
Chandler, A. Address
Cheever, S. Address
Cheney, J. Address
Colman, H. Address
Colman, H. Agricultural Addresses
Cook, Z. Address
Coxe, T. Address
Cushing, C. Address
Cutbush, E. Address
Darlington, W. Address
Darlington, W. Discourse
Davidge, F. Address
Davie, W. Address
Dearborn, H. Address
Denny, N. Address
Dickinson, S. Address
Dillingham, W. Address
Dix, J. Address
Dodge, A. Address
Douglas, S. Address
Drake, D. Discourse
Eastman, M. Discourse
Eldridge, J. Address
Eliot, T. Address
Ellsworth, H. Address
Emerson, G. Address
English, W. Address
Everett, E. Address
Fisher, S. Address
Fiske, O. Address
Foote, H. Address
Foster, E. Address
Francis, J. Address
Furman, G. Address
Gadsden, J. Oration
Gage, N. Address
Genet, E. Address
Gholson, T. Address
Gold, T. Address
Gold, T.R. Address
Goodwin, I. Address

Gowen, J. Address
Greeley, H. Address
Hall, J. Address
Hallowell, B. Address
Halsey, A. Address
Hammond, J. Address
Harris, T. Discourse
Hazard, T. Address
Hillard, G. Address
Holcomb, C. Address
Holley, M. Discourse
Holmes, J. Address
Horry, E. Address
Hosack, D. Discourse
Howes, F. Address
Humphreys, D. Discourse
Ingersoll, J. Address
Jackson, C. Address
Jenkins, J. Address
Johnson, B. Address
Jones, J. Address
Keim, G. Address
Kelley, E. Address
Kennedy, J. Address
Kersey, J. Lectures
King, D. Address
King, W. Address
Kirby, E. Address
Latrobe, J. Address
Lawrence, M. Address
Lee, Z. Address
Le Ray de Chaumont, J. Address
Le Ray de Chaumont, V. Address
Lincoln, L. Address
Lincoln, W. Address
Logan, G. Address
Lowell, J. Address
Lumpkin, J. Address
Lunt, G. Address
Lyford, S. Address
Lynch, J. Address
Madison, J. Address
Mapes, J. Address
Marsh, G. Address
Mason, C. Oration
Matlack, T. Oration
Mather, T. Address
Maxcy, V. Address
McCoun, W. Address
McNaughton, J. Address
Mease, J. Address
Meigs, H. Address
Minor, F. Address
Mitchill, S. Address
Morris, J. Address
Morse, O. Address
Moseley, E. Address
Moulton, J. Address
Muse, J. Address
Nichols, A. Address
Niles, J. Oration
Norton, J. Address
Paige, A. Address
Parsons, S. Address
Patterson, J. Address
Perry, G. Address
Peters, R. Address
Peters, R. Discourse
Phinney, E. Address
Pickering, T. Address
Pickering, T. Discourse
Plumer, W. Address
Quincy, J. Address

Rawle, W. Address
Rayner, K. Address
Ridgely, H. Address
Robbins, A. Address
Rose, R. Address
Rost, P. Oration
Ruffin, E. Address
Russell, C. Address
Russell, G. Address
Russell, J. Address
Saltonstall, L. Address
Sanborn, E. Address
Schroeder, J. Discourse
Seabrook, W. Address
Sedgwick, T. Address
Shepard, C. Address
Skinner, J. Address
Skinner, J. Christmas Gift
Smith, O. Address
Spencer, J. Address
Stevenson, A. Address
Stimson, E. Address
Stone, E. Address
Stout, Z. Address
Stuart, I. Address
Tallmadge, J. Address
Taylor, G. Address
Tilghman, W. Address
Tillinghast, J. Address
Tinelli, L. Address
Todd, C. Address
Townsend, J. Address
Tufts, G. Address
Tyson, J. Address
Vaux, R. Address
Wakeman, T. Lecture
Walker, L. Address
Wall, J. Address
Wall, J. Speech
Ward, M. Address
Watterston, G. Address
Wayland, F. Address
Webster, N. Address
Weston, E. Address
Whitaker, D. Address
Wigg, W. Address
Wilder, M. Address
Wright, J. Address

Agricultural, national
Carey, J. Claims Of Agriculture
Coxe, T. View Of The United States
Crèvecoeur, M. Letters From An American Farmer
Johnston, J. Notes On North America
Landreth, D. Commissioner Of Patents
McKillop, J. Harvest Prospects
National Convention of Farmers, Gardeners and Silk Culturists. Proceedings
Sketch Of The Present State Of Commerce, Agriculture
Spurrier, J. Practical Farmer
Swan, J. Address
United States. Patent Office. Report
Watson, E. Modern Agricultural Societies

Agriculture -- Far West
Alta California
Farnham, E. California
Forbes, A. California
Seyd, E. California And Its Resources
Trask, J. Geology Of Northern And Southern California
Trask, J. Geology Of The Coast Mountains

Agriculture -- Great Lakes
Birkbeck, M. Bemerkungen auf einer Reise

Ridgeway Farm And Coal Co. Report
Roberts, J. Pennsylvania Farmer
Struthers, T. 120,000 acres Of Good Arable Land
Sullivan Co. [Pa.]. Farm and Improvement Co.
To Persons, Inclining To Emigrate
Troost, G. Geological Survey
Tulloh, A. Large Tracts of Land
Tyson, P. State Agricultural Chemist
Weymouth Farm And Agricultural Company. Lands
Wolbert, C.J. Auctioneer. Bucks County Farms

Agriculture -- New England
Campbell, P. Travels
Colman, H. Letter
Colman, H. Report
Eliot, J. Field-Husbandry
Harris, T. Insects
Humphreys, D. Discourse
Jackson, C. Agricultural Survey
Johnson, S. Annual Report
Massachusetts. Agricultural Survey. Report
Massachusetts. General Court. Select Committee. Report
Massachusetts. Society For Promoting Agriculture. Inquiries
Massachusetts. State Board of Agriculture. Abstract Of Returns
Massachusetts. State Board of Agriculture. Agriculture
Massachusetts. State Board of Agriculture. Annual Report
New Hampshire. State Board Of Agriculture. Agricultural Repository

Agriculture -- Plains
Caird, J. Prairie Farming
Des Moines Navigation Company. Iowa Lands
Ferris, J. Great West
Iowa. State Agricultural Society. Annual Report
Parker, N. Iowa
Parker, N. Minnesota Handbook

Agriculture -- South East
Baudry Des Lozières, L. Second Voyage A La Louisiane
Cabell, N. Agriculture In Virginia
Chazotte, P. Culture Of Vines
Coxe, T. Number Of Farms
Debar, J. Distribution of Farm Lots
De Bow, J. Industrial Resources
Emmons, E. North-Carolina Geological Survey
For Sale ... Farm Lands
Forstall, E. Louisiana
Glen, J. South Carolina
Harper, L. Mississippi
Hilgard, E. Mississippi
Judd, T. Sugar
Louisiana. Agriculturists' And Mechanics' Association. State Fair
Memorial On The Practicality Of Growing Vineyards
Mitchell, E. Geology Of North-Carolina
Morse, J. Georgia Western Territory
North American Land Company. Observations
Olmsted, D. Geology of North-Carolina
Ruffin, E. Survey Of South-Carolina
Sacket, A. For Sale, 400,000 Acres
Safford, J., and R. Owen. Mineral And Agricultural Resources
Sibbald, G. Pine Lands Of Georgia
Sketch Of Western Virginia
Smith, J. Review Of East Tennessee
Southern Agriculturist
Stork, W. East-Florida
Tennessee. State Agricultural Bureau. Biennial Report
Tennessee. State Agricultural Bureau. Transactions
Tuomey, M. Agricultural Survey
Virginia. Society For Promoting Agriculture. Memoirs
Virginia. State Agricultural Society. Report
Virginia. United Agricultural Societies. Address
Virginia. United Agricultural Societies. Memorial
Wailes, B. Mississippi
Western Virginia Farm And Agricultural Company. Farm of 20 Acres

Williams, E. Virginia

Agriculture -- South West
De Cordova & Frazier. Texas Land Register
Edward, D. History Of Texas
Woodman, D. Texas Emigrants

Apples
Chapman, T. Cyder-Maker's Instructor
Putnam, H. Touches On Agriculture

Architecture
Allen, L. Rural Architecture
Bordley, J. Country Habitations
Downing, A. Country Houses
Downing, A. Cottage Residences
Hammond, J. Practical Architect
Johnson, S. Rural Economy
Ruskin, J. Sheepfolds

Bees
Bevan, E. Honey Bee
Boswell, P. Bees, Pigeons, Rabbits
Butler, F. Farmer's Manual
Colvin, R. Bee And Hive
Cottage Bee Keeper
Du Couëdic de Villeneuve, P. Bee-Hive
Honig und dessen Nutzen
Kelsey, W. Apiarian's Guide
Langstroth, L. Honey-Bee
Mahan, P. Bee And Hive
Miner, T. Bee Keeper's Manual
Phelps, E. Bee-Keeper's Chart
Repplogel, G. Preservation Of Bees
Smith, J. Honey Bee
Souder, D. Management Of Bees
Townley, E. Humanity To Honey Bees
Vanière, J. Bees
Weeks, J. Managing Bees
Wiestling, J. Bienen-Wärter

Beets
Child, D. Culture Of The Beet
Church, E. Beet Sugar
Pedder, J. Sugar from the Beet
Pedder, J. Beet Sugar Society
Ronaldson, J. Sugar Beet

Berries
Buchanan, R. Culture Of The Grape
Eastwood, B. Cranberry
Fessenden, T. American Kitchen Gardener
Pardee, R. Strawberry

Bookkeeping
Barnum, H. Farmer's Own Book
Hough, F. Farm Record

Botany
Darlington, W. Agricultural Botany
Darlington, W. American Weeds
Marcet, J. Vegetable Physiology
Rodgers, M. Scientific Agriculture

Businesses -- Fertilizer
Allen & Needles
American Guano Company
Hanson, J.
Leinau, G. A.
Philadelphia Chemical Manure Manufacturing Co.
Philadelphia Guano Co.
Street Sweeping & Fertilizing Co.

Dadd, G. Reformed Cattle Doctor
Dadd, G. Diseases Of Cattle
Davy, J. Herd Book
Downing, J. Horned Cattle
Flint, C. Milch Cows
Gardner, A. Value of Milk
Guénon, F. Milch Cows
Haines, B. & C.S., Livestock breeders. Catalogue
Heeringen, E. Milk-Sickness
Howard, S. Breeders Of Devon Cattle
Jefferson Co. [N.Y.]. Agricultural Society. Cattle Show
Lawrence, B. Complete Cattle-Keeper
Lawrence, J. Poultry, Cows, Swine
Mills, J. Treatise On Cattle
New Castle Co. [Del.]. Temperance Convention. Cattle And Hogs
Pennsylvania. Agricultural Society. Hints For American Husbandmen
Powel, J. Short Horned Cattle
Quincy, J. Soiling Of Cattle
Small, M. Cattle Doctor
Thomson, R. Food Of Animals
Valley Agricultural Society. Cattle Show

Celery
Roessle, T. Celery

Chemical agriculture
Chaptal de Chanteloupe, J. Chymistry
Davy, H. Agricultural Chemistry
Falkner, F. Farmer's Manual
Gray, A. Scientific And Practical Agriculture
Hall, L. Farmer's Railroad
Higgins, J. State Agricultural Chemist
Johnson, C. Encyclopaedia
Johnson, S. Annual Report
Johnston, J. Chemistry And Geology
Kent, J. Farmer's Hand-Book
Liebig. J. Chemistry
Liebig, J. Organic Chemistry
Liebig, J. Principles Of Agricultural Chemistry
Liebig, J. Complete Works On Chemistry
MacNeven, W. Application Of Chemistry
Nash, J. Progressive Farmer
Norton, J. Scientific Agriculture
Rodgers, M. Scientific Agriculture
Smith, J. Productive Farming
Solly, E. Lectures On Chemistry
Squarey, C. Agricultural Chemistry
Stockhardt, J. Chemical Field Lectures
Topham, J. Chemistry Made Easy
Tyson, P. State Agricultural Chemist
Ure, A. Dictionary Of Chemistry

Clay
Hoskyns, C. Talpa

Climatology
Blodget, L. Climate Of The Summer Of 1853
Blodget, L. Climatology Of The United States
Essay On The Climate

Commerce
Book Of Commerce
Bordley, J. Intimations
Bordley, J. Necessaries
Briggs, I. Statements And Remarks
Brissot de Warville, J. Commerce Of America With Europe
Brissot de Warville, J. Considerations On The Relative Situation
Brissot de Warville, J. De La France Et Des États-Unis
Carey, H. Harmony Of Interests
Carey, H. The Prospect
Chambon, -----. Commerce De L'Amérique
Chambon, -----. Traité Général
Chesapeak

Flint, J. Letters From America
Freeman, S. Emigrant's Hand Book
Lorain, J. Hints To Emigrants
Marshall, J. Farmers And Emigrants Complete Guide
Marshall, J. Farmer's And Emigrant's Hand-Book
Melish, J. Advice To Emigrants
Parker, N. Iowa
Parker, N. Minnesota Handbook
Ross, G. Auswanderers Handbuch
Seyd, E. California And Its Resources
Sibbald, G. Pine Lands Of Georgia
Sketch Of Western Virginia
Smith, J. Review Of East Tennessee
Thomason, D. Hints To Emigrants
To Persons, Inclining To Emigrate
Woodman, D. Texas Emigrants

Essays, letters
Agricola. Series Of Essays
Beatty, A. Essays On Practical Agriculture
Bordley, J. Essays And Notes On Husbandry
Bordley, J. Additions And Amendments
Buckminster, W. Practical Farmer
Buel, J. Farmer's Companion
Buel, J. Farmers' Instructer
Crêvecoeur, M. Letters From An American Farmer
Crêvecoeur, M. Lettres D'Un Cultivateur Américain
Downing, A. Rural Essays
Essex [Mass.]. Agricultural Society. Agricultural Essays
Everett, D. Farmer's Monitor
Lathrop, L. Farmer's Library
Lloyd, D. Economy Of Agriculture
Manwaring, C. Essays
Select Essays
Taylor, J. Arator
Washington, G. Letters

Fairs
Ball, L. Address
Delaware Co. [Pa.]. Institute Of Science. Report
Essex [Mass.]. Agricultural Society. Cattle Show
Essex [Mass.]. Agricultural Society. Annual Exhibition
Franklin Co. [Mass.]. Agricultural Society. Eleventh Exhibition
Hamilton Co. [Ohio]. Agricultural Society. Proceedings
Illinois. State Agricultural Society. State Fair
Jefferson Co. [N.Y.]. Agricultural Society. Cattle Show
Kentucky. State Agricultural Society. Exhibition
Louisiana. Agriculturists' And Mechanics' Association. State Fair
Maryland. State Agricultural Society. Premiums
Massachusetts. Horticultural Society. Annual Exhibition
New Jersey. State Agricultural Society. List Of Premiums
New York City. American Institute. Premiums
New York City. American Institute. Annual Fair
Norfolk [Mass.]. Cattle Show & Fair
Pennsylvania. Horticultural Society. Report
Pennsylvania. Poultry Society. Exhibition
Pennsylvania. State Agricultural Society. Annual Exhibition
Pennsylvania. State Agricultural Society. List Of Judges
Pennsylvania. State Agricultural Society. List Of Premiums
Pennsylvania. State Agricultural Society. Reports Of Judges
Philadelphia. Society for Promoting Agriculture. Catalogue
Philadelphia. Society for Promoting Agriculture. Proceedings Of The Annual Exhibition
Philadelphia. Society for Promoting Agriculture. Regulations
St. Louis. Agricultural & Mechanical Association. Annual Fair
Schenectady Co. [N.Y.]. Agricultural Society. Exhibition
United States. Agricultural Society. National Exhibition
Valley Agricultural Society. Cattle Show

Farriery
Badcock, J. Farriery
Badcock, J. Groom's Oracle
Badcock, J. Veterinary Surgeon
Barnum, H. American Farrier

Bartlet, J. Gentleman Farrier's Repository
Bentwright, J. American Horse Tamer
Bracken, H. Farriery Improved
Bracken, H. Taplin Improved
Buchan, W. Every Man His Own Doctor
Burdon, W. Gentleman's Pocket-Farrier
Carver, J. Epidemic Disease
Carver, J. Veterinary Science
Carver, W. Practical Horse Farrier
Cheap And Famous Farrier
Complete Farrier
Domestic Animal's Friend
Every Man His Own Farrier
Farrier's Magazine
Franklin, A. American Farrier
Hassel, C. Farrier
Jewett, P. New-England Farrier
Jones, J. Elastic Horse-Shoe
Knowlson, J. Complete Farrier
Markham, G. Experienced Farrier
Mason, R. Pocket Farrier
Miles, W. Horse-Shoeing
Mills, J. Modern System Of Farriery
New American Pocket Farrier
Overdurff, E. Book Of Farriery
Pocket Farrier
Richardson, J. New-England Farrier
Rupp, I. Farmer's Complete Farrier
Simmons, J. American Pocket Farrier
Taplin, W. Farriery
Taplin, W. Stable Directory
Thompson, J. Farriery
White, J. System Of Farriery

Fertilizers
Allen & Needles, Oil merchants. Guano
American Guano Company. Experiments
Bartlett, E. Guano
Bommer, G. New Method
Colman, H. Bone Manure
Dana, S. Manures
Dana, S. Muck Manual
Dumas, J. Manures
Emerson, G. Superphosphate of Lime
Falkner, F. Farmer's Manual
Guano
Hammond, J. Marl
Hannam, J. Waste Manures
Hanson, Joseph B., Agent. Phosphatic Guano
Heermance, H. Farmer's Mine
Hyatt, J. Lime And Marl
Johnson, S. Manures
Kirwan, R. Manures
Leinau, G. American Fertilizer
Logan, G. Experiments On Gypsum
Morfit, C. Manures
Peters, R. Plaister Of Paris
Peysson, A. Public Health And Agriculture
Philadelphia Guano Company. Phosphatic Guanos
Puvis, M. Use Of Lime
Redd, G. Neue Entdeckung
Reinhold, C. Farmer's Promotion Book
Reynolds, J. Application Of Manures
Robbins, L. Fertilizing Compounds
Robinson, S. Guano
Ruffin, E. Exhausting And Fertilizing Systems
Ruffin, E. Calcareous Manures
Teschemacher, J. Guano

Fish breeding
Fry, W. Artificial Fish-Breeding
Massachusetts. General Court. Senate ... Artificial Propagation of Fish
Sabine, L. Principal Fisheries

Downing, A. Landscape Gardening
Follen, C. Landscape Gardening
Kemp E. Lay Out A Garden
Kern, G. Landscape Gardening
Meehan, T. Ornamental Trees
Warder, J. Hedges And Evergreens

Livestock
Allen, R. Domestic Animals
Culley, G. Live Stock
Glenn, J. Breeding Of Animals
Williams, T. Domestic Animals

Machinery
Albany Agricultural Works. Annual Circular
Albany Agricultural Works. Annual Descriptive Catalogue
Atkins, J. Atkins' Automaton
Ball, L. Address
Boas & Spangler. Ketchum's Improved Combined Harvester
Bump, J. Atmospheric Attempering Churn
English, Goodrich & Dixon. Grain & Seed Drill
Evans, E. Tobacco Press
Evans, O. License
Evans, O. Patent Rights
Evans. O. Mill-Wright & Miller's Guide
Fessenden, T. Register Of Arts
Hubbard, M. Reapers And Mowers
Jefferson, T. Mould Board
Landreth Family, Nurserymen. Agricultural Warehouse
Manny, J. Letters Patent of Cyrus H. McCormick
Mauck, R. Patent
McCormick, C. Patent
Poellnitz, F. Essay On Agriculture
Seward, W. Reaper
Simpson, T. Keil-Maschine
Stabler, E. Reaping Machines
Thomas, J. Farm Implements
United States. Agricultural Society. Reapers, Mowers, &c
Williamson, P. Travels

Maple Sugar
Jones, E. Sugar Maple Tree
Remarks On The Manufacturing Of Maple Sugar
Rush, B. Sugar Maple-Tree
Society for Promoting the Manufacture of Sugar .... Constitution

Mensuration
Pedder, J. Farmers' Land-Measurer
Villee, H. Bauer als Landmesser

Milling
Boston And Roxbury Mill Corporation. The subscriber
Evans, O. License
Evans. O. Mill-Wright & Miller's Guide
Hughes, W. American Miller
Hughlett, W. Mills
Union Manufacturing Company. Steam Flouring Mill
United States. Citizens. Oliver Evans' Patent

Mineralogy
Beck, L. Mineralogy Of New-York
Safford, J., and R. Owen. Mineral And Agricultural Resources

Peaches
Casey, J. Farmers' And Gardeners' Hive

Pears
Field, T. Pear Culture

Periodicals -- Agriculture
Agricultural Museum
Agriculturist's And Manufacturer's Magazine
American Agriculturist

American Farmer
American Farmer's Magazine
American Journal of Agriculture and Science
American Medical And Philosophical Register
Archives Of Useful Knowledge
Bristol Co. [Mass.]. Agricultural Society. Transactions
California. State Agricultural Society. Transactions
Ceres
Christian's, Scholar's, and Farmer's Magazine
Cincinnatus
Connecticut. Society For Promoting Agriculture. Transactions
Connecticut. State Agricultural Society. Transactions
Cultivator
De Bow's Review
Delaware Register
Essex [Mass.] Agricultural Society. Transactions
Farmer and Gardener
Farmers' Cabinet
Farmer's Museum
Farmer's Register (1798)
Farmer's Register
Hampshire Co. [Mass.]. Agricultural Society. Transactions
Hingham [Mass.]. Transactions
Illinois. State Agricultural Society. Transactions
Madison Co. [N.Y.]. Transactions
Massachusetts. Secretary of the Commonwealth. Transactions
Massachusetts. Society For Promoting Agriculture. Repository
Massachusetts. Society For Promoting Agriculture. Papers
Massachusetts. Society For Promoting Agriculture. Transactions.
Medical and Agricultural Register
Michigan. State Agricultural Society. Transactions
New Castle Co. [Del.]. Agricultural Society. Transactions
New Hampshire. State Agricultural Society. Transactions
New York City. American Institute. Journal
New York City. American Institute. Transactions
New-York Farmer And Horticultural Repository
New York. Society for the Promotion of Agriculture .... Transactions
New York. State Agricultural Society. Proceedings
New York. State Board of Agriculture. Memoirs
Norfolk [Mass.]. Agricultural Society. Transactions
Pennsylvania. Agricultural Society. Memoirs
Pennsylvania. Agricultural Society. Proceedings
Pennsylvania Farm Journal
Pennsylvania Farmer
Pennsylvania. State Agricultural Society. Report Of Transactions
Philadelphia. Society for Promoting Agriculture. Memoirs
Plough Boy
Plymouth Co. [Mass.]. Agricultural Society. Transactions
Progressive Farmer
Quarterly Journal Of Agriculture
Religious Farmer
Rhode Island. Society For The Encouragement Of Domestic Industry. Transactions
Rural Magazine
Rural Visiter
Southern Agriculturist
Western Journal
Wisconsin. State Agricultural Society. Transactions

Periodicals -- Horticulture

American Pomological Society. Proceedings
American Pomologist
American Silk Grower
American Silk Society. Journal
Burlington Silk Record
Farmer and Gardener
Floral Magazine
Florist And Horticultural Journal
Gardener's Monthly
Hingham [Mass.]. Transactions
Hoffy's Orchardist's Companion
Horticulturist
Magazine Of Gardening And Botany
Massachusetts. Horticultural Society. Transactions

New Haven Co. [Conn.]. Horticultural Society. Transactions
New-York Farmer And Horticultural Repository
North-Western Fruit-Growers' Association. Transactions
Ohio. State Pomological Society. Transactions
Pennsylvania. Horticultural Society. Transactions
Silk Culturist
Silkworm

Periodicals -- Livestock
Farmers' Cabinet
Farrier's Magazine

Philosophy
Blake, J. Farm And The Fireside
Blake, J. Farmer's Every-Day Book
Blake, J. Modern Farmer
Bordley, J. Intending to retire
Hirzel, H. Rural Socrates

Pigs
Ellsworth, H. American Swine Breeder
Lawrence, J. Poultry, Cows, Swine
Leib, M. Hog-Teaching
New Castle Co. [Del.]. Temperance Convention. Cattle And Hogs
Youatt, W. Pig

Potatoes
Bosson, C. Potatoe
Massachusetts. Secretary of the Commonwealth. Potato Rot
Massachusetts. Society For Promoting Agriculture. Potatoes
Smee, A. Potatoe Plant

Poultry, birds
Bement, C. American Poulterer's Companion
Bennett, J. Poultry Book
Boswell, P. Bees, Pigeons, Rabbits
Boswell, P. Poultry-Yard
Browne, D. American Bird Fancier
Browne, D. American Poultry Yard
Cock, M. American Poultry Book
Cooper, J. Cocking
Eccaleobion
Lawrence, J. Poultry, Cows, Swine
Mann, J. Bird-Keeper's Manual
Pennsylvania. Poultry Society. Exhibition
Pennsylvania. State Poultry Society. Constitution
Poultry Breeder's Text Book

Premiums
American Philosophical Society. Premiums
Essex [Mass.]. Agricultural Society. Account Of Premiums
New York City. American Institute. Premiums
Pennsylvania. Horticultural Society. Premiums
Royal Society Of Arts. Premiums

Preserving
Arthur, Burnham & Co., can mfrs. Fresh Fruits

Rabbits
Boswell, P. Bees, Pigeons, Rabbits

Receipts
Barnum, H. Farmer's Own Book
Beman, D. Mysteries Of Trade
Bolmer, P. New Receipts
Bolmer, P. Neuen Rezepten
Curtis, S. Collection Of Recipes
41 Bewährte Recepte
Farmer's And Horsemen's True Guide
Fessenden, T. Register Of Arts
Frost, J. Omnibus
Hohman, J. Albertus Magnus
Hohman, J. Land-und Haus-Apotheke

Cattaraugus Co. [N.Y.]
Cayuga Co. [N.Y.]
Chester Co. [Pa.].
Chester County [Pa.]. Cabinet Of Natural Science
Columbian Institute For The Promotion Of Arts And Sciences
Delaware State
Delaware Co. [Pa.]. Institute of Science
Dorchester [Md.].
Downingtown [Pa.]. Society For ... Natural Knowledge
Essex Co. [N.J.]
Hunterdon Co. [N.J.]
Jefferson Co. [N.Y.]
Kent County [Del.]
Lancaster Co. [Pa.]
Lehigh Co. [Pa.]
Maryland
Maryland State
Mercer County [Pa.]
Monroe Co. [N.Y]
Montgomery Co. [Md.].
Montgomery Co. [N.Y.]
Montgomery Co. [Pa.]
New Castle Co. [Del.]
New Jersey State
New York City. American Institute
New York Society For The Promotion Of Agriculture
New York State
Niagara and Erie [N.Y.]
Oneida Co. [N.Y.]
Oswego Co. [N.Y.]
Otsego Co. [N.Y.]
Pennsylvania
Pennsylvania. Poultry Society
Pennsylvania State
Pennsylvania. State Poultry Society
Philadelphia Society for Promoting Agriculture
Prince George's Co. [Md.]
Queens Co. [N.Y.]
Rensselaer Co. [N.Y.]
St. Lawrence Co. [N.Y.]
Saratoga Co. [N.Y.]
Schenectady Co. [N.Y.]
Schoharie Valley [N.Y.]. Stock Growers' Association
Schuylkill C.[Pa.]
Somerset Co. [Pa.].
Susquehanna Co. [Pa.]

Societies, agricultural -- New England
Albany Co. [N.Y.]
Aquidneck [R.I.]
Barnstable Co. [Mass.]
Bennington Co. [Vt.]
Berkshire [Mass.]
Bristol Co. [Mass.]
Connecticut River Valley
Connecticut Society for Promoting Agriculture
Connecticut State
Essex [Mass.]
Franklin Co. [Mass.]
Grafton Co. [N.H.].
Hampshire Co. [Mass.]
Hampshire, Hampden And Franklin [Mass.]
Hartford Co. [Conn.]
Hillsborough [N.H.]
Hingham [Mass.]
Massachusetts Society For Promoting Agriculture
Middlesex Co. [Conn.]
Middlesex Co. [Mass.]
New Hampshire State
New Haven Co. [Conn.]
New London And Wincham Cos. [Conn.].
Norfolk [Mass.].
Norwich Co. [Conn.]

Statistics
Bigelow, J. Statistical Tables
Coxe, T. Arts And Manufactures
Donnant, D. Statistical Account
Illinois. General Assembly ... Agricultural Statistics
Jay, J. Statistical View
Pitkin, T. Statistical View

Sugar
Archbald, W. Fabrication pour le Sucre Brut
Child, D. Culture Of The Beet
Church, E. Beet Sugar
Degelos, P. Sugar
Forstall, E. Louisiana
Hyde, J. Chinese Sugar-Cane
Judd, T. Sugar
Lovering, J. Chinese Sugar Cane
McCulloh, R. Sugar
Olcott, H. Sorgho And Imphee
Orleans Territory. House Of Representatives. Culture Of Sugar
Pedder, J. Sugar from the Beet
Pedder, J. Beet Sugar Society
Rhode Island. Society For The Encouragement Of Domestic Industry. Chinese Sugar Cane
Silliman, B. Sugar Cane

Textbooks
Adams, D. Agricultural Reader
Bentz, L. Elements Of Agriculture
Gray, A. Scientific And Practical Agriculture
Olcott, H. Yale Agricultural Lectures
Waring, G. Elements Of Agriculture

Tobacco
Hand, F. Tobacco Plants
Jenifer, D. Convention Of Tobacco Planters
Nolte, Vincent & Co., Merchants. Statements of our exports
Tatham, W. Tobacco
Tobacco Convention. Proceedings
Watterston, G. Tobacco Plant

Treatises
Allen, R. American Farm Book
American Farmer's Guide
American Husbandry
Armstrong, J. Treatise On Agriculture
Beecher, H. Plain And Pleasant Talk
Binns, J. Treatise On Practical Farming
Bowler, M. Treatise On Agriculture
Burger, J. Economy Of Farming
Butler, F. Farmer's Manual
Casey, J. Farmers' And Gardeners' Hive
Cobbett, W. Cottage Economy
Colman, H. Agriculture And Rural Economy
Cooke, G. Complete English Farmer
Copeland, R. Country Life
Coulton, Miss. Farm Of Four Acres
Dabney, J. Address To Farmers
Denny, S. Farmer
Gaskins, J. Farmer's Guide
Gleanings from ... Husbandry
Mechi, J. How To Farm Profitably
Parkinson, R. Experienced Farmer
Pedder, J. Dialogues Between A Father And Son
Sinclair, J. Code Of Agriculture
Stephens, H. Book Of The Farm
Thaer, A. Principles Of Agriculture
Thompson, C. Notes On Farming
Todd, S. Young Farmer's Manual
Varlo, C. New System Of Husbandry
Wells, S. The Farm
Wiggins, F. Farmer's Instructor
Young, A. Rural Oeconomy

APPENDIX II -- CHRONOLOGY

Omitted from this Appendix are almanacs, which are arranged chronologically within place of publication in the alphabetical listing, and periodicals, which have wide and sometimes spasmodic ranges of dates. Titles are represented by key words only.

1622 Bonoeil, J. His Maiesties Graciovs Letter

1650 Williams, E. Virginia

1675 Fenwick, J. New Jersey

1684 Tryon, T. Country-Man's Companion

1685 Budd, T. Good Order Established

1725 Rawle, F. Inhabitants of Delaware To become Rich

1731 Hall, F. British Plantations

1748 Eliot, J. Field-Husbandry

1749 Eliot, J. Field-Husbandry

1751 Eliot, J. Field-Husbandry

1753 Eliot, J. Field-Husbandry

1754 Eliot, J. Field-Husbandry
Society for Encouraging Industry and Employing the Poor. Linen

1757 Postlethwayt, M. Britain's Commercial Interest

1758 Pullein, S. Culture of Silk
Royal Society Of Arts. Premiums

1759 Eliot, J. Field-Husbandry
Pullein, S. Preserving the Seeds

1760 Eliot, J. Field-Husbandry

1761 Glen, J. South Carolina

1762 Chapman, T. Cyder-Maker's Instructor

1764 Markham, G. Experienced Farrier
Royal Society Of Arts. Premiums

1765 Quincy, E. Hemp-Husbandry

1766 Marcandier. Hemp
Stork, W. East-Florida

1767 Mitchell, J. Great Britain And North America

1768 Chesapeak
Williamson, P. Travels

1769 Stork, W. East-Florida

1770 Boissier de Sauvages de la Croix, P. Silk-Worms
Cluny, A. American Traveller
Deigendesch, J. Ross-Artzney-Büchlein
Ellis, J. Bringing Over Seeds
Postlethwayt, M. History Of America

1771 Deigendesch, J. Ross-Artzney-Büchlein
Ottolenghe, J. Breeding Silk-Worms
Prince Family, Nurserymen. To Be Sosd
Wohl-eingerichtetes Vieh-Arzney-Buch

1772 Cooke, G. Complete English Farmer

1773 Ellis, J. Preserving Seeds

1774 Philadelphia. Committee Of The City And Liberties. Breed of Sheep

1775 American Husbandry
Bartlet, J. Gentleman Farrier's Repository

1776 Bordley, J. Necessaries
Paine, R. Common Salt
Young, A. Rural Oeconomy

1777 Select Essays

1778 Royal Society Of Arts. Premiums

1780 Matlack, T. Oration

1782 Chambon, -----. Commerce De L'Amérique
Crèvecoeur, M. Letters From An American Farmer
Hand, F. Tobacco Plants

1783 Ayscough, S. Remarks On The Letters From An American Farmer
Bartram's Botanic Garden. Catalogue
Bartram's Botanic Garden. Catalogue of American Trees
Chambon, -----. Traité Général
Cluny, A. Voyageur Americain
Crèvecoeur, M. Letters From An American Farmer
Lettre d'un Américain sur les productions

1784 Bordley, J. Courses Of Crops
Crèvecoeur, M. Lettres D'Un Cultivateur Américain
Smyth, J. Tour In The United States

1785 Philadelphia. Society for Promoting Agriculture. Address
Varlo, C. New System Of Husbandry

1786 Bowler, M. Treatise On Agriculture

1787 Coxe, T. Address
Crèvecoeur, M. Lettres D'Un Cultivateur Américain
Cutler, M. Explanation Of The Map
Ten Minutes Advice ... to purchase a Horse
Thompson, C. Notes On Farming

1788 Brissot de Warville, J. Considerations On The Relative Situation
Jefferson, T. Whale-Fishery
Philadelphia. Society for Promoting Agriculture. Laws
Philadelphia. Society for Promoting Agriculture. Lectures Upon Agriculture

1789 Bordley, J. Purport of a Letter on Sheep
Cutler, M. Description Du Sol
Philadelphia. Society for Promoting Agriculture. Laws

1790 Deane, S. New-England Farmer
Essay On The Culture Of Silk
Poellnitz, F. Essay On Agriculture
Queries, On the present state of Husbandry
Remarks On The Manufacturing Of Maple Sugar

1791 Brissot de Warville, J. De La France Et Des États-Unis
Deigendesch, J. Ross-Artzney-Büchlein
Kurzgefasstes Arzney-Büchlein
Remarks On The Manufacturing Of Maple Sugar
Smyth, J. Voyage Dans Les États-Unis
United States. Secretary of State. Cod and Whale Fisheries

1792 Bordley, J. Sketches on Rotations Of Crops
Rush, B. Sugar Maple-Tree
United States. Secretary of State. Cod And Whale Fisheries
Young, A. Rural Economy

1793 Bordley, J. Intimations

1793 Campbell, P. Travels
Coxe, T. Plan For encouraging Agriculture
Crèvecoeur, M. Letters From An American Farmer
Massachusetts. Society For Promoting Agriculture. Laws
Society for Promoting the Manufacture of Sugar .... Constitution
Spurrier, J. Practical Farmer

1794 Bordley, J. Intimations
Bordley, J. Outlines Of A Plan
Brissot de Warville, J. Commerce Of America With Europe
Coxe, T. View Of The United States
Kurzgefasstes Arzney-Büchlein
Neuer Erfahrner, Amerikanischer, Haus-und Stallarzt
Short ... Treatise On ... The Wine-Grapes
Taplin, W. Stable Directory

1795 American Farmer's Guide
Brissot de Warville, J. Commerce Of America
Cheap And Famous Farrier
Evans. O. Mill-Wright & Miller's Guide
Jewett, P. New-England Farrier
Kurzgefasstes Arzney-Büchlein
Mills, J. Treatise On Cattle
Moreau de Saint Méry, M. Chevaux En Amérique
Moreau de Saint Méry, M. Horses In America
New York. Society for the Promotion of Agriculture .... Plan for collecting

1796 American Philosophical Society. Premiums
Bordley, J. Sketches On Rotations Of Crops
Bracken, H. Farriery Improved
Child, S. Every Man His Own Brewer
Dabney, J. Address To Farmers
Hughlett, W. Mills
Massachusetts. Society For Promoting Agriculture. Rules
Mills, J. Modern System Of Farriery
North American Land Company. Observations

1797 Bordley, J. Intending to retire
Bordley, J. Queries selected from a paper
Deane, S. Newengland Farmer
Kurzgefasstes Arzney-Büchlein
Logan, G. Experiments On Gypsum
Logan, G. Agricultural Experiments
Markham, G. Experienced Farrier
Morse, J. Georgia Western Territory
Peters, R. Plaister Of Paris
Taplin, W. Farriery

1798 Bordley, J. Cattle pastured
Bordley, J. Country Habitations
Bracken, H. Farriery Improved
Essays On The Spirit of Legislation
Fennell, J. Salt Works
Lambert, J. Essay On Farming
Massachusetts. Society For Promoting Agriculture. Potatoes
Memorial On The Practicality Of Growing Vineyards

1799 Bordley, J. Essays And Notes On Husbandry
Bordley, J. Hemp
Everett, D. Farmer's Monitor
Lavenir, Auctioneer. Lots for Sale
Marshall, C. Gardening
Parkinson, R. Experienced Farmer
Peck, W. Slug Worm
Wohlerfahrne Baum-Gartner

1800 Back Lands And Improvements
Des Blinden Rosshandlers Arzneymittel
Bordley, J. Additions And Amendments
Essays On The Spirit of Legislation
Hirzel, H. Rural Socrates
Lithen, J., Nurseryman. Catalogue
Logan, G. Promoting Agriculture

Massachusetts. Society For Promoting Agriculture. Inquiries
Rose, J. Mineralisches Pferde-Pulver
Tatham, W. Tobacco
Tulloh, A. Large Tracts of Land
Washington, G. Letters

1801 Bordley, J. Essays And Notes On Husbandry
For Sale ... Farm Lands
Jones, T., and S. James. Land on Conemaugh
Moore, T. Error Of American Agriculture
Sibbald, G. Pine Lands Of Georgia
Washington, G. Letters

1802 Forsyth, W. Fruit Trees
Leib, M. Hog-Teaching
Smith, D., Nurseryman. Catalogue

1803 Baudry Des Lozières, L. Second Voyage A La Louisiane
Binns, J. Treatise On Practical Farming
Forsyth, W. Fruit Trees
Franklin, A. American Farrier
Gleanings from ... Husbandry
Kurz gefasstes, Ross-Arzney Büchlein
Moore, T. Ice-Houses
Washington, G. Letters

1804 Culley, G. Live Stock
Forsyth, W. Epitome Of ... Fruit Trees
Gardiner, J., and D. Hepburn. American Gardener
Krafft, M. American Distiller
MacMahon, B., Nurseryman. Catalogue
Roberts, J. Pennsylvania Farmer

1805 Brodin de la Jutais, P. L'Abondance
Complete Family Brewer
Donnant, D. Statistical Account
Eddowes, R. Wheat Moth
Jefferson, T. Mould Board
Kirk, C. Tracts of Land
Parkinson, R. Experienced Farmer's Tour
Parkinson, R. Tour In America
Ritter, C. Europäisches Pferde-Pulver

1806 Burke, B. The Horse
Johnson, S. Rural Economy
MacMahon, B., Nurseryman. American Gardener's Calendar
Moore, C. Foreign Carrying Trade
Orleans Territory. House Of Representatives. Culture Of Sugar

1807 Bartram's Botanic Garden. Catalogue Of Trees
Evans, E. Tobacco Press
Evans, O. Mill-Wright's & Miller's Guide
Jewett, P. New-England Farrier
Kirwan, R. Manures
Pocket Farrier
Souder, D. Management Of Bees
Thompson, J. Farriery

1808 Blockley and Merion [Pa.]. Agricultural Society. Rules & Regulations
Fessenden, T. Register Of Arts
Lasteyrie du Saillant, C. Cotton-Tree
Prince Family, Nurserymen. Catalogue Of Fruit
Vaniere, J. Bees
Vergilius Maro, P. Georgics

1809 Allaire, J. Voyages De M. Michaux
Complete Farrier
Essay On The Climate
Freitag, J. Deutsche Pferd-Arzt
Livingston, R. Essay On Sheep
M'Harry, S. Practical Distiller
Sicher ... Gauls-Doktor Büchlein

1810 American Medical and Philosophical Register. Prospectus
Antidote To The Merino-Mania
Commissioners For The Cultivation Of The Vine. Culture
Complete Farrier
Derby, E. Merino Sheep
Livingston, R. Essay On Sheep
Mackenzie, G. Sheep
Peters, R. Plaister Of Paris
Redd, G. Neue Entdeckung

1811 Bard, S. Guide For Young Shepherds
Daubenton, L. Advice To Shepherds
Engelman, D., Nurseryman. Flower Seeds
Landreth Family, Nurserymen. Catalogue
Tessier, A. Merinos And Other Sheep

1812 Pennsylvania Vine Company
Philadelphia. Society for Promoting Agriculture. Laws
Taplin, W. Stable Directory

1813 Downing, J. Horned Cattle
Hall, H. Distiller
Livingston, R. Essay On Sheep
MacMahon, B., Nurseryman. Catalogue
United States. Citizens. Oliver Evans' Patent

1814 Bakewell, R. Wool
Coxe, T. Arts And Manufactures
Mease, J. Diseases Of Domestic Animals
Nicholson, J. Farmer's Assistant
Taylor, J. Arator

1815 Bracken, H. Taplin Improved
Brownell, T. Address
Clater, F. Cattle Doctor
Coppinger, J. American Practical Brewer
Davy, H. Agricultural Chemistry
Holland Land Co. 230,000 Acres
Kurz gefasstes Ross Arzeney Büchlein
Nicholson, J. Farmer's Assistant

1816 Briggs, I. Statements And Remarks
Buchan, W. Every Man His Own Doctor
Coxe, T. Number Of Farms
Gold, T. Address
Humphreys, D. Discourse
Lasteyrie du Saillant, C. Pastel, Or Woad
Peters, R. Discourse
Pitkin, T. Statistical View
Rohlwes, J. Allgemeines Vieharzneibuch

1817 Birkbeck, M. Notes On A Journey
Carver, J. Epidemic Disease
Carver, J. Veterinary Science
Clater, F. Cattle Doctor
Coxe, T. Cotton Wool Cultivation
Coxe, W. Fruit Trees
Cutbush, E. Address
Evans, O. Patent Rights
Mease, J. Address
Rohlwes, J. Gauls-Doctor-Buch
Swan, J. Address
Taylor, J. Arator
Watterston, G. Tobacco Plant

1818 Birkbeck, M. Bemerkungen auf einer Reise
Birkbeck, M. Letters From Illinois
Birkbeck, M. Notes On A Journey
Boston And Roxbury Mill Corporation. The subscriber
Bristed, J. Resources
Carver, J. Age Of The Horse
Carver, W. Practical Horse Farrier
Cobbett, W. Year's Residence

Coxe, T. Cotton Culture
Domestic Animal's Friend
Ellsworth, H. Address
Evans, O. License
Evans, O. Mill-Wright's & Miller's Guide
Fearon, H. Sketches of America
Fernagus de Gelone. Manuel-Guide Des Voyageurs
Gardiner, J., and D. Hepburn. American Gardener
Hall, H. Distiller
Hohman, J. Land-und Haus-Apotheke
Homespun, P. Receipt Book
Le Ray de Chaumont, J. Address
Logan, G. Address
Lowell, J. Address
Madison, J. Address
Mease, J. Pattern Farm
New Family Receipt Book
Philadelphia. Society for Promoting Agriculture. Address
Philadelphia. Society for Promoting Agriculture. Memorial
Philadelphia. Society for Promoting Agriculture. Encouragement of Agriculture
Pickering, T. Address
Sinclair, J. Code Of Agriculture
Taylor J. Arator
Virginia. Society For Promoting Agriculture. Memoirs
Webster, N. Address
White, J. System Of Farriery

1819 Agricola. Series Of Essays
Armstrong, J. Treatise On Agriculture
Birkbeck, M. Extracts From A Supplementary Letter
Birkbeck, M. Lettres Sur Les Nouveaux Etablissemens
Butler, F. Farmer's Manual
Casey, J. Farmers' And Gardeners' Hive
Cobbett, W. Year's Residence
Curtis, S. Collection Of Recipes
Davie, W. Address
Davy, H. Agricultural Chemistry
DeWitt, S. Agricultural College
Fearon, H. Sketches of America
Fearon, H. Skizzen von America
Haines, C. Board Of Agriculture
Jefferson Co. [N.Y.]. Agricultural Society. Cattle Show
Krauss, J. Haus-und Kunst-Buch
Lincoln, L. Address
Lorain, J. Hints To Emigrants
MacMahon, B., Nurseryman. American Gardener's Calendar
Melish, J. Advice To Emigrants
New Family Receipt-Book
Parkes, S. Advantages Of Using Salt
Philadelphia. Society for Promoting Agriculture. Society was incorporated
Practical American Gardener
Quincy, J. Address
Rawle, W. Address
Sacket, A. For Sale, 400,000 Acres
Thomas, D. Western Country
Wiestling, J. Bienen-Wärter

1820 Brush, J. Wheat And Flour
Burr, R. Black River
Carver, W. Practical Horse Farrier
Courtauld, G. Address
Courtauld, G. Circular letter
Eaton, A. Geological Survey
Gold, T.R. Address
Gratz, S., J., and J. 4,830 Acres Land
Hohman, J. Verborgene Freund
Howitt, E. Tour Through The United States
Le Ray de Chaumont, J. Remarks
Maule, J. Esteemed Friend
Maxcy, V. Address
New York. State Board of Agriculture. Address
Nicholson, J. Farmer's Assistant
Nolte, V. & Co., Merchants. Statements of our exports

Pennsylvania. Society For The Encouragement Of American Manufactures. Memorial
Pickering, T. Discourse
Rose, R. Address
Simpson, T. Keil-Maschine
Tilghman, W. Address
Virginia. United Agricultural Societies. Address
Watson, E. Modern Agricultural Societies

1821 Beatson, A. New System Of Cultivation
Bigelow, L. Address
Broom, J. Address
Butler, F. Farmer's Manual
Casey, J. Flower Roots
Chazotte, P. Culture Of Vines
Cobbett, W. American Gardener
Essex [Mass.]. Agricultural Society. Cattle Show
Evans, O. Mill-Wright And Miller's Guide
Genet, E. Address
Moulton, J. Address
Nichols, A. Address
Philadelphia. Society for Promoting Agriculture. Cultivation Of Flax
Plumer, W. Address
Russell, J. Address
Ure, A. Dictionary Of Chemistry

1822 Abbot, A. Address
Biddle, N. Address
Burges, T. Address
Cobbett, W. Year's Residence
Colman, H. Address
Cropper, Benson & Co. Cultivation Of Cotton
Deane, S. New-England Farmer
Denny, N. Address
Eastman, M. Discourse
Eaton, A. Geological And Agricultural Survey
Flint, J. Letters From America
Gardiner Lyceum. Address
Jewett, P. New-England Farrier
New Hampshire. State Board Of Agriculture. Agricultural Repository
Robbins, A. Address
Schenectady Co. [N.Y.]. Agricultural Society. Exhibition
Thacher, J. American Orchardist

1823 Adlum, J. Memoir
Bates, I. Address
Burdon, W. Gentleman's Pocket-Farrier
Essex Co. [N.J.]. Agricultural Society. Constitution
Faux, W. Memorable Days In America
Gardiner Lyceum. Catalogue
Morris, J. Address
Pennsylvania. General Assembly .... Report Of The Committee on Agriculture
Peters, R. Address
Powel, J. Strictures Upon Arator's Attack
Prince Family, Nurserymen. Catalogue Of Fruit
Rohlwes, J. Allgemeines Vieharzneibuch
Stimson, E. Address
Van Ranst, C. Celebrated Horse
Van Rensselaer, J. Salt

1824 Adams, D. Agricultural Reader
Carey, M. Address
Denny, S. Farmer
Drowne, W. Compendium Of Agriculture
Eaton, A. Geological And Agricultural Survey
Gardiner Lyceum. Catalogue
Goodwin, I. Address
Hosack, D. Discourse
Howes, F. Address
Jewett, P. New England Farrier
Landreth Family, Nurserymen. Catalogue
Le Ray de Chaumont, J. Address
New York. State Horticultural Society. Prospectus
Parker, M. Arcana Of Arts And Sciences

Villee, H. Bauer als Landmesser

1825 Beman, D. Mysteries Of Trade
Briggs, C. Discourse
Cobbett, W. Woodlands
Gardiner Lyceum. Laws
Lathrop, L. Farmer's Library
Lorain, J. Nature And Reason Harmonized
Lyford, S. Address
MacNeven, W. Application Of Chemistry
Newtown [Pa.] ... Association for Detecting Horse Thieves. Constitution
Powel, J. Reply To Col. Pickering's Attack
Prince Family, Nurserymen. Annual Catalogue
Putnam, H. Touches On Agriculture
Rensselaer Institute. Constitution
Simmons, J., comp. American Pocket Farrier
Thacher, J. American Orchardist
Vaux, R. Address
Walker, L. Address

1826 Adlum, J. On Making Wine
Cobbett, W. Giardiniere Americano
Croll, M. Sammlung Auserlesenen Recepte
Dufour, J. Vine-Dresser's Guide
Essex [Mass.]. Agricultural Society. Account Of Premiums
Jewett, P. New-England Farrier
Landreth Family, Nurserymen. Catalogue
Lathrop, L. Farmer's Library
Mitchill, S. Address
Rensselaer Institute. Constitution
Schmidt, D. Haus-Arzneybuch
Sharpless, J. Bombyx Mori
Smith, D., Nurseryman. Catalogue
Troost, G. Geological Survey

1827 Balmer, D. Sammlung von Neuen Recepten
Carey, M. Address
Carter, N. Address
Gadsden, J. Oration
Mitchell, E. Geology Of North-Carolina
Olmsted, D. Geology of North-Carolina
Pennsylvania. Agricultural Society. Hints For American Husbandmen
Pennsylvania. Horticultural Society. Constitution
Pennsylvania. State Convention. Journal
Sketch Of The Present State Of Commerce, Agriculture
Squibb, R. Gardener's Calendar
Tibbits, G. Creating Home Markets
Tillinghast, J. Address
Tufts, G. Address

1828 Adlum, J. Memoir
Bartram's Botanic Garden. Periodical Catalogue
Breck, S. Address
Cobbett, W. Year's Residence
Drake, D. Discourse
Fessenden, T. New American Gardener
Green, R. Ornamental Flowers
Hazzi, J. Rearing Of Silk-Worms
Hohman, J. Verborgene Freund
Holley, M. Discourse
Horry, E. Address
Keagy, J. Woodside Institute
Kecht, J. Weinbau in Gärten
Kersey, J. Lectures
Lathrop, L. Farmer's Library
Little Eagle ... Tract
Mason, R. Pocket Farrier
New York. Horticultural Society. Proceedings
North Carolina. General Assembly .... Cotton And Woollen
Pennsylvania. Society For Promoting The Culture Of The Mulberry. Meeting
Pennsylvania. Society For Promoting The Culture Of The Mulberry. Directions
Prince, W. Treatise On Horticulture
Richardson, J. New-England Farrier

Russell, J., Nurseryman. Catalogue
Schroeder, J. Discourse
Seabrook, W. Address
Townsend, J. Address
Vernon, W. Mulberry Tree
Wilson, W. Kitchen-Garden

1829 Burdon, W. Gentleman's Pocket-Farrier
Cobbett, W. Emigrant's Guide
Dearborn, H. Address
Du Couëdic de Villeneuve, P. Bee-Hive
Francis, J. Address
Le Ray de Chaumont, J. Address
Loubat, A. Vine Dresser's Guide
Lynch, J. Address
Manwaring, C. Essays
Massachusetts. Horticultural Society. [Constitution]
Muskingum [Ohio]. Agricultural And Manufacturing Society. Report
Pascalis-Ouvière, F. Silk-worm Nurseries
Prince Family, Nurserymen. Annual Catalogue
Schmidt, D. Haus-Arzneybuch
Tibbits, G. Creating Home Markets

1830 Allen, S. Address
Archbald, W. Fabrication pour le Sucre Brut
Badcock, J. Veterinary Surgeon
Beardsley, H. Domestic Instructor
Burges, T. Address
Cobbett, W. Emigrant's Guide
Cook, Z. Address
Degelos, P. Sugar
Fellenberg Institution
Fessenden, T. New American Gardener
Halsey, A. Address
Homergue, J., and P. Du Ponceau. Essays On American Silk
Honig und dessen Nutzen
Lawrence, J. Horse
Le Ray de Chaumont, V. Address
Marcet, J. Vegetable Physiology
Mauck, R. Patent
Miller, J. Der verborgene Arzt
New York City. American Institute. Annual Fair
Pennsylvania. Horticultural Society. Report
Pennsylvania. Society For Promoting The Culture Of The Mulberry. Address
Phinney, E. Address
Powel, J. Short Horned Cattle
Prince, W.R. & W. Treatise On The Vine
Rafinesque, C. American Manual Of The Grape Vines
Repplogel, G. Preservation Of Bees
Sedgwick, T. Address
Stout, Z. Address
Wolbert, C.J. Auctioneer. Bucks County Farms

1831 Badcock, J. Groom's Oracle
Bartram's Botanic Garden. Periodical Catalogue
Bolmer, P. New Receipts
Bolmer, P. Neuen Rezepten
Cheney, J. Address
Cobb, J. Growth Of The Mulberry Tree
Dickinson, S. Address
Everett, E. Address
Fellenberg Institution
Fiske, O. Address
Fowler, J. Journal Of A Tour
Historical Review ... Of The Silk Culture
Ismar, F. Fellenberg's Institution
Kirby, E. Address
Pennsylvania. Horticultural Society. Constitution
Prince, W.R. & W. Pomological Manual
Smith, J. Honey Bee
To Persons, Inclining To Emigrate
Ward, M. Address

1832 Barnum, H. American Farrier
Barnum, H. Farmer's Own Book
Colman, H. Address
Du Ponceau, P. Silk
Essex [Mass.]. Agricultural Society. Mulberry Trees
Friends Of Domestic Industry. Cotton
Harris, T. Discourse
Hibbert and Buist, Nurserymen. Flower Garden Directory
Jones, E. Sugar Maple Tree
Lawrence, B. Complete Cattle-Keeper
Lawrence, J. Poultry, Cows, Swine
Lawrence, M. Address
Lloyd, D. Economy Of Agriculture
MacMahon, B., Nurseryman. American Gardener's Calendar
Mease, J. Cordage
New York City. American Institute. Charter
New York. State Agricultural Society. Convention
Porter, G. Silk Manufacture
Prince, W.R. & W. Pomological Manual
Ruffin, E. Calcareous Manures
Steuart, H. Planter's Guide
Towar, A. Amerikanische Pferdearzt
Zünd, J. Pferde-und Vieharzney-Kunde

1833 Allegheny College. Report Of Trustees
Badcock, J. Veterinary Surgeon
Cobb, J. Growth Of The Mulberry Tree
Cobbett, W. Cottage Economy
Colman, H. Address
Columbian Horticultural Society. Constitution
Farmer's Year Book
Hasty Defence Of The Farmers & Distillers
Kennedy, J. Address
Kenrick, W. New American Orchardist
Marcet, J. Vegetable Physiology
Maryland. Horticultural Society. Constitution
Pennsylvania. General Assembly .... Bericht der Committee über Ackerbau
Pennsylvania. General Assembly .... Report Of The Committee On Agriculture
Pennsylvania. General Assembly .... Report Of The Committee On Education
Perry, G. Address
Silliman, B. Sugar Cane
Whitaker, D. Address

1834 Baldwin, H. Address
Book Of Commerce
41 Bewährte Recepte
Evans, O. Mill-Wright And Miller's Guide
Farmer's And Horsemen's True Guide
Fisher, S. European Vine
Latrobe, J. Address
Mahan, J. Italian Mulberry
Pittsburgh. Horticultural Society. Constitution
Watterston, G. Address

1835 Bridgeman, T. Young Gardener's Assistant
Chaptal de Chanteloupe, J. Chymistry
Davidge, F. Address
Dearborn, H. Address
Hickey, W. Flower Garden
Kenrick, W. Silk Grower's Guide
Kenrick, W. New American Orchardist
Moseley, E. Address
Pitkin, T. Statistical View
Prince Family, Nurserymen. Annual Catalogue
Prince Family, Nurserymen. Catalogue Of Green House Plants
Roberts, E. Mulberry Tree
Ruffin, E. Calcareous Manures
Shirreff, P. Tour Through North America
Wakeman, T. Lecture
Woodman, D. Texas Emigrants

1836 Bartram's Botanic Garden. Periodical Catalogue
Bridgeman, T. Florist's Guide

Comstock, F. Culture Of Silk
Edward, D. History Of Texas
Evans, O. Mill-Wright And Miller's Guide
Gibbons, W. Culture Of Silk
Judd, T. Sugar
King, D. Address
Landreth Family, Nurserymen. Catalogue
Maryland. General Assembly ... Report
Pedder, J. Sugar from the Beet
Pedder, J. Beet Sugar Society
Puvis, M. Use Of Lime
Weeks, J. Managing Bees
Weston, E. Address

1837 Americanisches Noth-und Hülfs-Büchlein
Betton, S. Address
Book of Commerce
Browne, P. Indian Corn
Browne, P. Veterinary Art
Church, E. Beet Sugar
Colman, H. Letter
Colman, H. Report
Du Ponceau, P. Silk Bill
Hohman, J. Verborgene Schatz und Haus-Freund
Illinois In
Lincoln, W. Address
McNaughton, J. Address
Muse, J. Address
Pennsylvania. Horticultural Society. Report
Sketch Of Western Virginia
Strong, J. Silk Worm
Tinelli, L. Mulberry
Weeks, J. Managing Bees

1838 Badcock, J. Veterinary Surgeon
Bigelow, J. Statistical Tables
Busby, J. Principal Vineyards
Colman, H. Spring Wheat
Colman, H. Bone Manure
Ellsworth, H. Upper Wabash
Gage, N. Address
Gaskins, J. Farmer's Guide
Hill, D. & C. A., Auctioneers. Hot House Plants
Julien, S. Silk Worms
Kenrick, W. Americanischen Seidebauers Anweisung
Lelievre, J. Jardinier De La Louisiane
Manning, R. Book Of Fruits
Massachusetts. General Court. Select Committee. Report
New Haven Co. [Conn.]. Horticultural Society. Report
Pennsylvania. Horticultural Society. Constitution
Roberts, E. Mulberry Tree
Sayers, E. Flower Garden Companion
Schroeder, J. Discourse
Shepard, C. Address

1839 Buel, J. Address
Buist, R. American Flower Garden Directory
Clarke, J. Treatise On The Mulberry Tree
Cobb, J. Growth Of The Mulberry Tree
Colman, H. Report
Comstock, F. Culture Of Silk
Dennis, J. Silk Manual
Forbes, A. California
Hohman, J. Albertus Magnus
Homergue, J. Silk Culturist's Manual
Kenrick, W. Silk Grower's Guide
Lee, Z. Address
MacMahon, B., Nurseryman. American Gardener's Calendar
Pennsylvania. Horticultural Society. Premiums
Pennsylvania. Silk Convention. Proceedings
Rafinesque, C. American Manual Of The Mulberry Trees
Roberts, E. Mulberry Tree
Sayers, E. Fruit Garden Companion

Todd, C. Address
Washington, G. Letters
Whitmarsh, S. Mulberry Tree

1840 Armstrong, J. Treatise On Agriculture
Bridgeman, T. Florist's Guide
Bridgeman, T. Kitchen Gardener's Instructor
Buckminster, W. Practical Farmer
Child, D. Culture Of The Beet
Clarke, T. Manual for Producing Silk
Colman, H. Agricultural Addresses
Colman, H. Third Report
Ellsworth, H. American Swine Breeder
Hoare, C. Grape Vine
Hohman, J. Verborgene Freund
Jackson, C. Agricultural Survey
Jenifer, D. Convention Of Tobacco Planters
Latrobe, J. Manual Labor School
Mason, C. Oration
Massachusetts. Agricultural Survey. Report
Overdurff, E. Book Of Farriery
Pedder, J. Dialogues Between A Father And Son
Pennsylvania. Horticultural Society. Catalogue Of The Library
Pennsylvania. Horticultural Society. Constitution
Philadelphia. Society for Promoting Agriculture. Charter
Ronaldson, J. Sugar Beet
Tinelli, L. Address
Tobacco Convention. Proceedings
Weeks, J. Managing Bees
Wiggins, F. Farmer's Instructor
Winter, G. Pferde-Arzt

1841 Booth, J. Geological Survey Of The State Of Delaware
Boswell, P. Poultry-Yard
Clarke, J. Sericon
Colman, H. Report
Curtis, J. Corn Laws
Darlington, W. True Grasses
Downing, A. Lardscape Gardening
Dudgeon, T. Nire Years Residence
Gentle, A. Every Man His Own Gardener
Harris, T. Insects
Illinois. General Assembly ... Agricultural Statistics
Liebig, J. Organic Chemistry
Niles, J. Oration
Rowe, A., Nurseryman. Catalogue
Skinner, J. Christmas Gift
Tallmadge, J. Address
Winter, G. Pferde-Arzt

1842 Allardice, R. Agricultural Tour
Beck, L. Mineralogy Of New-York
Boswell, P. Bees, Pigeons, Rabbits
Buel, J. Farmer's Companion
Cobbett, W. American Gardener
Dana, S. Muck Manual
Downing, A. Cottage Residences
Glenn, J. Breeding Of Animals
Gray, A. Scientific And Practical Agriculture
Harris. T. Insects
Johnston, J. Chemistry And Geology
Leib, I. Wohlerfahrner Pferde-Arzt
Monroe Co. [N.Y.]. Agricultural Society. Annual Report
Morse, O. Address
Pedder, J. Farmers' Land-Measurer
Plymouth Co. [Mass.]. Agricultural Society. Constitution
Prince Family, Nurserymen. Catalogue Of Bulbous
Ruffin, E. Calcareous Manures
Rupp, I. Farmer's Complete Farrier
Smith, J. Review Of East Tennessee
Squarey, C. Agricultural Chemistry

1843 Barnard, D. Address

Bevan, E. Honey Bee
Bommer, G. New Method
Burger, J. Economy Of Farming
Cock, M. American Poultry Book
Dana, S. Manures
Falkner, F. Farmer's Manual
Hassel, C. Farrier
Heeringen, E. Milk-Sickness
Heermance, H. Farmer's Mine
Hohman, J. Verborgene Freund
Liebig. J. Chemistry
Loudon, J. Gardening For Ladies
Mather, T. Address
McCoun, W. Address
Parsons & Co., Nurserymen. Catalogue
Ruffin, E. Survey Of South-Carolina
Saltonstall, L. Address
Smith, J. Productive Farming
Tallmadge, J. Address
Townley, E. Humanity To Honey Bees

1844 Barbour, I. Silk Culture
Beatty, A. Essays On Practical Agriculture
Bridgeman, T. Young Gardener's Assistant
Buist, R. Rose Manual
Clater, F. Cattle Doctor
Complete Florist
Dana, S. Manures
Darlington, W. Address
Downing, A. Cottage Residences
Downing, A. Landscape Gardening
Eccaleobion
Hamilton Co. [Ohio]. Agricultural Society. Proceedings
Hannam, J. Waste Manures
Johnson, C. Encyclopaedia
Manning, R. New England Fruit Book
National Convention of Silk Growers. Proceedings
New York City. American Institute. Annual Report
Pennsylvania. Horticultural Society. By-Laws
Prince Family, Nurserymen. Catalogue Of Fruit
Queens Co. [N.Y.]. Agricultural Society. Annual Report
Seabrook, W. Cotton
Skinner, J. Address
Small, M. Cattle Doctor
Tuomey, M. Agricultural Survey
United States Practical Receipt Book

1845 Bartlett, E. Guano
Bement, C. American Poulterer's Companion
Bommer, G. New Method
Breck, C., Nurseryman. New Seed Store
Buist, R. American Flower Garden Directory
Cotton
Downing, A. Fruits And Fruit Trees
Eliot, T. Address
Ellwanger & Barry, Nurserymen. Catalogue
Fitch, A. Wheat-Fly
Forstall, E. Louisiana
Furman, G. Address
Gilroy, C. History of Silk
Hohman, J. Albertus Magnus
Johnston, J. Chemistry And Geology
Lunt, G. Address
Mapes, J. Address
Marshall, J. Farmer's And Emigrant's Hand-Book
Meigs, H. Address
Morrell, L. American Shepherd
Muse, J. Address
New American Pocket Farrier
New York City. American Institute. Report
Rost, P. Oration
Stuart, I. Address
Tallmadge, J. Address

Teschemacher, J. Guano
Topham, J. Chemistry Made Easy
Ure, A. Recent Improvements
Winter & Co., Nurserymen. Catalogue

1846 Bosson, C. Potatoe
Chandler, A. Address
Chester Co. [Pa.]. Horticultural Society. Constitution
Cincinnati Horticultural Society. Charter
Darlington, W. Address
Dean, A. Claims Of Long Lake
Every Man His Own Gardener
Fitch, A. Hessian Fly
Gardner, D. Farmer's Dictionary
Guénon, F. Milch Cows
Hammond, J. Marl
Hohman, J. Verborgene Freund
Louisiana. Agriculturists' And Mechanics' Association. State Fair
Meigs, H. Address
Morrell, L. American Shepherd
National Convention of Farmers, Gardeners and Silk Culturists. Proceedings
New Castle Co. [Del.]. Temperance Convention. Cattle And Hogs
New York City. American Institute. Annual Report
New York City. American Institute. Premiums
New York. Legislature .... Report
Olmsted, D. Memoir Of Eli Whitney
Prince Family, Nurserymen. Catalogue Of Roses
Prince, W.R. Manual Of Roses
Rensselaer Institute. Established Especially
South Carolina. State Agricultural Society. Proceedings
Stewart, J. Stable Economy
Stone, E. Address
Thomas, J. Fruit Culturist
Thomson, R. Food Of Animals
Williams, E. Wheat Trade
Youatt, W. Horse

1847 Allen, J. Culture of the Grape
Alta California
Bridgeman, T. Kitchen Gardener's Instructor
Buist, R. Family Kitchen Gardener
Burchard, N. Address
Cole, S. American Veterinarian
Complete Florist
Darlington, W. Agricultural Botany
Darlington, W. Discourse
Delaware Co. [Pa.]. Institute Of Science. Report
Dillingham, W. Address
Dodge, A. Address
Downing, A. Fruits And Fruit Trees
Essex [Mass.]. Agricultural Society. Agricultural Essays
Fitch, A. Hessian Fly
Guénon, F. Milch Cows
Hohman, J. Verborgene Schatz und Haus-Freund
Horse: Its Habits
Jenkins, J. Address
Johnson, G. Modern Gardening
Kitchen And Fruit Gardener
Liebig, J. Chemistry
Manning, R. New England Book Of Fruit
New York City. American Institute. Annual Report
Rensselaer Institute. Polytechnic School
Smee, A. Potatoe Plant
Springfield, R. Horse And His Rider
Stephens, H. Book Of The Farm
Stevenson, A. Address
Tallmadge, J. Address
Washington, G. Letters
Youatt, W. Pig

1848 Albany Agricultural Works. Annual Descriptive Catalogue
Allen, J. Practical Treatise
Allen, R. Domestic Animals

Bacon, L. Address
Bentz, L. Elements Of Agriculture
Bowen, J. Address
Busby, J. Grapes And Wine
Delaware Co. [Pa.]. Institute of Science. Report
Dumas, J. Manures
Hoare, C. Grape Vine
Hyatt, J. Lime And Marl
Johnson, B. Address
Mann, J. Bird-Keeper's Manual
Marsh, G. Address
Massachusetts. Horticultural Society. Annual Exhibition
McCulloh, R. Sugar
Morfit, C. Manures
National Convention of Fruit Growers. Proceedings
New York City. American Institute. Annual Report
Norton, J. Address
Ohio. State Board Of Agriculture. Annual Report
Pittsburgh. Horticultural Society. Constitution
Rodgers, M. Scientific Agriculture
Spencer, J. Address
Summer, T. Cotton Plant
Tallmadge, J. Address
Thomason, D. Hints To Emigrants

1849 Allen & Needles, Oil merchants. Guano
Allen, L. Address
Allen, R. American Farm Book
Cole, S. American Fruit Book
Delaware. State Agricultural Society. Proceedings
Delaware Co. [Pa.]. Institute Of Science. Report
Downing, A. Landscape Gardening
Emerson, G. Address
Emmons, E. Agriculture Of New-York
Essex [Mass.]. Agricultural Society. Constitution
Hammond, J. Address
Ingersoll, J. Address
Jaques, G. Fruit Trees
Kelsey, W. Apiarian's Guide
Miner, T. Bee Keeper's Manual
Mt. Airy [Pa.]. Agricultural Institute. Catalogue
New York City. American Institute. Annual Report
Ohio. State Board of Agriculture. Annual Report
Parsons, S. Address
Wilder, M. Address
Williams, T. Domestic Animals

1850 Boutwell, G. Address
Browne, D. American Bird Fancier
Browne, D. American Poultry Yard
Buist, R. Family Kitchen Gardener
Cass, L. Address
Cushing, C. Address
Dadd, G. Veterinary Reform
Delaware Co. [Pa.]. Institute Of Science. Report
Downing, A. Country Houses
Fisher, S. Address
Higgins, J. State Agricultural Chemist
Howitt, W. Country Year-Book
Jackson, C. Address
James, C. Cotton
Johnston, J. Chemistry And Geology
New York City. American Institute. Annual Report
New York. Legislature .... Agricultural College
Norfolk [Mass.]. Cattle Show & Fair
Norton, J. Scientific Agriculture
Ohio. State Board of Agriculture. Annual Report
Pedder, J. Farmers' Land-Measurer
Pennsylvania. Horticultural Society. Catalogue Of The Library
Philadelphia. Society for Promoting Agriculture. Proceedings Of The Annual Exhibition
Philadelphia. Society for Promoting Agriculture. Reports And Letters
Russell, C. Address
Sanborn, E. Address

Solly, E. Lectures On Chemistry
Trumbull Co. [Ohio]. Agricultural Society. Transactions

1851 Barry, P. Fruit Garden
Breck, J. Flower-Garden
Browne, D. American Bird Fancier
Carey, H. Harmony Of Interests
Carey, H. The Prospect
Cottage Bee Keeper
Dadd, G. Reformed Cattle Doctor
Delaware Co. [Pa.]. Institute Of Science. Report
Emerson, G. Address
Evans, G. Dairyman's Manual
Foster, E. Address
Freeman, S. Emigrant's Hand Book
Gardner, A. Value of Milk
Hovey, C. Fruits Of America
Jenks, F., Merchant. Circular
Johnston, J. Notes On North America
Leuchars, R. Hot-Houses
Lindley, J. Theory Of Horticulture
Lumpkin, J. Address
MacMahon, B., Nurseryman. American Gardener's Calendar
Maryland. Horticultural Society. Constitution
Neill, P. Fruit, Flower, And Kitchen Garden
Norton, J. Scientific Agriculture
Ohio. State Board of Agriculture. Annual Report
Pennsylvania. State Agricultural Society. List Of Premiums
Pennsylvania. State Agricultural Society. Proceedings
Peysson, A. Public Health And Agriculture
Philadelphia. Society for Promoting Agriculture. Proceedings Of The Annual Exhibition
Prince George's Co. [Md.]. Agricultural Society. Proceedings
Ross, G. Auswanderers Handbuch
Ruskin, J. Sheepfolds
Russell, G. Address
Thomas, J. American Fruit Culturist
Waring, W. Fruit Grower's Hand Book
Wright, J. Address

1852 American Rose Culturist
Ball, L. Address
Barry, P. Fruit Garden
Bonynge, F. Future Wealth Of America
Bowen, E. Pictorial Sketch-Book Of Pennsylvania
Brincklé, W. Remarks On Entomology
Buchanan, R. Culture Of The Grape
Buist, R. American Flower-Garden Directory
Dana, S. Manures
Delaware Co. [Pa.]. Institute Of Science. Report
Douglas, S. Address
Essex [Mass.]. Agricultural Society. Cattle Show
Fessenden, T. American Kitchen Gardener
Gowen, J. Address
Hallowell, B. Address
Higgins, J. State Agricultural Chemist
Indiana. State Board Of Agriculture. Report
Johnson, L. Flower Gardener
Lindley, J. Theory of Horticulture
Massachusetts. Secretary of the Commonwealth. Potato Rot
Morris, Paschall & Co., Nurserymen. Catalogue
Ohio. State Board of Agriculture. Annual Report
Pennsylvania. State Agricultural Society. Proceedings
Pennsylvania. State Agricultural Society. Reports Of Judges
Pests Of The Farm
Reid, W., Nurseryman. Catalogue
Reid, W. Wholesale Price List
Robinson, S. Guano
Ruffin, E. Exhausting And Fertilizing Systems
Russell, G. Address
Scott, T. Address On Education
Stephens, H. Book Of The Farm
Thaer, A. Principles Of Agriculture
Thomas, J. American Fruit Culturist

Tubbs, E.M. Gardener
United States. Patent Office. Report
Wayland, F. Address

1853 Allen, J. Practical Treatise
Allen, L. Rural Architecture
Blake, J. Farm And The Fireside
Blodget, L. Climate Of The Summer Of
Booth, J. Analysis Of Soils
Bowen, E. Pictorial Sketch-Book
Bowie, W. Address
Browne, P. Trichologia Mammalium
Buist, R. Select Catalogue
Byrn, M. Complete Practical Distiller
De Bow, J. Industrial Resources
Delaware Co. [Pa.]. Institute Of Science. Report
Downing, A. Country Houses
Downing, A. Rural Essays
Emerson, G. Address
Evans, O. Mill-Wright And Miller's Guide
Every Man His Own Farrier
Gowen, J. Address
Greeley, H. Address
Hazard, T. Address
Higgins, J. State Agricultural Chemist
Hohman, J. Verborgene Freund
Holcomb, C. Address
Hovey, C. Fruits Of America
King, W. Address
Langstroth, L. Honey-Bee
Lebanon Co. [Pa.]. Horse Company. Charter
Meehan, T. Ornamental Trees
Nash, J. Progressive Farmer
Nefflen, J. Yield Of The Milch-Cow
Pedder, J. Farmers' Land-Measurer
Pennsylvania. Poultry Society. Exhibition
Pennsylvania. State Poultry Society. Constitution
Philadelphia. Society for Promoting Agriculture. Proceedings Of The Annual Exhibition
Ruffin, E. Address
Ruffin, E. Agricultural Education
Sabine, L. Principal Fisheries
Stockhardt, J. Chemical Field Lectures
United States. Patent Office. Report
Ure, A. Dictionary Of Arts, Manufactures
Wigg, W. Address

1854 Allen, R. American Farm Book
Black, J. Address
Blake, J. Modern Farmer
Book Of Commerce
Bowen, E. Off-Hand Sketches
Bowen E. Pictorial Sketch-Book
Bradford Co. [Pa.] Agricultural Society. First Annual Report
Buel, J. Farmers' Instructer
Buist, R. American Flower-Garden Directory
Curtis & Lincoln. Catalogue
Eley, J. American Florist
Elliott, F. Fruit Book
Ellwanger & Barry, Nurserymen. Catalogue
Freeoff, P.R., Nurseryman. Select List
Fry, W. Artificial Fish-Breeding
Gholson, T. Address
Guénon, F. Milch Cows
Hall, J. Address
Higgins, J. State Agricultural Chemist
Higgins, J., and C. Bickell. Sea Island Cotton
Holcomb, C. Address
Holmes, J. Address
Hoskyns, C. Talpa
Jones, J. Address
Keim, G. Address
Lehigh Co. [Pa.] Agricultural Society. Report
Martin, W. Training The Grape Vine

Massachusetts. State Board of Agriculture. Agriculture
Massachusetts. State Board of Agriculture. Annual Report
Ohio. State Board of Agriculture. Annual Report
Pardee, R. Strawberry
Pennsylvania. Polytechnic College. Announcement
Philadelphia. Society for Promoting Agriculture. Minutes
Poultry Breeder's Text Book
Randall, H. Sheep Husbandry
Rayner, K. Address
Seward, W. Reaper
Stabler, E. Reaping Machines
Thomas, J. Farm Implements
Trask, J. Geology Of The Coast Mountains
Union Manufacturing Company. Steam Flouring Mill
United States. Patent Office. Report
Ure, A. Dictionary Of Arts, Manufactures
Wailes, B. Mississippi
Waring, G. Elements Of Agriculture
Wright, J. Address

1855 Allen, I. Address
Arthur, Burnham & Co., can mfrs. Fresh Fruits
Atkins, J. Atkins' Automaton
Batsto Farm And Agricultural Company. Investment Full of Interest
Central Illinois Farm, Coal And Lumber Co.
Christy, D. Cotton Is King
Davy, J. Herd Book
Debar, J. Distribution of Farm Lots
De Cordova & Frazier. Texas Land Register
Emerson, G. Address
Essex [Mass.]. Agricultural Society. Cattle Show
Fitch, A. Insects
Griscom, D., Nurseryman. Catalogue
Haines, B. & C.S., Livestock breeders. Catalogue
Hohman, J. Verborgener Freund
Illinois Central Rail Road Company. Farm And Wood Lands
Illinois Land Company. Prairie Farm Lands
Johnson, C. Encyclopaedia
Kern, G. Landscape Gardening
Landis, C. Practical Hints
Landreth Family, Nurserymen. Catalogue
Liebig, J. Principles Of Agricultural Chemistry
Manny, J. Letters Patent of Cyrus H. McCormick
Marshall, J. Farmers And Emigrants Complete Guide
Massachusetts. State Board of Agriculture. Annual Report
McCormick, C. Patent
McKillop, J. Harvest Prospects
Minor, F. Address
Norton, J. Scientific Agriculture
Ohio. State Board of Agriculture. Annual Report
Overman, C. Osage Orange Hedge
Parker, N. Iowa
Pennsylvania Farm And Land Association. Full Account
Pennsylvania. State Agricultural Society. List Of Judges
Philadelphia. Society for Promoting Agriculture. Catalogue
Ridgway Farm And Coal Co. Farm Within The Reach
Struthers, T. 120,000 acres Of Good Arable Land
Sullivan Co. [Pa.]. Farm And Improvement Company. Constitution
Western Virginia Farm And Agricultural Company. Farm of 20 Acres
Wright, J. Address
Youatt, W. Sheep

1856 Albany Agricultural Works. Annual Circular
Bement, C. American Poulterer's Companion
Bennett, J. Poultry Book
Biddle, J. Address
Buchanan, R. Culture Of The Grape
Byrn, M. Complete Practical Brewer
Cheever, S. Address
Cherry Hill Nursery. Catalogue
Dadd, G. Modern Horse Doctor
Des Moines Navigation Company. Iowa Lands
Downing, A. Country Houses

Eastwood, B. Cranberry
Farnham, E. California
Ferris, J. Great West
Ferris, W. L., Nurseryman. Catalogue
Fisher, S. Native Grapes
Guernsey, O., and J. Willard. History Of Rock County
Hall, L. Farmer's Railroad
Higgins, J. State Chemist
Hohman, J. Long Lost Friend
Howard, S. Breeders Of Devon Cattle
Hughes, W. American Miller
Jones, J. Elastic Horse-Shoe
Kennicott, J. & C., Nurserymen. Catalogue
Landreth Family, Nurserymen. Agricultural Warehouse
Leinau, G. American Fertilizer
Liebig, J. Complete Works On Chemistry
Miles, W. Horse-Shoeing
Munn, B. Practical Land Drainer
New Jersey. State Agricultural Society. List Of Premiums
Ohio. State Board Of Agriculture. Annual Report
Parker, N. Iowa Handbook
Pennsylvania. Farmers' High School. Memorial
Pennsylvania. State Agricultural Society. List Of Premiums
Philadelphia Guano Company. Phosphatic Guanos
Reemelin, C. Vine-Dresser's Manual
Reinhold, C. Farmer's Promotion Book
Ridgway Farm And Coal Co. Bauerey und Heymath
Ridgway Farm And Coal Co. Farm And Home
Ridgeway Farm And Coal Co. Report
Smith, O. Address
Sullivan Co. [Pa.]. Farm and Improvement Co.
Trask, J. Geology Of Northern And Southern California
Tyson, J. Address
United States. Agricultural Society. National Exhibition
United States. Patent Office. Report
Virginia. State Agricultural Society. Report
Wells, D. Year-Book of Agriculture
Weymouth Farm And Agricultural Company. Lands
Wilder, M. Address
Wisconsin. State Agricultural Society. Report
Yale College. School of Science. Appeal
Yale College. School Of Science. Proposed Plan

1857 American Emigration Land Company. 50,000 Acres Of Land
Bache, A. Address
Badcock, J. Farriery
Blake, J. Farmer's Every-Day Book
Blodget, L. Climatology Of The United States
Boas & Spangler. Ketchum's Improved Combined Harvester
Colman, H. Agriculture And Rural Economy
Cook, G. Agricultural Resources
Dadd, G. Horse
Downing, A. Fruits And Fruit Trees
Dreer, Henry A., Nurseryman. Catalogue
Emerson, G. Address
Fisher, S. Address
Flint, C. Grasses
Frost, J. Omnibus
Gerhard, F. Illinois
Granniss, G., Land agent. For Sale
Harper, L. Mississippi
Hooper, E. Western Fruit Book
Hyde, J. Chinese Sugar-Cane
Illinois Central Rail Road Co. Farming And Wood Lands
Kelley, E. Address
Koogle, J.D. Farmer's Own Book
Langstroth, L. Honey-Bee
Linsley, D. Morgan Horses
MacMahon, B., Nurseryman. American Gardener's Calendar
Manley, J. Cattaraugus County
Maryland. State Agricultural Society. Premiums
Massachusetts. General Court. Senate ... Artificial Propagation of Fish
Michigan. State Agricultural College. Agricultural College

New Jersey. State Agricultural Society. List Of Premiums
New York. State Agricultural Society. Dedication
Ohio. State Board Of Agriculture. Annual Report
Olcott, H. Sorgho And Imphee
Parker, N. Minnesota Handbook
Patterson, J. Address
Pennsylvania. Farmers' High School. Address
Pennsylvania. State Agricultural Society. List of Premiums
Philadelphia. Society for Promoting Agriculture. Proceedings
Pleasantville [Pa.]. Horse Company. Constitution
Rhode Island. Society For The Encouragement Of Domestic Industry. Chinese Sugar Cane
Robbins, L. Fertilizing Compounds
Russell, R. North America
Safford, J., and R. Owen. Mineral And Agricultural Resources
Stevenson, A. Address
Street Sweeping & Fertilizing Co.
Thomas, M., & Sons, Auctioneers. Catalogue
Thomas, M., & Sons, Auctioneers. For Sale: Springbrook
Turner, J. Cotton Planter's Manual
United States. Agricultural Society. Reapers, Mowers, &c
United States. Agricultural Society. National Exhibition
United States. Patent Office. Report
Washington [D.C.]. Horticultural Society. Constitution
White, S. Training Colts
Wisconsin. State Agricultural Society. Report

1858 Bentwright, J. American Horse Tamer
Cabell, N. Agriculture In Virginia
Chester Co. [Pa.]. Agricultural Society. Annual Report
Downing, A. Fruits And Fruit Trees
Eldridge, J. Address
Emmons, E. North-Carolina Geological Survey
English, W. Address
English, Goodrich & Dixon. Grain & Seed Drill
Field, T. Pear Culture
Flint, C. Milch Cows
Hammond, J. Practical Architect
Haven, C. Vine Growing Resources
Higgins, J. State Chemist
Hilgard, E. Mississippi
Iowa. State Agricultural Society. Annual Report
Johnson, S. Annual Report
Kemp E. Lay Out A Garden
Knowlson, J. Complete Farrier
Landreth, D. Commissioner Of Patents
Lawson, W. Orchard and Garden
Lee, Z. Address
Lovering, J. Chinese Sugar Cane
Mahan, P. Bee And Hive
Maryland. Agricultural College. Report
Massachusetts. School of Agriculture. List Of Officers
Massachusetts. Society For Promoting Agriculture. Transactions.
Ohio. State Board of Agriculture. Annual Report
Olcott, H. Sorgho And Imphee
Pennsylvania. Farmers' High School. Address
Pennsylvania. Farmers' High School. Annual Report
Pennsylvania. State Agricultural Society. Officers, Charter
Phelps, E. Bee-Keeper's Chart
Philadelphia. Society for Promoting Agriculture. Charter
Reemelin, C. Vine-Dresser's Manual
Reynolds, J. Application Of Manures
Seyd, E. California And Its Resources
Swallow, G. Grape Culture In Missouri
Tennessee. State Agricultural Bureau. Biennial Report
Thomas, J. Annual Register
United States. Patent Office. Report
Walden, J. Soil Culture
Warder, J. Hedges And Evergreens
Wells, S. The Farm

1859 American Guano Company. Experiments
Beecher, H. Plain And Pleasant Talk
Bump, J. Atmospheric Attempering Churn

Caird, J. Prairie Farming
Cheltenham Nurseries. Catalogue
Cherry Hill Nursery. Catalogue
Cincinnati. Horticultural Society. Brief History
Colvin, R. Bee And Hive
Cooper, J. Cocking
Copeland, R. Country Life
Dadd, G. Diseases Of Cattle
Dana, S. Muck Manual
Darlington, W. American Weeds
Dix, J. Address
Downing, A. Landscape Gardening
Emerson, G. Superphosphate of Lime
Fisher, S. Address
Follen, C. Landscape Gardening
Foote, H. Address
Hanson, J., Agent. Phosphatic Guano
Herbert, H. Horse-Keepers
Hubbard, M. Reapers And Mowers
Illinois. State Agricultural Society. State Fair
Jay, J. Statistical View
Johnson, S. Manures
Johnston, J. Chemistry And Geology
Kent, J. Farmer's Hand-Book
Kentucky. State Agricultural Society. Exhibition
Massachusetts. State Board of Agriculture. Abstract Of Returns
Massachusetts. State Board of Agriculture. Annual Report
Mechi, J. How To Farm Profitably
New York. State Agricultural College. Charter Ordinances
Ohio. State Board of Agriculture. Annual Report
Pennsylvania. Farmers' High School. Annual Report
Pennsylvania. Farmers' High School. Catalogue
Pennsylvania. State Agricultural Society. Annual Exhibition
Quincy, J. Soiling Of Cattle
Ridgely, H. Address
St. Louis [Mo.]. Vine & Fruit Growers' Association. Views And Proceedings
Storke, E. Farm And Gardens
Taylor, G. Address
United States. Agricultural Society. National Exhibition
Wall, J. Address
Watson, A. American Home Garden

1860 Allen, R. American Farm Book
Allen, S. Fibrilia
Biddle, C. Address
Carey, J. Claims Of Agriculture
Chorlton, W. American Grape Grower's Guide
Coulton, Miss. Farm Of Four Acres
Eastwood, B. Cranberry
Essex [Mass.]. Agricultural Society. Annual Exhibition
Fisher, S. Address
Fitch, A. Pernicious Species
Flint, C. Grasses
Franklin Co. [Mass.]. Agricultural Society. Eleventh Exhibition
Guano
Hilgard, E. Mississippi
Hillard, G. Address
Hough, F. Farm Record
Iowa. State Agricultural Society. Annual Report
Johnson, L. Flower Gardener
Klippart, J. Wheat Plant
Maryland. Agricultural College. Circular
New Jersey. State Agricultural Society. Annual Report
Ohio. State Board Of Agriculture. Annual Report
Olcott, H. Yale Agricultural Lectures
Paige, A. Address
Pennsylvania. Farmers' High School. Catalogue
Philadelphia. Society for Promoting Agriculture. Catalogue
Philadelphia. Society for Promoting Agriculture. Regulations
Randall, H. Sheep Husbandry
Roessle, T. Celery
St. Louis. Agricultural & Mechanical Association. Annual Fair
Schenck, P. Gardener's Text-Book

Schoharie Valley [N.Y.]. Stock Growers' Association. By-Laws
Soles, A. Book On The Horse
Sorsby, N. Horizontal Plowing
Spangler, A. Grape Question
Spangler, A. Year Book
Stephens, H. Book Of The Farm
Tennessee. State Agricultural Bureau. Transactions
Thomas, J. American Fruit Culturist
Todd, S. Young Farmer's Manual
Trimble, C. Speech
Tyson, P. State Agricultural Chemist
United States. Patent Office. Report
Valley Agricultural Society. Cattle Show
Wall, J. Address
Wall, J. Speech